THERMOPHYSICAL PROPERTIES OF MATERIALS AND DEVICES

THERMOPHYSICAL PROPERTIES OF MATERIALS AND DEVICES

IVth National Conference on Thermophysical Properties—NCTP'07

Kollam, Kerala, India 20 – 23 September 2007

EDITORS

P. Predeep
National Institute of Technology
Calicut, Kerala, India

S. Prasanth
A. S. Prasad
Sree Narayana College
Kollam, Kerala, India

All papers have been peer reviewed

SPONSORING ORGANIZATIONS

Thermophysical Society of India, Guwahati
Condensed Matter Physics Laboratory, Sree Narayana College, Kollam
Centre for Advanced Scientific Research, Kollam
Amrita Vishwa Vidyapeetham, Kollam
Department of Science and Technology, New Delhi
Board of Research in Nuclear Sciences, Mumbai
Kerala State Council for Science, Technology and Environment, Trivandrum
Indian National Science Academy, New Delhi

AMERICAN INSTITUTE OF PHYSICS

Melville, New York, 2008

AIP CONFERENCE PROCEEDINGS ■ VOLUME 1004

Editor and Conference Chair:

P. Predeep
Professor of Physics
National Institute of Technology
Calicut, Kerala 673 601
India

E-mail: ppredeep@gmail.com

Co-Editors:

S. Prasanth
A. S. Prasad

Condensed Matter Physics Laboratory
Sree Narayana College
Kollam, Kerala 691 001
India

E-mail: prasanth.kl@gmail.com
asp.physics@gmail.com

L.C. Catalog Card No. 2008924870
ISBN 978-0-7354-0523-3
ISSN 0094-243X
Printed in the United States of America

CONTENTS

NANO MATERIALS

DEVICES AND APPLICATIONS

POSTER PRESENTATIONS

Preface

This collection of research papers is from the National Conference on Thermophysical Properties (NCTP), which is the 4th in a series of biannual conferences on thermophysical properties organized by the prestigious Thermophysical Society of India (TPSI) during September 2007 at Kollam, Kerala. The first Asian thermophysical conference was held in the Guwahati University isecond in the Rajasthan University, Jaipur, and the third in the University of Goa. The fourth one was at Kollam, a rich historical tapestry woven in a verdant waterscape and one of the most sought after tourist destinations in India. The event turned out to be a unique forum for the exchange of ideas and newest technical information on thermophysical properties for both the novice and the expert, and for the experts to share experience with peers. The next NCTP is scheduled to be at Baroda, MS University, Gujarat, India.

In this context I would like to thank all the sponsors of the conference as we l as the office bears of TPSI, especially its founder president Prof.M.N.Bora of the Guwahati University. I also kindly remembers the timely help and support (including financial one) offered by Prof. N.S.Saxena of the Rajasthan University

Also thanks are due to all those who helped in making 4th NCTP a grand success and also the publication of this made possible. Besides the funding agencies and TPSI, but for my research scholars and colleagues, this would not have been a success and I would like to record my gratitude to them.

P. Predeep
Editor

INVITED PRESENTATIONS

Room temperature giant magneto-impedance in polycrystalline $La_{0.7}Ba_{0.25}Sr_{0.05}MnO_3$ manganite

Soma Das and T. K. Dey*

Thermophysical Measurements Laboratory
Cryogenic Engineering Centre
Indian Institute of Technology,
Kharagpur 721302, India

Abstract: The magneto-impedance (MI) effect of polycrystalline $La_{0.7}Ba_{0.25}Sr_{0.05}MnO_3$ manganites measured at 300K gives a value of ~51% at 500 Oe under the working frequency of 50KHz. The field dependence of MI nearly saturates at ~55% in a magnetic field of 1KOe. MI decreases sharply with increase in frequency upto 200 KHz and saturates at higher frequencies. MI is linear between ±20 and ±250 Oe and varies from 2% to 45% depending upon the working frequency. The present investigation reveals that the polycrystalline $La_{0.7}Ba_{0.25}Sr_{0.05}MnO_3$ manganite can be used as a potential magnetic sensor for near room temperature application.

Keywords: Magneto-impedance, Manganites.
PACS: 75.50.y, 07.55.w

* Corresponding author
E mail: tapasdey@hijli.iitkgp.ernet.in

INTRODUCTION

The discovery of giant magneto-impedance (GMI) in amorphous and nanocrystalline wires[1-3], ribbons[4-5], films[6-8], multilayers[9] has been the starting point of a large number of experimental and theoretical studies because of their important technological applications in magnetic recording and sensor devices. GMI effect in a material is obtained because of the variation of the transverse magnetic permeability of the material under an applied magnetic field. At higher frequencies, a large increase in circumferential permeability μ_φ of the material (i.e. the transverse permeability for the circumferential field generated with ac current) is obtained because of ferromagnetic resonance and this large circumferential permeability at resonance strongly decreases the skin depth (or magnetic penetration depth) δ as: $\delta = \sqrt{\frac{2\rho}{\omega\mu_\varphi}}$ (ρ is the resistivity, $\omega = 2\pi f$); consequently a large increase in impedance results. Application of an axial magnetic field leads to a substantial decrease in transverse permeability and, therefore increases the penetration depth. Consequently, the sample displays giant magneto-impedance (GMI).

Since last decade, doped lanthanum manganite perovskites are considered as a potential sensor materials because of their strong magnetic field dependence of their

CP1004, *Thermophysical Properties of Materials and Devices-NCTP '07*,
edited by P. Predeep, S. Prasanth, and A. S. Prasad

electrical resistivity.[10-11] Although less explored compared to their colossal dc magneto-resistance (CMR), ac negative GMI effect has also been reported in these materials.[12-19] However there exists wide differences in ac behavior of manganites in high frequency range and the explanation of ac behavior is not yet well understood.

In the present communication, we report our recent results on the ac magneto-transport properties of polycrystalline $La_{0.7}Ba_{0.25}Sr_{0.05}MnO_3$ compound in the frequency range from 50KHz to 15MHz. Our results reveal strong dc field variation of the impedance (both real and imaginary components) in the ferromagnetic metallic state below Currie temperature (T_C). Both these components decrease sharply with field up to 1KOe and tend to saturate at higher fields. Frequency dependence of the MI shows maximum at 1MHz, which is more prominent at higher temperatures near T_C. The maximum MI observed at 300K under 1KOe is -55% and -65% at 50 KHz and 15 MHz respectively.

EXPERIMENTAL TECHNIQUE

Powder samples of $La_{0.7}Ba_{0.25}Sr_{0.05}MnO_3$ were prepared by thermolysis of an aqueous, polymeric bond precursor solution starting from water-soluble coordinated complexes of the composing metal ions. La_2O_3, $Ba(NO_3)_2$, $Sr(NO_3)_2$, and $Mn(CH3COO)_2$ tetra-hydrate (99% purity) have been used as starting materials. Basic details of the preparation method have been reported earlier.[20] AC susceptibility of all the samples was measured using a home-made susceptometer operating at 111Hz and a driving field of 0.10 Gauss. Curie temperature (T_C) was deduced from the minimum of $\delta\chi / \delta T$ vs. T curves.

The sample of approximate cylindrical in shape is used for measurement of impedance. A signal coil of 20 turns was wound over the midpoint of the sample, and the impedance of the coil with and without the sample was measured with a Wayne Kerr Model 6520A Precison Impedance Analyzer. The biasing dc field up to 4KOe was applied parallel to the exciting ac field. The resistive and reactive components of impedance were measured between 50 KHz and 15 MHz using excitation voltage of 0.5V (ac field of 0.9 Oe).

RESULTS AND DISCUSSIONS

The polycrystalline $La_{0.7}Ba_{0.25}Sr_{0.05}MnO_3$ sample prepared by chemical route was phase identified by X-Ray diffraction (figure 1a). The structural parameters are obtained by refining the experimental data using Rietveld FULLPROF program (using rhombohedral lattice structure a=b=5.5203Å, c=13.5535Å and with space group $R\bar{3}C$). The tolerance factor t, defined as: $t = (r_A + r_0)/\sqrt{2}(r_B + r_0)$, r_A, r_B, r_O are ionic radii of A, B and oxygen for ABO_3 type compounds, comes out to be 0.873 for this compound, which is higher than that for $La_{0.7}Sr_{0.3}MnO_3$ (0.853). This increase in tolerance factor due to presence of Ba in this compound signifies the release of structural strain in $La_{0.7}Ba_{0.25}Sr_{0.05}MnO_3$ system[21-22]. The temperature dependence of ac susceptibility gives Curie temperature (T_C) to be 343K.

Figure 1(b)-(c) shows the frequency dependence of the resistive(R) and reactive part of the impedance of $La_{0.7}Ba_{0.25}Sr_{0.05}MnO_3$ at 300K and at different field from 0 to 4KOe. As the frequency increases above a critical value (~1MHz), a sharp increase in both resistive and reactive components and therefore increase in impedance Z $(=\sqrt{(R^2+X^2)})$ with frequency is observed. This critical frequency corresponds to the condition $d/\delta_m=1$ where $2d$ and δ_m are the thickness of the sample and the penetration depth

respectively. A substantial decrease in both resistive (R) and reactive (X) component of impedance (Z) with field is clearly shown (figure 1) and which is responsible to exhibit large MI of the system.

In general, the MI effect is explained in terms of screening of electromagnetic field in magnetic metallic system according to classical electrodynamics[14,16-17,23]. In the ferromagnetic metallic state of manganites, charge transport occurs because of e_g electrons of Mn^{4+} ions, obtained with the divalent substitution at Lanthanum site, by means of double-exchange interaction[24]. With the increase of frequency, the penetration depth δ decreases and the charge carriers therefore prefer to flow near the surface of the material like conduction electrons in metal leading to a high value of impedance. In case of soft ferromagnet, high value of permeability μ_φ decreases the penetration depth (δ), with corresponding increase in impedance. Under large dc magnetic field, because of nearly complete alignment of spins parallel to the biasing field, the material becomes magnetically homogeneous and the permeability drops to low value giving high value of penetration depth and consequently large MI effect results.

Figure 2 shows the field dependence of R, X, Z at 300K for a few selected frequencies (viz., 50 KHz, 100 KHz, 1 MHz and 15 MHz). As shown in fig 2, all the three quantities become field independent above 1KOe, while below 1KOe a sharp change is observed at all the frequencies. Therefore, the MI as well as magneto-resistance (MR=$\Delta R/R$) and magneto-reactance (MX= $\Delta X/X$) of this system is expected to give significant variation at lower field and at lower frequencies (<50 KHz). It may be noted that, out of the resistive and reactive components, the frequency dependence of the resistive part increases from 4% at 100 KHz to 55% at 1MHz and 75% at 15MHz. The magneto-impedance due to resistive part i.e. $\Delta R/R$, which is responsible for exhibiting CMR properties in doped lanthanum manganites, is seen here to give high frequency CMR for the present case. In general, large magneto-resistance of polycrystalline bulk manganites are observed at low temperature below T_C which is explained in terms of spin polarized tunneling of charge carriers across grain boundary[25-27]. However, the effect of high frequency in controlling CMR is still to be explored. On the other hand, the reactive part of magneto-impedance ($\Delta X/X$) in the present case does not exhibit strong frequency dependence (figure 2). $\Delta X/X$ reaches a maximum of ~66% at 1MHz satisfying the condition $d/\delta_m=1$ and saturates at higher frequencies. All the three quantities, viz., $\Delta R/R$, $\Delta X/X$ and $\Delta Z/Z$, saturates above 1KOe and for the entire frequency range used for the present study. This critical field of 1KOe denotes the field at which complete spin alignment of the grain boundaries takes place in $La_{0.7}Ba_{0.25}Sr_{0.05}MnO_3$ compound. Nevertheless, it also corresponds to the minimum value of transverse permeability and hence maximum penetration depth.

Figure 2, also indicates that the behavior of MI (*ΔZ/Z*) is mainly dominated by MX (*ΔX/X*) in low frequency region, while for higher frequencies, both MR and MX influences the overall behavior of MI of the system.

Figure 3(a) shows the frequency dependence of MI at three selected dc fields. A sharp change in MI with frequency upto a critical value of 200 KHz is clearly evident and beyond which MI is fairly frequency

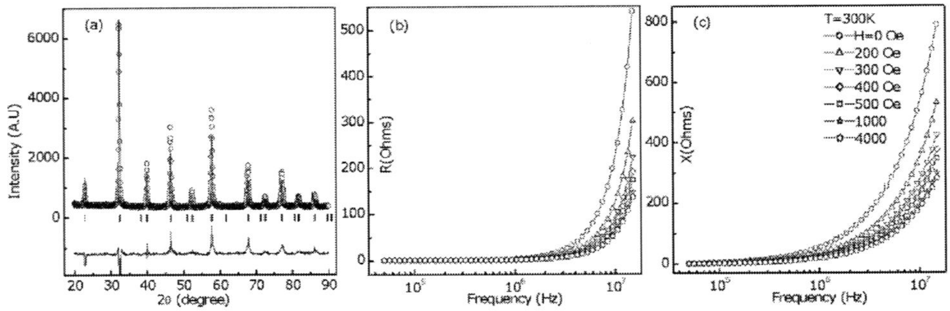

FIGURE 1: (a) Rietveld plot of XRD data for $La_{0.7}Ba_{0.25}Sr_{0.05}MnO_3$ compound. Points are the observed profile and solid line is the calculated one. Vertical line below it indicates the position of allowed Bragg reflection. The line at the bottom gives the error between the observed and calculated profiles. Figs. (b) & (c) shows the measured frequency dependence of resistance (R) and reactance (X) under different dc fields up to 4KOe respectively.

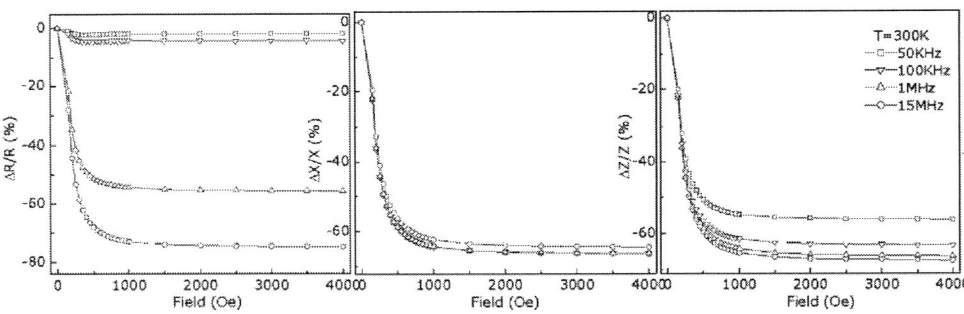

FIGURE 2: *ΔR/R*, *ΔX/X* and *ΔZ/Z* as a function of magnetic field at 50KHz, 100KHz, 1MHz and 15MHz respectively for $La_{0.7}Ba_{0.25}Sr_{0.05}MnO_3$ compound at 300K.

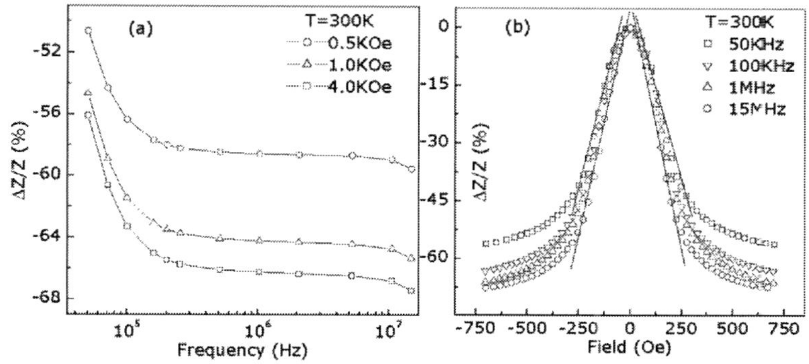

FIGURE 3: (a) Frequency dependence of *ΔZ/Z* at 300K in a dc field of 0.5, 1 and 4KOe for $La_{0.7}Ba_{0.25}Sr_{0.05}MnO_3$ compound (b) Linear variation of *ΔZ/Z* between ±20 and ±250 Oe at different frequencies.

independent. It may be noted that in a dc field of 0.5KOe, MI increases from 51% to 59% as the frequency is increased from 50 KHz to 15 MHz. However, the corresponding change in MI is between ~55% and ~65% in a field of 1KOe. This implies that $La_{0.7}Ba_{0.25}Sr_{0.05}MnO_3$ has potential for low field (0.5KOe) and low frequency (50 KHz) magnetic sensor for near room temperature application. Figure 3(b) shows the MI curve of the same sample at ±500Oe for four different frequencies. A linear increase in MI from ~2% to ~45% is clearly displayed between ±20 and ±250Oe. This linear change in MI with fields for frequencies between 50 KHz and 15 MHz obtained in $La_{0.7}Ba_{0.25}Sr_{0.05}MnO_3$ compound gives added advantage of this material as magnetic sensor.

CONCLUSIONS

In conclusion, the study of the transport properties of $La_{0.7}Ba_{0.25}Sr_{0.05}MnO_3$ manganites by ac impedance spectroscopy reveals that both resistive and reactive part of impedance is sensitive to frequency and magnetic field. The magneto-resistance is seen to give a large change from 4% to 75% as frequency is varied between 100 KHz and 15MHz. Magneto-inductance does not however, show such strong frequency sensitivity. It reaches maximum at 1MHz and saturates at higher frequencies. Magneto-impedance is governed primarily by magneto-inductance at low frequencies and is influenced by both MR and MX at higher frequencies. MR, MX and MI show sharp increase up to 1KOe and saturate in higher dc fields (>1KOe). This field (1KOe) for the present compound denotes the field required for complete alignment of spins across the grain boundaries. It also corresponds to the situation for minimum transverse permeability and hence maximum penetration depth. $La_{0.7}Ba_{0.25}Sr_{0.05}MnO_3$ displays a magneto-impedance of 51% at 500 Oe field and 50KHz at room temperature and shows a linear variation of MI from 2% to 45% for a field change of ±20 to ±250Oe depending on the working frequency range used. All these features suggest that polycrystalline $La_{0.7}Ba_{0.25}Sr_{0.05}MnO_3$ manganite has potential for application as a sensitive and inexpensive magnetic sensor for room temperature applications.

ACKNOWLEDGEMENTS

One of the authors (TKD) is grateful to Council of Scientific & Industrial Research, New Delhi (India) for the financial assistance in the form of a research project.

REFERENCES

1. L. V. Panina, K. Mohri, Appl. Phys. Lett. **65**, 1189 (1994).
2. J. Velázquez, M. Vázquez, D.-X. Chen, and A. Hernando, Phys. Rev. B **50**, 16737 (1994).
3. L. V. Panina, K. Mohri, T. Uchiyama, K. Bushida, M. Noda, IEEE Trans. Magn. **31**, 1249 (1995).
4. R. L. Sommer, C. L. Chien, Appl. Phys. Lett. **67**, 857 (1995).
5. F. L. A. Machado, C. S. Martins and S. M. Rezende, Phys. Rev. B **51**, 3926 (1995).
6. R. L. Sommer, C. L. Chien, Appl. Phys. Lett. **67**, 3346 (1995).
7. N. Jiang, K. Yamakawa, N. Honda, K. Ouchi, IEEE Transaction on Magn. **34**, 1339 (1998).
8. M. Takezawa, H. Kikuchi, M. Yamaguchi, K. I. Arai, IEEE Trans. Magn. **36**, 3664 (2000).
9. M. N. Baibich, J. M. Broto, A. Fert, F. Nguyen Van Dau, F. Petroff , P. Eitenne, G. Creuzet, A. Friederich, and J. Chazelas, Phys. Rev. Lett. **61**, 2472 (1988).
10. J. M. D. Coey, M. Viret, von S. Molnar, Adv. Phys. **48**, 167 (1999).
11. M. B. Salamon, M. Jaime, Rev. Mod. Phys. 73, 583 (2001).
12. S. E. Lofland, S. M.Bhagat,S. D. Tyagi, Y. M. Mukovskii, S. G. Karabashev, A. M. Balbashov, J. Appl. Phys. **80**, 3592 (1996).
13. S. K. Ghatak, B. Kaviraj, T. K. Dey, J. Appl. Phys. **101**, 023910 (2007).
14. J Hu, H Qin, J. Mag. Mag. Mat. **234**, 419 (2001).
15. J Hu, H Qin, H Niu, L Zhu, J Chen, W Xiao, Y Pei, J. Mag. Mag. Mat. **261**, 105 (2003).
16. H Qin, J Hu, J Chen, Y Wang, Z wang, J. Appl. Phys. **91**, 10003 (2002).
17. J Hu, H Qin, Sol. Stat. Comm. **116**, 159 (2000).
18. J Hu, H Qin, Mat. Sc & Engg. B **77**, 280 (2000).
19. G M B Castro, A R Rodrigues, F L A Machado, A E P de Araujo, R F Jardim, A K Nigam, J Alloys & Comp. **369**, 108 (2004).
20. A Ray, T K Dey, J. Mag. Mag. Mat. **266**, 268 (2003).
21. J B Goodenough, J. Appl. Phys. **81**, 5330 (1997).
22. J Vergara, R J Ortega-Hertogs, V Madurga, F Sapina, Z El-Fadli, E Martinez, A Beltran, K V Rao, Phys. Rev. B **60**, 1127 (1999).
23. A Gupta, J Z Sun, J. Mag. Mag. Mat. **200**, 24 (1999).
24. C Zener, Phys. Rev. **81**, 440 (1951).
25. P Raychaudhuri, T K Nath, A K Nigam, R Pinto, J. Appl. Phys. **84**, 2048 (1998).
26. P Raychaudhuri, K Sheshadri, P Taneja, S Bandyopadhyay, P Ayyub, R Pinto, S Chaudhuri, S B Roy, Phys. Rev. B **59**, 13919 (1999).
27. S Das, T K Dey, Sol. Stat. Comm. **134**, 837 (2005).

Thermal and Mechanical Characterization of Aniline-Formaldehyde Copolymer

Narendra S. Saxena* and Neeraj Jain

Semi-conductor & Polymer Science Laboratory, 5, Vigyan Bhawan, Deptt. of Physics, University of Rajasthan, Jaipur- 302004, India.
*Corresponding E-mail: n_s_saxena@rediffmail.com

Abstract. Copolymer of aniline and formaldehyde has been synthesized by chemical oxidative polymerization using $FeCl_3$ as an oxidant in aqueous acidic medium. Thermal transport properties like effective thermal conductivity (λ_e), effective thermal diffusivity (χ_e) and dynamic mechanical properties such as modulus (E′), loss tangent (tanδ) together with their temperature dependence have been investigated by transient plane source (TPS) method and dynamic mechanical analyzer (DMA) respectively. It has been found that both λ_e and χ_e increase with increase in temperature initially from room temperature to 80 °C and then decrease in the temperature range from 80 to 110 °C. For further increase in temperature λ_e & χ_e increase slightly and then attains a saturation value. The variation of modulus and loss tangent shows similar response and indicates a phase transformation at about 115 °C. This behavior of thermal transport properties is explained in terms of variation in mean free path of phonons arising due to structural changes in the material. Differential scanning calorimetry (DSC) was also performed to confirm the above said phase transformation. The results on phase transformation are in good agreement both by TPS and DMA.

Keywords: Aniline-formaldehyde copolymer; transient plane source; effective thermal conductivity; effective thermal diffusivity; modulus and loss tangent
PACS: 12.15.Cj, 82.35.Jk

INTRODUCTION

Conductive polymers have recently attracted a great deal of attention because of their many applications in lightweight batteries, solar cells, electro-chromic switching, variable transmission window, sensors and nonlinear optical materials [1,2]. Within the group of conductive polymers, polyaniline (PANI), its derivatives and copolymers have emerged as the most promising materials for technical applications due to its typical electrical, electrochemical, environmental and thermal stability [3-4]. Although PANI system possesses many unusual and attractive chemical aspects, including dopability by both oxidation and derivatization with a wide variety of substituents yet there doped form did not show solubility or melting, which made them highly unprocessible and gave them poor mechanical properties. To overcome this problem considerable efforts have been devoted in developing new methods to synthesize polyaniline with better mechanical properties, while maintaining its high conductivity.

CP1004, *Thermophysical Properties of Materials and Devices-NCTP '07*,
edited by P. Predeep, S. Prasanth, and A. S. Prasad

One of the methods employed to improve the processability of PANI is the preparation of composites or blends by the use of an insulating polymer as the main matrix. This has been aimed to combine the good mechanical properties of the insulating polymer with the electrical properties of PANI. The methods reported in the literature for the preparation of composites or blends of PANI are the electrochemical polymerization of PANI [5-6]. Electrochemically synthesized polyaniline is difficult to process due to its low solubility while chemically synthesized polyaniline suffers from low molecular weight, which results in low mechanical strength [7]

To improve the mechanical properties of polyaniline a new strategy has been employed by introducing cross-linking of polyaniline in formaldehyde matrix. This allows the synthesis of high molecular weight conductive copolymers. In the present study copolymer of aniline and formaldehyde has been synthesized by chemical oxidative polymerization of aniline-formaldehyde condensate (AFC). The investigation of thermal transport properties like effective thermal conductivity (λ_e) and effective thermal diffusivity (χ_e) is of great importance as it gives valuable information regarding processing conditions and possibility of different applications. The glass transition temperature T_g or softening temperature is important in polymers, since processing temperature and final application temperature are highly dependent on these properties. Also, conductive polymers can't be melt processed hence T_g or softening temperature becomes more important [8-11].

For aniline formaldehyde copolymer thermal transport properties λ_e and χ_e and their temperature dependence in the temperature range (30-140 °C) have been studied using transient plane source (TPS) method. Differential scanning calorimetry (DSC) has also been performed for the same sample in the above temperature range at a heating rate 10 °C /min.

Materials like metals and ceramics show time-invariant deformation at constant temperature and hence any measurement of strain will lead to a constant value of modulus. However materials like polymers exhibit time-dependent deformation therefore dynamic mechanical analyzer (DMA) is used for the measurement of mechanical properties.

MATERIAL PREPARATION

Copolymer of aniline and formaldehyde has been synthesized by oxidizing aniline – formaldehyde condensate (AFC) with the help of $FeCl_3$ in acidic medium at room temperature. AFC has been prepared by allowing reaction of aniline and formaldehyde as described elsewhere [12].

In this particular synthesis, 0.1 mole of aniline was added to 37% formaldehyde solution as per stiochiometry ratio. Condensation was carried for 24 hours until the colour of solution turns to deep red. Acidic solution of $FeCl_3$ (0.1 mole) was added to oxidize the AFC by continuous stirring. Stirred mixture was neutralized with 10 % NaOH solution, resulting in an insoluble precipitate of copolymer of aniline formaldehyde. The precipitate resulting from the reaction was washed with distilled water until the filtrate was free from alkali and dried under dynamic vacuum for 6 hrs.

THERMAL AND MECHANICAL CHARACTERIZATION

The powder so obtained has been used for fabrication of pellets. Twin Pallets (2mm thickness, 12mm diameter) have been prepared from the powder material with a load of 5 ton. The sample holder containing these samples is placed in a furnace that has a sensitivity of 1 °C. In addition, a sensitive thermometer is kept just above the sample pieces inside the furnace to monitor the temperature of the sample. After achieving isothermal condition in the sample, a constant current pulse of width 10 s and height 0.0736Amp is passed through the heating element. The measurements reported in this paper have been performed with a TPS element [13], which is made of 10μm-thick nickel foil with an insulating layer made of 50μm-thick kapton on each side of the metal pattern. The detail theory for TPS measurements is given elsewhere [13]. The analysis of the data was performed as obtained by Gustafsson [14]. The TPS element had a resistance of about 3.26 Ω at room temperature and a temperature coefficient of resistance (TCR) about $4.6*10^{-3}$ °C^{-1}. Because of the change in average temperature of the sensor, the potential across it changes. A digital multimeter is used to record the transient potential difference across the terminals and the current through the TPS sensor is supplied by a digital power supply. The current in the circuit is adjusted according to the nature of the sample material. Multiple readings at appropriate intervals are taken to ensure the accuracy of the results. The TPS method used here is capable of recording the temperature of the sample through the TPS sensor itself.

Mechanical characterizations of the sample have been carried out with Tritec 2000 Dynamic mechanical analyzer (DMA). Single cantilever bending method was applied to investigate storage modulus (E'), loss modulus (E") and mechanical loss factor (tanδ) and the sample has been scanned over a temperature range 30 °C to 130 °C. Frequency of the oscillations has been fixed at 1 Hz and the strain amplitude at 0.005 mm which is well within the linear viscoelastic region. The heating rate is 2 °C/min for temperature dependence and this has been confirmed by time-temperature scan. The sample dimensions for the experiment were 4.60mm by 11mm with a thickness of 2.42 mm.

RESULT AND DISCUSSION

The value of the λ_e and χ_e for aniline-formaldehyde copolymer in the temperature range 30-140 °C have been plotted in Figs. 1-2. From these figures it can be observed that the values of λ_e and χ_e increase from 30-60 °C and after this it decreases sharply between 60-100 °C. Now for further increase of temperature beyond 100 °C the values of λ_e and χ_e continuously increase till 140 °C. However, the increment is very small in the range 110-140 °C.

The initial increasing behaviour of λ_e and χ_e can be explained on the basis of decrease in moisture content in the sample with the increase of temperature. The moisture content in the sample is expected because of oxidation of sample carried out by $FeCl_3$ during the synthesis of copolymer. With the increase of temperature, the sample becomes more and more dry and the intermolecular distance decreases owing to loss of water molecules, which in turn increases the value of λ_e and χ_e.

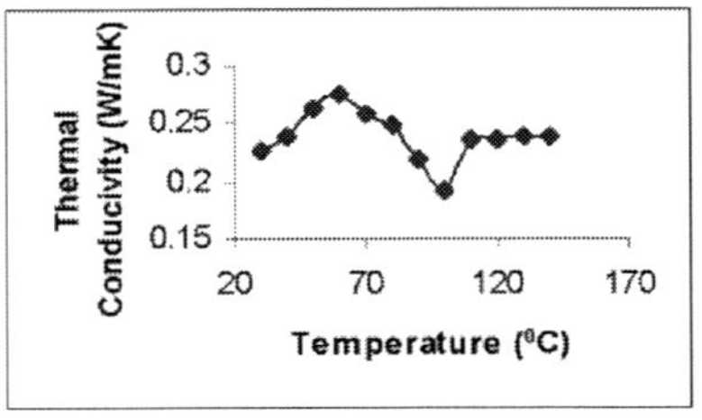

FIGURE 1: Effective thermal conductivity vs temperature.

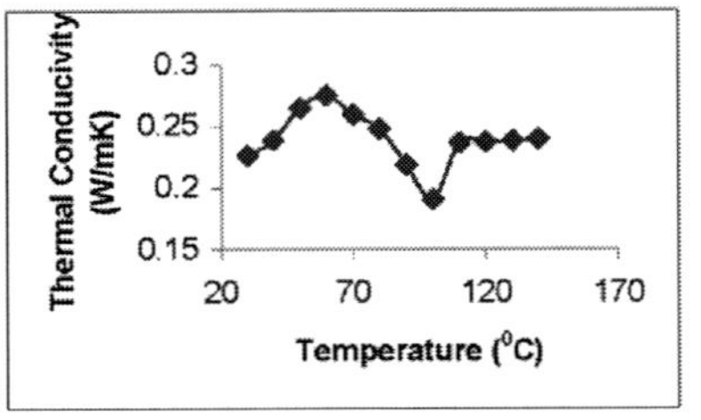

FIGURE 2. Effective thermal diffusivity vs temperature.

A sudden decrement is observed in the value of λ_e and χ_e from 60-100 °C. In this temperature range energy absorbed by the sample is used in bringing some structural change. This structural change in the sample is also experimentally confirmed by Differential Scanning Calorimetry (DSC). A typical DSC thermogram at heating rate of 10 °C/min of aniline-formaldehyde copolymer is shown in fig.3. The characteristic endothermic phenomenon is evident in DSC thermogram in the temperature range as mentioned above. This change in the structure brings disorderness and enhances the phonon scattering in the sample. Therefore, small decrement in phonon mean free path is expected, thereby decreasing the value of λ_e and χ_e. The basic matrix of polyaniline is amorphous in nature [1of polyaniline and formaldehyde shows glass transition in this temperature range and confirms its amorphous nature.

Beyond temperature 110°C, the sample starts attaining a stable ordered configuration reducing the phonon scattering in the sample. Thus a slight increase in phonon mean free path will be there which in turn brings about an increase in λ_e and χ_e. For further increase of temperature there is no change in the value of λ_e and χ_e. This indicates that a permanent change in the structure has occurred and the increase of temperature is not causing any additional change into it.

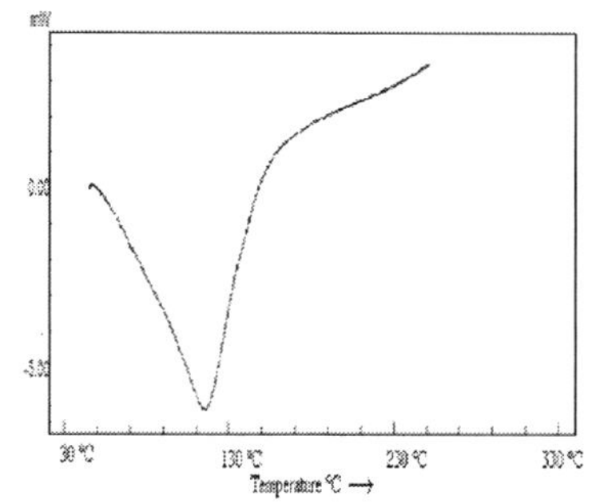

FIGURE 3. DSC thermogram of aniline-formaldehyde copolymer at the heating rate of 10 °C

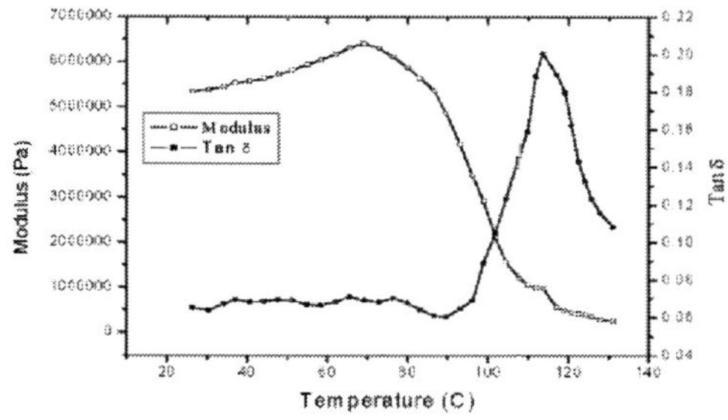

FIGURE 4. Modulus and loss tangent vs temperature.

Figure 4 shows the temperature dependant variation in modulus (E′) and loss tangent (tanδ) over a temperature range 25 °C to 130 °C. As the temperature increases from room temperature to 70 °C the value of modulus increases, which shows the increased stiffness of the sample. This is due to solid-state transition. At the room temperature the molecules are tightly compressed and with rise in temperature the material warms and expands, hence the free volume increases, which results in the movements of localized bonds (bending and stretching) and side chains. This contributes to the increase in the strength of the sample. The observed increment in sample stiffness also confirms the presence of moisture in the sample. Now as the temperature increases the moisture content is decreased and the system starts acquiring compact configuration resulting into more stiffness. Above 70 °C modulus values decreases rapidly with increase in temperature, which indicates the transition of aniline formaldehyde copolymer from amorphous state towards the rubbery state. After 115 °C the rate of decrease of modulus is very slow and suggests a permanent change in the structure, as explained above, and the system attains a stable configuration.

From the temperature scan of tanδ from room temperature to 75 °C it is clear that the value of loss tangent is almost constant indicating the uniform storage capacity of the polymer material. The decreasing values of loss tangent from 75 °C to 90 °C can be attributed to the decrease of moisture content, which results in increase in compactness of the material leading to a higher storage capacity and lower values of tanδ. The values of tanδ increase sharply above 90 °C and a peak is observed at about 115 °C. The peak value in the loss tangent curve corresponds to the phase transition temperature, which is in good agreement with the values observed by DSC and TPS. In the transition zone, the relaxed elastic modulus slumps due to irreversible flow and the set up fails to measure the exact values of tanδ beyond 130 °C.

CONCLUSION

Variation in λ_e and χ_e of the aniline formaldehyde copolymer in the temperature range from room temperature to 140 °C is suggestive of some kind of structural change in the material from 60 °C to 110 °C. A parallel study by dynamic mechanical analyzer confirms the transition of aniline formaldehyde copolymer from amorphous state to rubbery state. The softening temperature or phase transformation region is in good agreement by both the experiments and DSC also confirms this phase transition. The thermal response of the copolymer at higher temperatures shows stable configuration suggesting sintered material is better option then the as prepared copolymer.

ACKNOWLEDGEMENTS

Authors are thankful to Dr. Arun Pratap, MSU Baroda for his kind assistance in carrying out DSC analysis of aniline-formaldehyde copolymer. Besides this, authors are also thankful to Ms. Vinodini Shaktawat and Mr. Vibhav K Saraswat who helped in various ways during the course of this work.

REFERENCES

1. J.R. Fried, *Polymer science and Technology*, New Delhi: Prentica Hall India, (1999).
2. H.S. Nalwa, *Hand book of Organic Conductive Molecules and Polymers*, John Wiley and sons Ltd (1997) pp 2.
3. A.G. MacDiarmid, J.C. Chaing and A.F. Richter, *Synth. Met.* **18,** 285-290 (1987).
4. E.M. Genies, A. Boyle, M. Lapkowski and C. Tsintavis, *Synth. Met.* **36**, 139-182 (1990).
5. E.E.M. SaCak and M. Karakiqla, *Polym. Int.* **39**, 153-159 (1996).
6. F. Lawrence, Rozsnyai and M.S. Wrighton, *Chem. Mater.* **8**, 309-311 (1996).
7. M. Abe, A. Ohtani, Y. Umemoto, S. Akizuki, M. Ezoe, H. Higuchi, K. Nakamoto, A. Okuno and Y. Noda, *J Chem Soc, Chem Commun* **22**, 1736-1738 (1989).
8. K. Traore, W.T.K. Stevenson, B.J. McCormick, R.C. Dorey, S. Wen and D. Meyers, *Synth. Met.* **40**, 137-153 (1991).
9. R.H. Baughman, J.F. Wolf, H. Eckhardt and L.W. Schacklette, *Synth. Met.* **25**, 121-137 (1988).
10. L. Ding, X. Wang and R.V. Gregory, *Synth. Met.* **104**, 73-78 (1999).
11. C. Yang, Z. Fang, J. Lin, W. Lin and H. Zhou, *Thermochim. Acta.* **352**, 159-164 (2000).
12. P. Kuo, B. Wey and W. Cheu, *J. Appl. Polym. Sci.* **48,** 1953-1961 (1993).
13. G.P. Joshi, N.S. Saxena, T.P. Sharma, V. Dixit and S.C.K. Mishra, *J. Phys. Chem. Solids* **64**, 2391-2396 (2003).
14. S.E. Gustafsson, *Rev. Sci. Instrum.* **62**, 797-804 (1991).
15. S.C.K. Mishra, R. Bhattacharya and R.J. Angelucci, *Indian Inst. Sci.* **8**, 563-567 (2001).

Applications of Frequency Modulated Thermal Wave Imaging For Non-destructive Characterization

Ravibabu Mulaveesala[a] and Suneet Tuli[b]

[a]*PDPM-Indian Institute of Information Technology, Design and Manufacturing, IT Bhawan, Jabalpur Engineering College Campus, Gokalpur, Jabalpur-482 011.*

[b]*Centre for Applied Research in Electronics, Indian Institute of Technology Delhi, New Delhi, India-110016*

Abstract: A novel technique, frequency modulated thermal wave imaging (FMTWI) has recently been proposed and demonstrated for non-destructive subsurface defect detection in solids. Its theory and mathematical analysis is provided in this paper illustrating, how the technique simultaneously combines advantages of conventional pulse based thermography as well as lock-in thermography (LT). The specimen is heated for launching thermal waves into the sample, not at a single frequency (Lock-in) or at all frequencies (pulse), but in a desired range of frequencies. While peak power requirement is reduced, phase images obtained retain known advantages. Experimental results are also presented in support.

Keywords: Frequency modulation, Thermal wave, Thermography, Fourier transform.
PACS: 42.65.Ky, 42.79.Nv, 06.30.Ft, 06.20.Fn

INTRODUCTION

In recent years, thermal wave physics has been playing an important role in the area of nondestructive evaluation (NDE) of subsurface defects in metallic, insulating and composite materials. It is achieved by observing, recording and analyzing the thermal response at the material surface to a heat stimulus, and is broadly known as active thermography in contrast to passive thermography where no heat stimulus is applied. Significant work has been done and understanding achieved in the propagation of thermal waves in materials and their reflection from subsurface features [1].

Presently, three different approaches of active thermal nondestructive techniques are predominantly in use: Pulse Thermography (PT), lock in Thermography (LT) and Pulse Phase Thermography (PPT) [1-3]. In pulse thermography, the tested material is warmed or cooled with a short duration energy pulse (example optical, induction heating and ultra sonic heating etc.) and a measurement of the temporal evolution of the surface temperature is performed with an infrared (IR) camera. The surface temperature gradients on the sample help to localize the defects inside the material. However the surface temperature gradients are caused not only because of hidden defects but also affected by local variations of emissivity on the material surface as well as due to non-uniform heating. Lock-in thermography uses periodic sinusoidal thermal excitation in order to derive information of reflected thermal wave phase and magnitude [4] even at considerably low peak powers. The phase angle has the

CP1004, *Thermophysical Properties of Materials and Devices-NCTP '07*,
edited by P. Predeep, S. Prasanth, and A. S. Prasad

advantage of being less sensitive to local variations of illumination or of surface emissivity. Because of its mono-frequency excitation, the depth resolution of the experiment in a one run gets fixed (i.e. fixed 'thermal wavelength'). To detect defects located at various depths in the test sample, repetition of the test at various frequencies becomes a time consuming process. The experimental arrangement for pulse phase thermography [2,3] is similar to that of pulse thermography, but the fundamental idea of processing the captured image sequence during the experiment, is different. In PPT, the extraction of various frequency components in the captured sequence is performed by Fourier transform on each pixel of the thermal image sequence [3]. The phase images thus obtained show all the merits of the phase images obtained with LT (less sensitive to surface in-homogeneities and variations in illumination). Theoretically, the short duration excitation pulse in PPT does launch a large number of frequency components into the test samples, but the higher order components usually do not have sufficient energy to propagate deep into the sample. To overcome some of the traditional limitations of conventional thermal wave imaging techniques (resolution, peak power, depth of penetration), the present work focuses on non-stationary forms of thermal excitation technique: frequency modulated thermal wave imaging [5-8]. A physical insight based on mathematical analysis of the underlying thermal wave process is provided.

THEORY OF THERMAL WAVES

Thermal waves produced in homogeneous semi infinite solids are theoretically examined in this section. The model for studying thermal waves is based on the equation of heat flow due to conduction, which in a one dimensional case (in the absence of heat sources and sinks) is given as follows [9],

$$\frac{\partial^2 T(x,t)}{\partial x^2} = \frac{1}{\alpha}\frac{\partial T(x,t)}{\partial t} \tag{1}$$

where $T(x,t)$ is the temperature, $\alpha = k/\rho c$ is the thermal diffusivity, k is the thermal conductivity, ρ is the density and c is the specific heat, x is space coordinate and t is time.

Thermal conductivity and the thermal diffusivity are assumed independent of the temperature, this approximation being valid when the incident radiation intensity is not strong.

The solution for the $T(x,t)$ is obtained after taking into account the boundary conditions at the irradiated sample surface (x=0) and the opposite face (x=d), with $d \to \infty$ in case of semi-infinite solid. In thermal wave imaging a common technique to generate thermal waves is through absorption of incident intensity modulated light beam at the surface x=0.

Sinusoidal modulated surface heating is achieved by a sinusoidal modulated heat flux Q incident over the sample surface,

$$Q(x=0,t) = Q_0 e^{j2\pi ft} \tag{2}$$

The heat flux causes a similar modulation in the surface temperature T at the sample surface (x=0),

$$T(x=0,t) = T_\infty + T_0\,\mathrm{e}^{2\pi jft} \qquad (3)$$

where T_∞ is obtained from boundary condition at $x = d \rightarrow \infty$

1(a) 1(b)

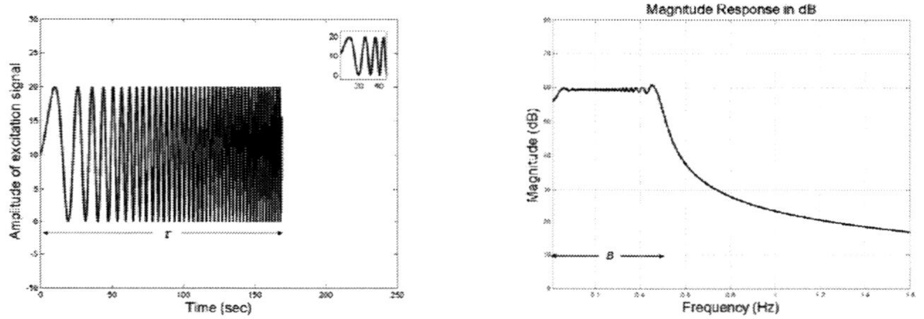

FIGURE 1.(a)Typical frequency modulated intensity profile of the halogen lamps for duration τ, and (b) it's frequency response magnitude with bandwidth B. Inset in (a) shows magnified view of the initial time varying signal.

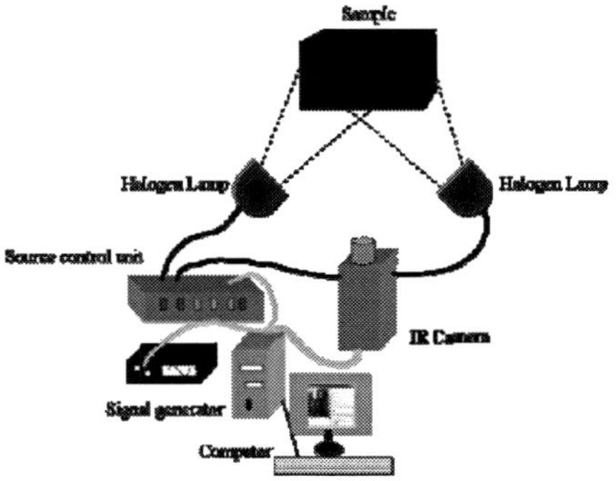

FIGURE 2.Schematic diagram experimental arrangement of the frequency modulated thermal wave imaging.

$$\lim_{x\to\infty} T(x,t) = T_\infty \qquad (4)$$

In this analysis, we neglect T_∞ and consider only the dynamic variations in temperature so that

$$T(x=0,t) = T_0\,\mathrm{e}^{2\pi jft} \qquad (5)$$

The expression for $T(x,t)$ can further be written in terms of real and imaginary parts,

$$T(x,t) = T_0 e^{-x\sqrt{\frac{\pi f}{\alpha}}}[\cos(\kappa) + j\sin(\kappa)] \quad (6); \quad \text{Here} \quad \kappa = 2\pi ft - x\sqrt{\frac{\pi f}{\alpha}} \qquad (7)$$

Eq.(6) shows the resultant temperature distribution caused by the wave like propagation of the heat flow into the bulk of the sample. Damping of the thermal waves as they propagate into the sample depends not only on the thermal properties of the sample but also on the frequency of incident heating. From the above eq.(7), a thermal diffusion length μ can be written as follows

$$\mu = \sqrt{\frac{\alpha}{\pi f}} \tag{8}$$

Knowledge of the thermal properties of the sample and modulation frequency of the incident thermal excitation thus enables the thermal diffusion length to be calculated, which is a measure of the depth probed by the thermal waves within the sample.

THERMAL WAVES IN FMTWI

Unlike the case of sinusoidal intensity modulated surface heating of the sample, the proposed technique in this novel approach utilizes frequency modulated waves instead. Frequency modulated surface heat cycle of duration τ with a bandwidth *B* is obtained by means of linear frequency modulated heat flux (Figure 1(a) & (b)) incident on the sample surface,

$$Q(x=0,t)=Q_0 e^{j2\pi(ft+\frac{Bt^2}{2\tau})} \tag{9}$$

This results in a corresponding frequency modulated surface temperature as shown below

$$T(x=0,t)=T_\infty + T_0\,\mathrm{e}^{2\pi j(ft+\frac{Bt^2}{2\tau})} \tag{10}$$

With

$$T_\infty = \lim_{x\to\infty} T(x,t) \tag{11}$$

Again, neglecting T_∞ and considering only the dynamic variations in temperature as in the previous single frequency sinusoidal excited thermal wave analysis, eq.(10) reduces to

$$T(x=0,t)=T_0\,\mathrm{e}^{2\pi j(ft+\frac{Bt^2}{2\tau})} \tag{12}$$

Next, a complex exponential solution is assumed for *T(x,t)* of the form

$$T(x,t)=X(x)e^{2\pi j(ft+\frac{Bt^2}{2\tau})} \tag{13}$$

The function *X(x)* will be determined so that the differential equation and it's boundary conditions will be satisfied.
Substituting eq.(13) in eq.(1)

$$X''e^{2\pi j(ft+\frac{Bt^2}{2\tau})} = \frac{2\pi j}{\alpha}(f+\frac{Bt}{\tau})Xe^{2\pi j(ft+\frac{Bt^2}{2\tau})}$$

..(14)

The assumed solution is constructed such that the time dependence would cancel out in eq.(14). Thus

$$X'' - \frac{2\pi j}{\alpha}(f+\frac{Bt}{\tau})X = 0 \qquad (15)$$

Since eq.(15) is of second order, it requires two boundary conditions. At x=0,

$$T(0,t) = X(0)e^{2\pi j(ft+\frac{Bt^2}{2\tau})} = T_0 e^{2\pi j(ft+\frac{Bt^2}{2\tau})}$$

..(16)

or

$$X(0) = T_0 \qquad (17)$$

Similarly

$$\lim_{x\to\infty} X(x) = 0 \qquad (18)$$

The solution is now obtained by applying the usual methods for ordinary differential equations with constant coefficients. Thus

$$X(x) = Ae^{x\sqrt{\frac{2\pi j}{\alpha}(f+\frac{Bt}{\tau})}} + Be^{x\sqrt{\frac{2\pi j}{\alpha}(f+\frac{Bt}{\tau})}}$$

..(19)

In eq.(19), $e^{x\sqrt{\frac{2\pi j}{\alpha}(f+\frac{Bt}{\tau})}}$ becomes infinite as x tends to infinity. Consequently B in eq.(19), must be taken to be zero, which therefore reduces to

$$X(x) = Ae^{-x\sqrt{\frac{2\pi j}{\alpha}(f-\frac{Bt}{\tau})}} \qquad (20)$$

The value of A in eq.(20) comes from eq.(17)

$$X(0) = T_0 = A \qquad (21)$$

Therefore the final solutions for the X is

$$X(x) = T_0 e^{-x\sqrt{\frac{2\pi j}{\alpha}(f+\frac{Bt}{\tau})}} \qquad (22)$$

By substituting $\sqrt{j} = \frac{1}{\sqrt{2}} + j\frac{1}{\sqrt{2}}$ in the above equation

X(x) becomes

$$X(x) = T_0 e^{-x\sqrt{\frac{2\pi}{\alpha}(f-\frac{Bt}{\tau})}(\frac{1}{\sqrt{2}}+j\frac{1}{\sqrt{2}})} \qquad (23)$$

$$= T_0 e^{-x\sqrt{\frac{\pi}{\alpha}(f+\frac{Bt}{\tau})}} e^{-jx\sqrt{\frac{\pi}{\alpha}(f+\frac{Bt}{\tau})}} \qquad (24)$$

This may be combined so that eq.(13) now gives the solution for *T(x,t)*

$$T(x,t) = T_0 e^{-x\sqrt{\frac{\pi}{\alpha}(f+\frac{Bt}{\tau})}} e^{-jx\sqrt{\frac{\pi}{\alpha}(f+\frac{Bt}{\tau})}} e^{2\pi j(ft+\frac{Bt^2}{2\tau})} \quad ..(25)$$

This expression for $T(x,t)$ can now be written in terms of real and imaginary parts.

$$T(x,t) = T_0 e^{-x\sqrt{\frac{\pi}{\alpha}(f+\frac{Bt}{\tau})}} e^{j\left(2\pi(ft+\frac{Bt^2}{2\tau})-x\sqrt{\frac{\pi}{\alpha}(f+\frac{Bt}{\tau})}\right)} \quad ..(26)$$

$$= T_0 e^{-x\sqrt{\frac{\pi}{\alpha}(f+\frac{Bt}{\tau})}} [Cos(\kappa) + j\sin(\kappa)] \quad (27)$$

Where

$$\kappa = 2\pi(ft + \frac{Bt^2}{2\tau}) - x\sqrt{(\frac{\pi}{\alpha}(f + \frac{Bt}{\tau})} \quad (28)$$

The thermal diffusion length μ_{fm} in the case of frequency modulated heating can thus be written as follows

$$\mu_{fm} = \sqrt{\frac{\alpha}{\pi(f + \frac{Bt}{\tau})}} \quad (29)$$

Here B/τ is the modulation factor, which controls the slope of frequency variation with time of a linear frequency modulated signal. So from the above equation it is clear that the thermal diffusion length depends on the bandwidth (Figure 1(b)) of the frequency modulated thermal excitation over the sample surface. This helps to scan the entire depth of the sample with time during one frequency modulated cycle. If the modulation factor B/τ is zero then the diffusion length formula becomes as in (eq.8), the special case of single frequency sinusoidal intensity modulated thermography (LT). The thermal wave length (λ) in case of frequency modulated thermal wave imaging is given by

$$\lambda = 2\pi\mu = 2\pi\sqrt{\frac{\alpha}{\pi(f + \frac{Bt}{\tau})}} \quad (30)$$

It is clear from eq.(30) that the wavelength in FMTWI is not fixed as in the case of lock-in thermography, but varies as a function of frequency during the experiment. The wavelength will vary with time leading to variation in the depth resolution for detection of defects at different depth.

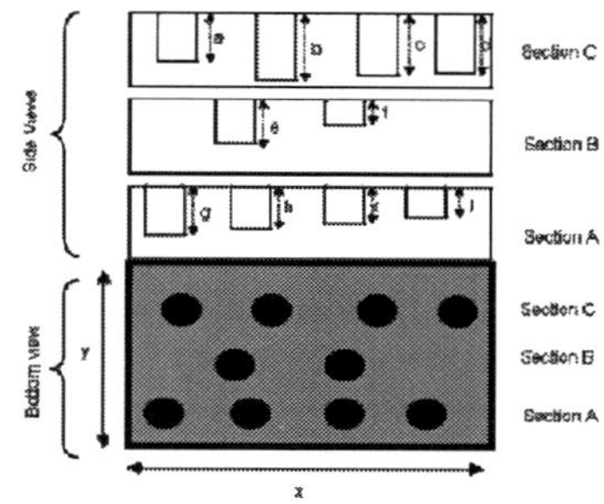

FIGURE 3. Dimensional diagram of the mild steel test sample.(All dimensions are in mm. Sample dimensions: x=210, y=112, and thickness is 11.2. Depth of blind holes (each 20 mm dia): a=9.1, b=11.1, c=10.2, d= 10.1, e=7.0, f=5.4, g=8.3, h=7.1, i=6.3, j=5.4.)

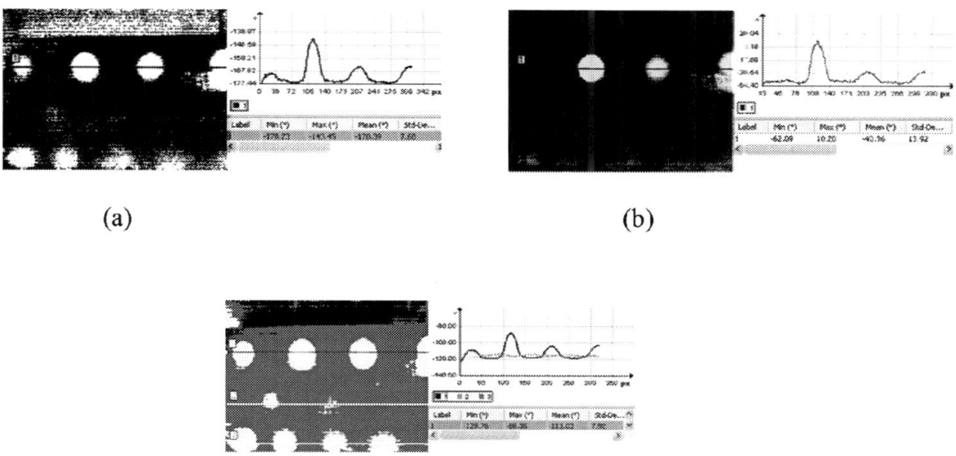

(a) (b)

(c)

FIGURE 4. Phase variations obtained over defects, located at different depths from the top surface at different frequencies. Phase images obtained at (a) 0.05 Hz, (b) 0.112 Hz, and (c) 0.214 Hz.

EXPERIMENTAL VERIFICATION FOR FMTWI

Experiments for FMTWI were carried out on a mild-steel sample using a CEDIP IR system [10] (3-5 μm). A linear frequency modulated signal of 169 seconds duration (τ) with a linear increase in frequency (B) from 0.01 Hz to 0.5 Hz (Figure 1(a) & (b)) is used to excite the halogen lamps (inset in Figure 1(a) shows the initial time varying signal). The intensity variations from the heat sources are assumed as similar to the excitation signal from experimental point of view. A schematic of the experimental arrangement and details of the test sample (mild steel) used are shown in Figure 2 and Figure 3, respectively. The phase and magnitude images of FMTWI were obtained through ALTAIR software [10]. Figure 4(a) to 4(c) shows the phase image of FMTWI at different modulation frequencies ((a) 0.05 Hz, (b) 0.112 Hz, (c) 0.214 Hz). The horizontal lines running through the images are markers that serve to highlight the relative variation of phase over the defects. Measurements were made over only one frequency-modulated cycle for FMTWI. It can be also be seen from Figure 4(a) to 4(c) that FMTWI can scan the entire sample thickness by utilizing thermal waves whose diffusion length changes with frequency (eq.(29)), depending on the modulation bandwidth of the incident frequency modulated surface heating. The measurement times are much shorter for FMTWI when compared to LT (if repeated at frequencies with in the required bandwidth). For a fixed test frequency (LT) the thermal diffusion length (eq.(8)) gets fixed and limits the depth resolution of the test. However in FMTWI the frequency varies with time causing variable depth probing. In this regard it can be compared to Pulse Phase thermography [3], in which even though comparatively much wider range of frequencies are probed, simultaneously in the test sample, the higher order harmonics may not have sufficient energy to cause the thermal wave to propagate deep enough. Further as compared to PT or PPT, considerably less peak power is required from the heat sources.

CONCLUSION

A theoretical basis has been provided for describing the proposed novel technique for frequency modulated thermal wave imaging. Experimental support of the proposed method has been given and the results show the capability of FMTWI for detection of the subsurface defects. Variations in phase images over the defective regions provide relative defect depth estimation.

REFERENCES

1. X.P.V. Maldague, *Theory and Practice of Infrared Thermography for Non- destructive Testing*, (New Jersey, Wiley-Interscience, 2001).
2. J.R. Lhota, S.M. Shepard, B.A. Rubadeux and T. Ahmed, Review of Progress in Quantitative Nondestructive Evaluation, **20A**, 492(2000).
3. X.P.V. Maldague, F. Galmiche and A. Ziadi, Infrared Physics & Technology, **43**, 175(2002).
4. D.Wu, and G. Busse, Revu Générale de Thermique, **37**, 693(1998).
5. R. Mulaveesala, and S. Tuli, Insight, **47**, No.4, 206(2005).
6. R. Mulaveesala, and S. Tuli, Materials Evaluation, **64**, No.10(2005).
7. S. Tuli and R. Mulaveesala, Quantitative Infrared Thermography Proceedings, H.6.1(2004).
8. S. Tuli and R. Mulaveesala, Journal of Quantitative Infrared Thermography, **2**, No.1, 41(2005).
9. Carlslaw HS and Jaeger JC, *Conduction of Heat in Solids*, (London: Oxford University Press, 1959).
10. CEDIP infrared systems, *www.cedip-infrared.com.*

Corresponding author:
[1]ravibabucareiitd@yahoo.co.in

Gamma Radiation Monitoring Through Thin Film of ClAlPc Doped With TiO_2

M. S. Roy, A. K. Gautam, M. Kumar, N. Prasad, Y. Janu, Y.S.Deol, R. K. Mishra, G.R. Choudhary and A. K. Sadh

Radiation processing and Device Group, Defence Laboratory, Jodhpur-342 011, Rajasthan, India

Abstract. Chloroaluminum phthalocyanine (ClAlPc) synthesized by adopting focused microwave synthesis approach was doped with nanocrystalline TiO_2 (5% by weight) and developed into the thin film sandwiched device having ITO/ClAlPc:TiO_2/Ag Schottky configuration by spin coating technique covering $1cm^2$ as an active area. The so fabricated device having initial dark current of the order of 0.2 5m A was exposed to variable dose of gamma radiation ranging from 1cGy to 10 Gy at a dose rate of 1 Gy/hour. The experimental observation reveals the generation of localized traps leading to structural disorder within the solid material. Doping with TiO_2 enhances the surface area of the film which in tern improves sensitivity of device to wider dose rage. Exposure of the device to variable dose of gamma radiation imparts decrease in forward bias current and capacitance characteristics with increase in radiation dose. Also, absorbance characteristics of the Al Pc: TiO_2 was analyzed before & after exposure to radiation which reveals that absorbance decreases with radiation dose leading to decrease in optical band gap.

Keywords: Schottky device, gamma radiation, ideality factor, structural disorder, absorbance
PACS: 29.40.Wk Solid-state detectors

INTRODUCTION

The quest towards the search of suitable sensors for developing radiation dosimeter has been the subject of extensive investigation in today's world scenario.

The p-type metal phthalocyanines have recently been identified as radiation sensing element [1-3] that can be employed for developing sensitive, compact and relatively very cost effective radiation dosimeter for gamma radiation monitoring.

The mechanism of interaction with gamma rays involves the electronic excitation leading to ionization. Exposure to radiation imparts carrier's removal with subsequent increase in structural defects.

In the present communication, chloroaluminum pthalocyanine (ClAlPc) doped with TiO_2, is taken as a semi conducting sensing element. The ITO/ClAlPc:TiO_2 /Ag device structure was fabricated and its optical and I-V characteristics were analyzed before and after exposure to radiation. It has been observed that exposure to gamma radiation imparts an increase in localized state of traps leading to structural disorder with increasing radiation dose.

CP1004, *Thermophysical Properties of Materials and Devices-NCTP '07*,
edited by P. Predeep, S. Prasanth, and A. S. Prasad

EXPERIMENTAL PROCEDURE

Synthesis of Chloroaluminum phthalocyanine (ClAlPc)

The material was synthesized through microwave-assisted synthesis process using focused microwave based SYNTHEWAVE-402 (make Prolabo, France). In the procedure, the fine grounded phthalic anhydride (18 mmol), urea (92 mmol), ammonium chloride (17 mmol), ammonium molybdate (5 mg as catalyst) and aluminum chloride (0.6 mmol) were taken in a Quartz cylindrical reaction vesicle (cap. 250 ml) and then irradiated with microwave (Power 300W) under the given temperature profile. The product obtained within 10 minutes reaction time was washed with hot water and then dried in vacuum oven for one hour. The re-precipitation through concentrated H_2SO_4 leads to the purest form of the material. Further, purification was carried out by soxlet extraction with acetone for 3 hour and then pure M-Pc was obtained approximately in 70% yield. The absorption spectra of M-Pcs, recorded with UV-Vis spectrophotometer reflect two absorption bands as reported for other metal-phthalocyanines. The structure of M-Pc is given in fig. (1).

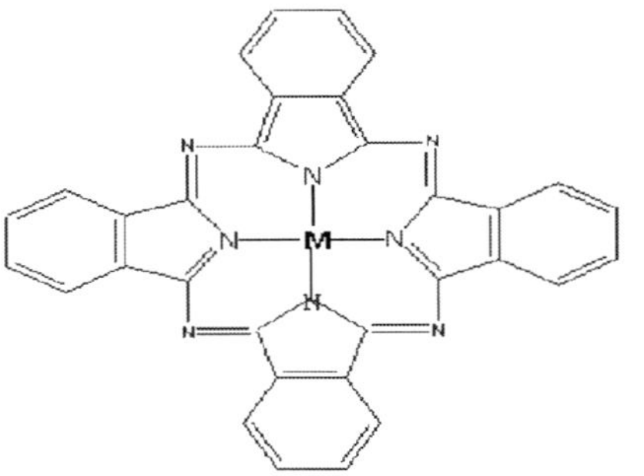

M = Al

FIGURE 1. Structure of Metal-phthalocyanine

Fabrication of thin film device

ClAlPc was blended thoroughly with TiO_2 by grinding in mortar and then device was fabricated by developing the thin-film of Al-Pc employing spin coating technique over indium tin oxide (ITO) coated glass substrate using DMF as solvent. The thickness of the film was 1000 nm. The Silver (Ag) contact was then made over the top of the organic film covering 1cm^2 active area of the device. The structure of fabricated device i.e. ITO/ClAl-Pc/Metal is as follows:

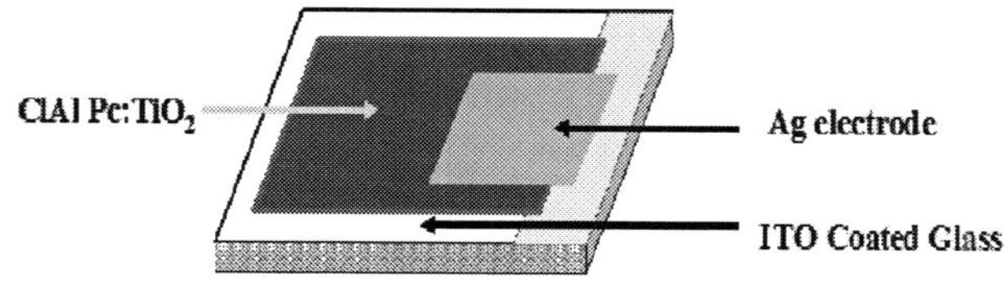

FIGURE 2. Structure of Device ITO/Al $Pc:TiO_2$/Ag

Characterization of the Device

The characterization of the device was carried out by recording J-V characteristics with biasing from 0 to 5 volts using HP semiconductor Analyzer (4145B) before and after exposure to variable doses of gamma radiation. The respective capacitance characteristics were also recorded with an impedance analyzer HP (41492) at frequency range from 1 to 100 Hz before and after exposure to radiation of dose 10 Gy. The calibrated gamma radiation source Co-60 (12.42 Ci) was used for irradiating the device. The dose rate of radiation was 1 Gy/Hr at a distance of 40.18 cm from the calibrated gamma source. The influence over optical properties has also been analyzed by evaluating a change in optical band gap and shifts in absorption peaks.

RESULT AND DISCUSSION

The optical properties of the film play a key role in analyzing the effects of radiation over the device performance. In the present device, it has been investigated that radiation imparts a decrease in degree of absorbance with a slight shifts in absorption peaks. The characteristics B and Q bands are well seen in the spectrum recorded in the form of thin film at 360 nm and 670 nm respectively. Exposure to radiation imparts a shift of B band from 360 to 365 and blue shifts in Q band i.e. from 670 nm to 660 nm, respectively after a cumulative dose of 10 Gy.

Further, the plot in between incident energy hυ hand absorbed energy $\alpha\ (h\upsilon)^2$ reveals that radiation imparts reduction in the extent of absorbed energy from 3.5×10^{-2} eV-cm^{-1} to 2.8×10^{-2} eV-cm^{-}1, though the inter-band transition remains the direct allowed transition.

However, the respective absorption band edge at lower energy side shifts from 695 nm to 705 nm causing a slight reduction in optical band gap from 1.8 to 1.7.

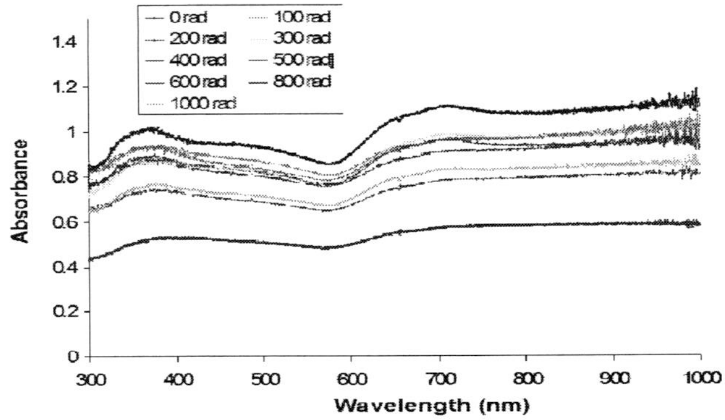

FIGURE 3:. Absorbance characteristics of $ClAlPc:TiO_2$ thin film

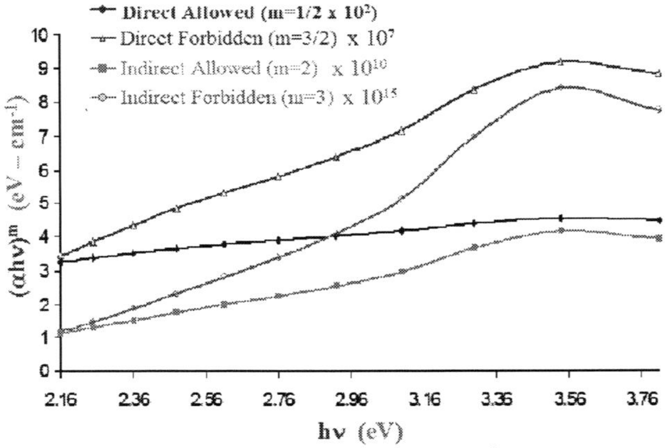

FIGURE 4: Incident energy hυ versus Absorbed energy $(\alpha h\upsilon)^2$ before irradiation with gamma radiation

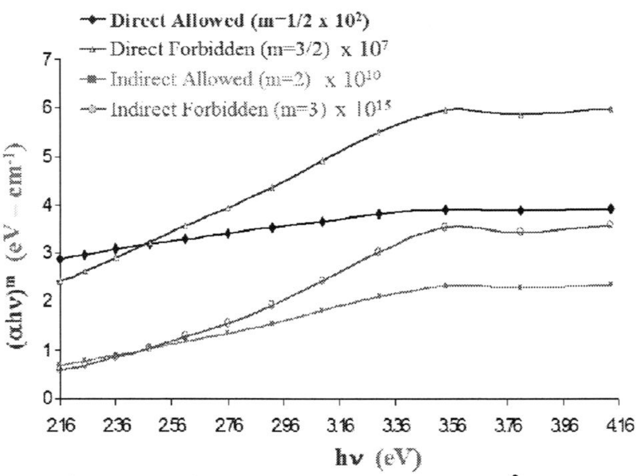

FIGURE 5:. Incident energy hυ versus Absorbed energy $\alpha\ (h\upsilon)^2$ after irradiation with gamma radiation (1000 rads)

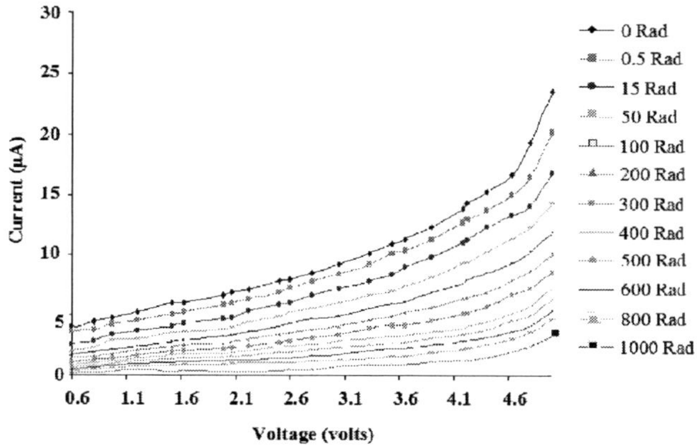

FIGURE 6:. I-V Characteristics of ITO/ClAl Pc: TiO_2/Ag

The J-V characteristics Fig.6 reveal that the rectification ratio of the device is getting decreased with exposure to accelerating doses of gamma radiation. The rectification is seen due to the p-type conductivity of ClAlPc:TiO_2 layer which form Schottky barrier with lower work function electrode Ag (4.2 eV) and ohmic contact with high work function ITO electrode (4.8 eV).

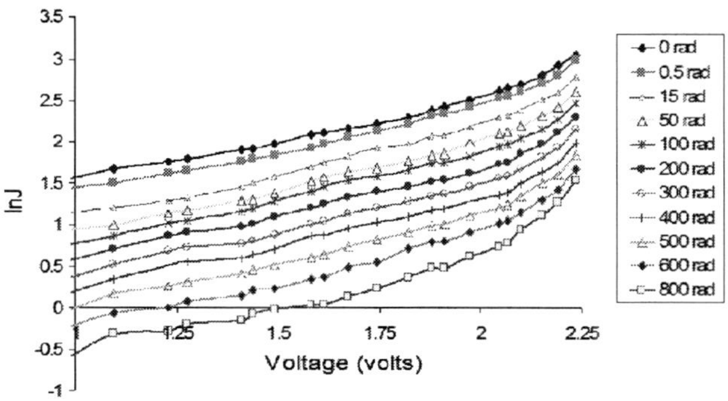

FIGURE 7: ln J vs V characteristics

ln J vs V characteristics shown in Fig.7 reveals two regions i) at < 1V attributed to electrode limited and ii) > 1 V responsible for bulk limited conduction. The slope of the curve provides the value of ideality factor that increases with increase in radiation dose. Also, the slope of this plot yields filed lowering coefficient **β** which is directly related with the relative permittivity ε_r of the material through the relation:

$$\varepsilon_r = q^3/ 4\pi m\ \beta^2$$

where q is electrical charge, m =1 represents pool franckel mechanism and m=4, reveals the Schottky conduction mechanism and **β** is field lowering coefficient. It is clear from this relation that relative permittivity increases with increasing radiation dose and simultaneously the value of **β** is decreases. The value of m in this device corresponds to 4 that mean Schottky conduction mechanism is predominating.
To understand the mechanism in terms of space charge limited conduction (SCLC), J-V characteristics is re-plotted in between lnJ and ln V as shown in Fig.8.

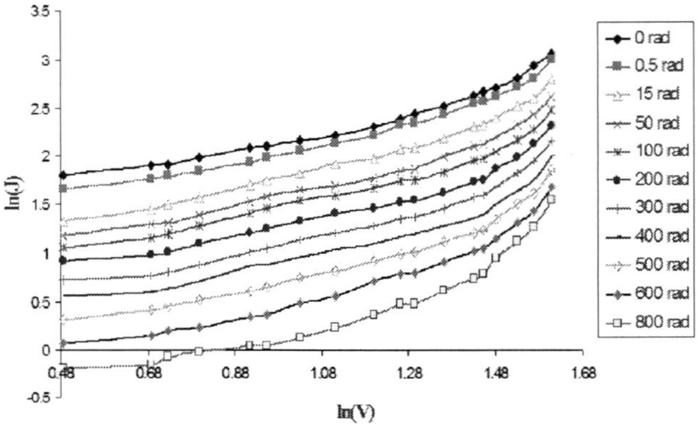

FIGURE 8: ln J versus ln V characteristics

It appears from the plot that before irradiation only two segments is seen i.e. i) below 1.5V and ii) beyond 1.5 V but after irradiation it becomes almost three segments plot i.e. i) below 1.5 ii) above 1.5 to 2.0V and iii) beyond 2.0V.
This observation signifies that how the radiation imparts acceleration of localized traps through a step wise conduction process i.e. i) thermally generated free carriers following Ohmic conduction to ii) smooth shifting towards SCLC by accumulated charges iii) trap free SCLC by shallow level traps and finally iv) trap predominated SCLC. It can be explained by the relation:

$$J = qp_o\mu_p V/d + 9/8\, \varepsilon\mu_p V^2/d^3$$

(Ohmic) (SCLC)

where q is the electrical charge, p_o is density of thermally generated free carriers, μ is the mobility, d is the thickness of semiconductor and ε is the dielectric constant of the material [4-5] .

Once it becomes trap dominated conduction process with increasing radiation dose then the level of traps and its distribution determines the generated disorder in the material. The power dependent relation $J \alpha V^m$ reveals the types of trap distribution, if the value of m = 2 or above then it is considered as exponentially distributed traps and if it is greater than 2 then single level trap distribution is predicted. The value of m can be obtained from the slope of ln J vs ln V plot. In the present device, the value of m is less than 2 and it decreases with increasing radiation. It does mean that, single level traps predominates the SCLC conduction process by accelerating disorder.

To ascertain the application of device as dosimeter, current vs dose relation is plotted and that shows an almost linear relation in between current versus dose relation.

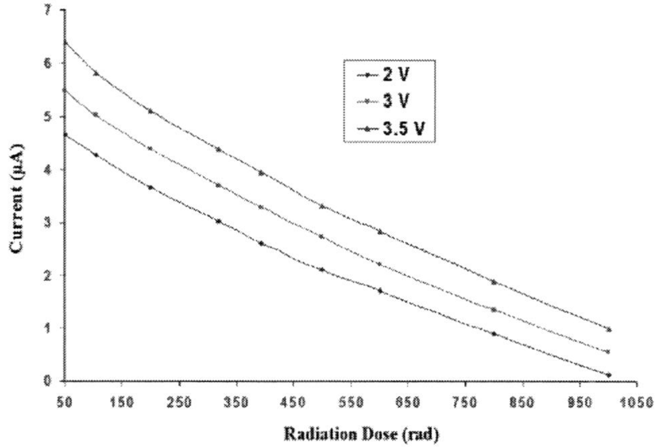

FIGURE 9: Variation of current with absorbed dose at different biasing voltages

CONCLUSION

The device having Schottky configuration was fabricated and tested for its response to variable doses of gamma radiation ranging from 1cGy to 1000 cGy in terms of changes in its optical and electrical characteristics. It has been found that exposure to radiation imparts decrease absorbance with a shifts in optical band gap. Through the detail analysis of J-V characteristics it is found that exposure to radiation causes a decrease in forward bias current due to removal of carriers and accelerated generation of traps.

REFERENCES

1. J. Arsaf, A.Al-Mohamad, *Radiation Measurement*, 2000, pp. 32 153-158
2. M.S. Roy, Manish Kumar, Pratibha Jaiswal and G.D.Sharma, *Radiation Measurement*, 2004, pp. 38 205-209
3. M. S. Roy, Anil Gautam, Manish Kumar, Proceeding of "International Workshop on Physics of Semiconductor Device (IWPSD-2005)" 2005, Vol II pp. 935
4. Z. Chiguvare, J. Parisi and D.Dyakonov, *Journal of Applied Physics*, 2003, Vol. 94, pp. 2440-2448
5. G.D. Sharma, Shailendra Sharma and M.S.Roy, *Materials Science and Engineering B*, 2004, Vol. 110, 135-142

ORAL PRESENTATIONS—THERMAL

Effect of Thermal Annealing on Structure, Optical and Electrical Properties of Hot Wall Deposited $CuInSe_2$ Polycrystalline Thin Films

S.Agilan[a,b*], D.Mangalaraj[b], Sa.K.Narayandass[b], G.Mohan Rao[c]

[a] *Department of physics, Coimbatore Institute of Technology, Coimbatore, India*
[b] *Department of Physics, Bharathiar University, Coimbatore, India*
[c] *Department of Instrumentation, Indian Institute of Science, Bangalore, India.*

Abstract. $CuInSe_2$ chalcopyrite thin films were deposited on glass substrates by hot wall deposition method. As grown films are polycrystalline in nature showing diffraction peaks corresponding to $CuInSe_2$ chalcopyrite. After annealing a narrowing and an increasing of intensity of these diffraction peaks are observed, due to an improvement in the crystallanity and an increase in the grain size in the films. The annealing process of $CuInSe_2$ films in vacuum does not affect considerably [Cu]/[In] ratio but leads to a decrease of Selenium. The optical absorption coefficient is slightly influenced by the annealing temperature. Variation of band gap energy of $CuInSe_2$ thin films with annealing temperature were studied. In the high temperature range (T>385K) the conductivity would be dominated by the transport of holes through the thermal excitation of carriers. In the low temperature range (T<385K) the activation energy is much less due to the possible onset of grain boundary scattering.

Keywords: Hot wall deposition, $CuInSe_2$ thin films, Annealling, Structure, Optical Electrical conductivity.
*Corresponding Author; e-mail: agils70@yahoo.co.in (S.Agilan) Ph: (0)999 490 999 3
PACS: 68.60Dv

INTRODUCTION

$CuInSe_2$ is a promising material for thin film solar cells because of its extraordinary radiation stability[1]. $CuInSe_2$ is a semiconducting compound which belongs to I-III-VI_2 chalcopyrite family. $CuInSe_2$ films possess certain exceptional material characteristics like band gap, absorption coefficient and minority carrier diffusion length which are particularly suitable for photovoltaic applications. Copper indium di selenide thin films have been deposited by several techniques like sputtering [2], molecular beam epitaxy [3] and vacuum evaporation [4] and so on.

In the present study stoichiometric $CuInSe_2$ thin films have been deposited on the well cleaned glass plates, by Hot Wall Deposition technique with the use of optimized quartz tube of length 11cm and diameter 1 cm. Deposited films have been annealed at different temperatures and the effect of annealing on structure, optical and electrical properties have been studied.

CP1004, *Thermophysical Properties of Materials and Devices-NCTP '07*,
edited by P. Predeep, S. Prasanth, and A. S. Prasad

EXPERIMENTAL TECHNIQUES

The synthesized bulk $CuInSe_2$ compound was evaporated and deposited onto chemically-cleaned glass substrates by a hot wall deposition technique. A schematic diagram of the hot wall deposition system is shown in Fig. 1. The substrate was held at a distance of less than 1mm exactly above the open end of the quartz tube acting almost as a lid closing the tube with the help of a substrate holder cum heater. $CuInSe_2$ thin films were deposited onto well-cleaned glass substrates with the source and wall temperatures around 1270 K. Heat treatment experiments have been performed using a one zone furnace and pyrex tubes evacuated to a pressure better than that $3x10^{-5}$ Torr then sealed off.

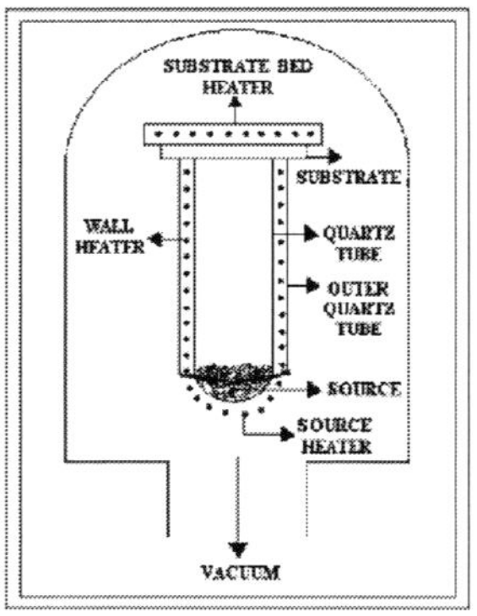

FIGURE 1. Schematic diagram of Hot Wall Deposition system.

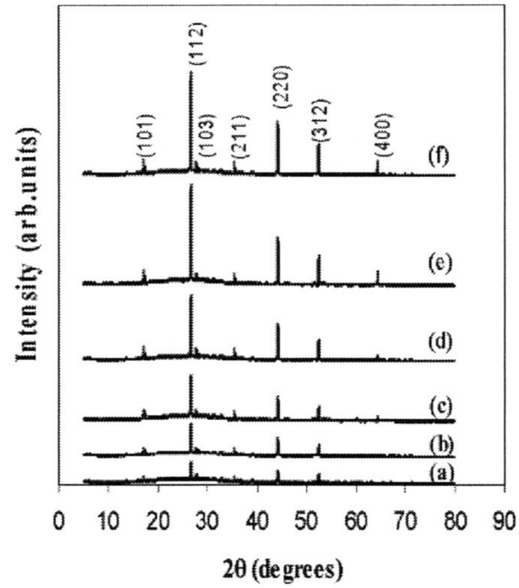

FIGURE 2. The X-ray diffraction pattern of **(a)** as deposited $CuInSe_2$ thin film and annealed films at **(b)** 300°C, **(c)** 350°C, **(d)** 400°C, **(e)** 450°C and **(f)** 500°C.

The samples were annealed for a period of 20minutes at various temperatures viz. 300°C, 350°C, 400°C, 450°C and 500°C. The structure of the films was studied using an X-ray diffractometer with CuK_α radiation (Philips model PW 3710). Optical transmittance spectrum was recorded by UV-VIS-NIR Spectrophotometer Model JASCO V-570. The electrical resistance of $CuInSe_2$ thin films were measured under vacuum of the order of 10^{-3} Torr in the temperature range 300 to 435 K using a Keithley 614 electrometer as an ohm-meter.

RESULT AND DISCUSSION

Effect of annealing on structural properties of hot wall deposited $CuInSe_2$ thin films:

The X-ray diffraction patterns of as deposited and annealed films with thickness of 410nm are given in Fig. 2. From this figure it can be seen that the as grown films are polycrystalline in nature showing diffraction peaks (101),(112),(103),(211), (220, 204) and (312, 116) corresponding to $CuInSe_2$ chalcopyrite. Chalcopyrite phase of the $CuInSe_2$ is identified and is in good agreement with ASTM value of X-ray powder file data [5].

FIGURE 3. SEM micrograph of the $CuInSe_2$ films annealed at 500°C.

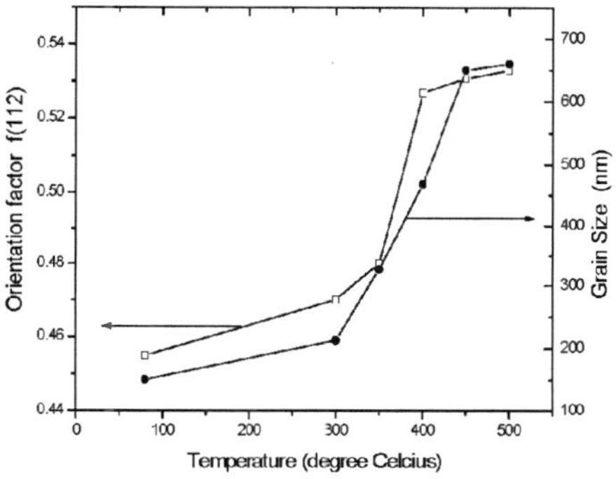

FIGURE 4 The effective grain size and preferred orientation factor f(112) as a function of annealing temperature.

TABLE 1. The [Cu]/[In] ratio and [Se]/[M] ratio before and after annealing.

Composition before annealing - atomic %			Cu/In before annealing	Composition after annealing - atomic %			Cu/In after annealing	[Se]/[M] before annealing	[Se]/[M] after annealing
Cu	In	Se		Cu	In	Se			
21.76	48.35	29.89	0.45	21.85	48.51	29.64	0.45	0.426	0.421
22.34	34.90	42.76	0.64	22.41	34.99	42.60	0.64	0.747	0.742
23.73	30.82	45.45	0.77	24.57	30.13	45.30	0.815	0.833	0.828
25.27	21.97	52.76	1.15	27.63	22.81	49.56	1.211	1.116	0.983
25.97	18.79	55.24	1.382	28.92	20.71	50.37	1.396	1.234	1.015
26.13	18.02	55.85	1.45	30.18	20.65	49.17	1.462	1.265	0.967
26.47	16.85	56.68	1.57	30.27	19.12	50.61	1.583	1.308	1.025

The (112) peak is significantly faster than the two doublets (220, 204) and (312, 116), indicating a relatively preferred structural orientation in the (112) plane. SEM micrograph of $CuInSe_2$ thin films annealed at 500°C is shown in Fig.3. It is visible that grains have grown well during the annealing process.

The magnitude of the preferred orientation factor f for a given plane (peak) relative to other planes (peaks) in a material was estimated according to the method given by earlier researchers[6-8]. The effective grain size and preferred orientation factor f(112) as a function of annealing temperature are shown in Fig.4. An effective grain size up to 660nm and f(112) factor up to 0.533 are observed for the annealed $CuInSe_2$ films.

Besides these, the effect of annealing on composition of a number of as prepared, Cu and In rich films have been studied. Table 1 shows the values of [Cu]/[In] ratio and [Se]/{[Cu]+[In]} ratio before and after annealing at 500°C. The ratio of selenium content[Se] to total concentration of copper and indium {[Cu]+[In]=[M} show a decrease in after annealing films exhibiting [Se]/[M] values greater than 0.828, due un-doubtly to selenium evaporation. Evidence for this conclusion is shown both by means of [Cu]/[In] ratio which remains practically unchanged over the whole composition range investigated and the slight red deposits observed on the inner wall of vacuum sealed tubes. The similar effect on CSVT deposited $CuInSe_2$ films was reported by Zouaoui et al.[9].

Effect of annealing on optical properties of hot wall deposited $CuInSe_2$ thin films:

Transmittance spectra of un-annealed and annealed $CuInSe_2$ thin films at different temperatures are shown in Fig.5. The spectra are characterized by considerable changes in the region 1000-1350 nm. The appearance of a steep edge in the transmission spectra after the thermal treatment indicates improvement of the crystallanity of the layers. The presence of a single slope in the curves in Fig. 6 for

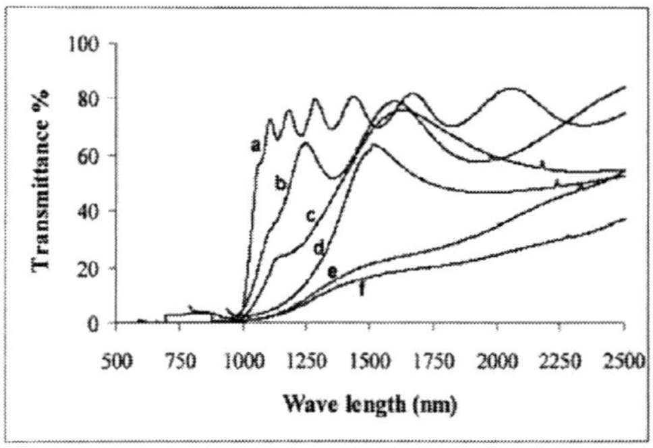

FIGURE 5. Transmittance spectra of **(a)** as deposited $CuInSe_2$ thin film and annealed films at **(b)** 300°C, **(c)** 350°C, **(d)** 400°C, **(e)** 450°C and **(f)** 500°C.

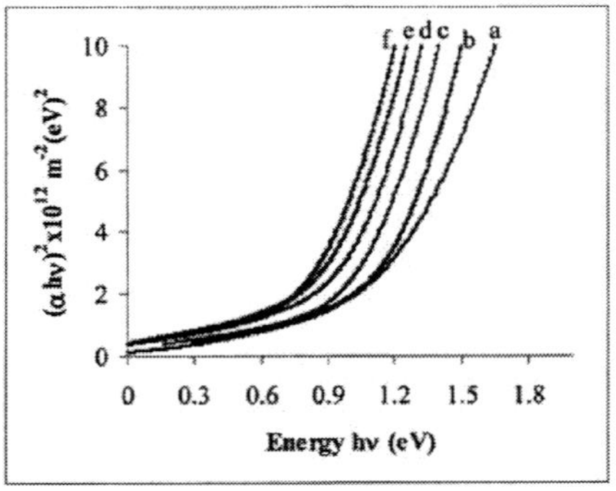

FIGURE 6. Plot of $(\alpha h\nu)^2$ versus hν for **(a)** as deposited $CuInSe_2$ thin film and annealed films at **(b)** 300°C, **(c)** 350°C, **(d)** 400°C, **(e)** 450°C and **(f)** 500°C.

the plot of $(\alpha h\nu)^2$ versus hν of $CuInSe_2$ thin films suggests that the transition type in all the films is direct allowed. The absorption coefficient is slightly influenced by the annealing temperature. Figure 7 shows the variation of band gap energy of $CuInSe_2$ thin films with annealing temperature. A similar trend was observed by Kazmerski et al[10] and it was attributed to various grain boundaries and compositional effects on the light absorbance characteristics of the materials. The band gap values for the annealed samples are in fair agreement with earlier reported values [11,12].

Effect of annealing on electrical properties of hot wall deposited $CuInSe_2$ thin films:

The effect of heat treatment on the electrical conductivity of the as prepared $CuInSe_2$ thin films were studied. Figure 8 shows the plot of log σ versus 1000/T for the as prepared and annealed $CuInSe_2$ thin films of thickness around 760nm. The plots are showing change in the slope near T= 385K. This suggests that two different conduction mechanisms are operative in the film. According to Kundu et al[11], in the high temperature range (T>385K) the activation energy is high and the conductivity

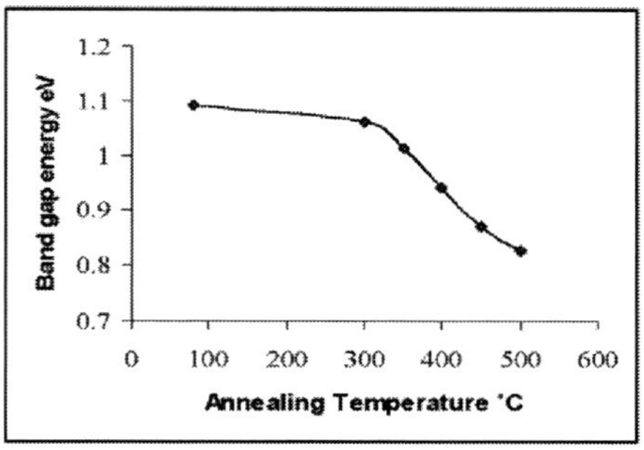

FIGURE 7. Variation of optical band gap energy of $CuInSe_2$ thin films with annealing temperature.

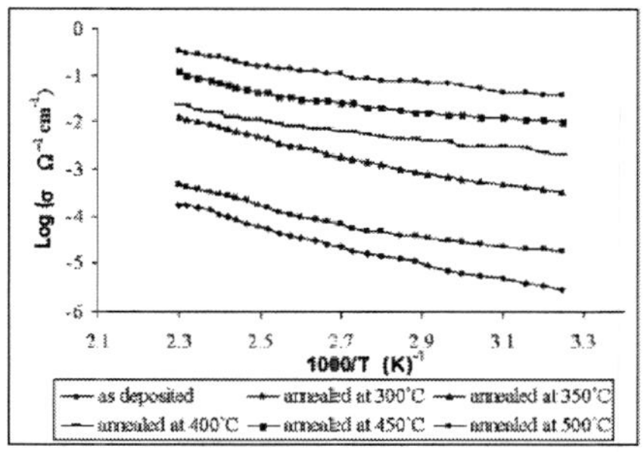

FIGURE 8. Plot of Logσ versus 1000/T for the as deposited and annealed $CuInSe_2$ thin films.

Annealing Temperature °C	Activation energy (eV)	
	Region I	Region II
As deposited	0.26	0.15
300	0.23	0.13
350	0.19	0.12
400	0.16	0.08
450	0.14	0.06
500	0.12	0.04

TABLE 2. Activation energy for conduction of as deposited and annealed $CuInSe_2$ thin films for the two regions of temperature.

would be dominated by the transport of holes through the thermal excitation of carriers from the acceptor levels to the valance band. In the low temperature range (T<385K) the activation energy is much less due to the possible onset of grain boundary scattering.

The activation energy for conduction in two different regions as a function of annealing temperature is given in Table 2. It is observed that the activation energy for conduction decreases as the annealing temperature increases and the conductivity of the film gets increased as the annealing temperature increases [12]. Tuttle et al[13] have stated that selenium interstitials act like deep acceptor levels while copper vacancies behave as shallow acceptor levels; as the temperature is lowered the deeper levels become progressively frozen out. Hence the conductivity is less when the temperature is lowered.

CONCLUSION

The annealing process of $CuInSe_2$ films in vacuum does not affect considerably [Cu]/[In] ratio but leads to a decrease of selenium. The variation of optical band gap energy of $CuInSe_2$ thin films with annealing temperature are studied. The activation energy for conduction decreases as the annealing temperature increases and the conductivity of the film gets increased as the annealing temperature increases.

ACKNOWLEDGEMENT

One of the authors (S.A) expresses his gratitude to "Tamil Nadu State Council for Science and Technology-Chennai" for the award of "Young Scientist Fellowship" to him.

REFERENCES

1. A.V.Postnikov, M.V.Yakushev, Thinsolid films 451-452 (2004) 141
2. J.Diekoszewsk, JJ.Loferski and R.Beaulieu, Solar Energy Materials, 2 (1980) 363
3. F.R.White, A.H.Clark, M.C.Graf, S.P. Grindle and L.L.Kazmerski, J. Vac.Sci. Technol., 16 (1979) 287.
4. R.A.Mickelson, B.J.Stanbery, J.E.Avery, W.S.Chen and W.E.Devancy , in Proc. 19th IEEE Photovoltaic Specialists Conf., 1987, P.744.
5. ASTM X-ray Powder file data 23-209.
6. F.J.Pern, R.Noufi, A.Mason and A.Fraz, Thin Solid Films, 202 (1991) 299.
7. F.K.Lotgering, J.Inorg. Nucl. Chem., 9 (1959) 113.
8. J.U. Knickerbocker and D.A.Payne, Ferroelectrics, 37 (1981) 733.
9. A.Zouaoui, M.Lachab, M.L.Hidalgo,A.Chaffa, C.Llinares and N.Kesri, Thin Solid Films 339 (1999) 10.
10. L.L.Kazmerski, M.Hallerdt, P.J.Ireland, R.A.Mickelsen and W.S.Chen, Journal of Vacuum Science and Technology A, 1 (1983) 395.
11. S.N.Kundu, D.Bhattacharya, S.Chaudhuri and A.K.Pal, Materials Chemistry and Physics 57 (1999) 207.
12. C.Guillen and J.Herrero, Journal of Applied Physics 71(11) (1992) 5479.
13. J.R.Tuttle, D.Albin, J.Goral, C.Kennedy and R.Noufi, Solar Cells 24 (1988) 67.

Inter Linkage of Structure and Thermo-Optic Behavior in Ge-Te-In System

Achamma George, D. Sushama and P. Predeep

Condensed Matter Physics Laboratory, Sree Narayana College, Kollam- 691001, Kerala, India.

* *Corresponding author: ppredeep@gmail.com*

Abstract. Bulk $Ge_{20}Te_{80-x}In_x$(x=0, 6,10 &16) glasses are prepared by the conventional melt quenching technique. The crystallization mechanism was studied using differential scanning calorimetry (DSC) performed at different heating rates [10,15&20°C/min]under non-isothermal conditions. The optical studies are performed on the given glass system, in the wave length range 400-900nm. From the optical absorption data the optical band gap energy (E_g) has been evaluated and studied as a function of the composition. The average heat of atomization and related parameters such as the effective coordination number and number of constraints number of these glasses are also calculated.

Keywords: Chalcogenide glasses, crystallization, heat of atomization, optical band gap.
PACS: 78.55.Qr

INTRODUCTION

Chalcogenide glasses have extensive applications as electronic and optoelectronic device materials [1]. The study of glassy material is currently being strongly driven by the application of calorimetric techniques; through DSC, it is possible to penetrate the glass forming mechanisms, determining kinetic parameters which describe the phenomena of nucleation and subsequent crystalline growth, from amorphous materials.

The present work deals with the differential scanning calorimetric studies on the glass transition and the crystallization behaviour of Ge-Te-In glasses of different compositions under non-isothermal conditions. Besides, the influence of Indium (In) on the optical and physical properties of Ge Te In glass system is investigated. Also the variations of the optical gap with composition and average heat of atomization are discussed. The study of the optical absorption and particularly the absorption edge is a useful method for the investigation of optically induced transitions [2] and for the provision of information about the band structure and energy gap in both crystalline and non-crystalline materials

EXPERIMENTAL

$Ge_{20}Te_{80-x}In_x$(x=0, 6, 10 & 16) glasses are prepared by the conventional melt quenching method. High-purity (99.999%) Ge, Te and In in appropriate atomic

CP1004, *Thermophysical Properties of Materials and Devices-NCTP '07*,
edited by P. Predeep, S. Prasanth, and A. S. Prasad

percentage proportions are weighed in quartz ampoules (10mm diameter). The contents of the ampoule are sealed under a vacuum of 10^{-4} Torr. The ampoules are kept in an electric furnace and slowly heated to1000°C. They are kept at this temperature for nearly 24hours for homogeneous mixing of the constituents. Then the ampoules are quenched in ice water to obtain the glassy samples. The amorphous nature of the samples is confirmed by X-ray diffraction technique. Absorption measurements in the wavelength range 400 - 900 nm are carried out using a double beam spectro photometer [Varian D M S].

RESULT AND DISCUSSION

Differential Scanning Calorimetric measurements

The DSC curves of the $Ge_{20}Te_{80-x}In_x$(x=0, 6, 10 & 16) glasses at a heating rate of 10^0C/min are shown in figure1. From the figure it is clear that the glass transition temperature increases up to 6 at% of indium there after as the indium content increases the value of T_g decreases. The decreasing trend of T_g with increasing In impurity can be explained by considering the structural changes due to the introduction of In atom, since T_g represents the rigidity of the glass network, it can be expected to scale with the average coordination number and the bond energy. The crystallization temperature T_c is also maximum at 6% of In in the system. It is found that in the given system, the presence of cross-link chains also increase the crystallization temperature T_c.

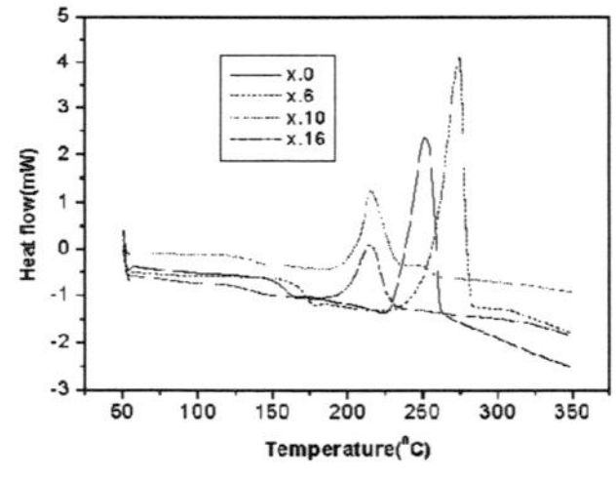

FIGURE 1. DSC traces of $Ge_{20}Te_{80-x}In_x$(x=0, 6, 10 & 16) system at a heating rate of 10^0C/min

The kinetic resistance to crystallization is higher for larger difference between T_c and T_g. The thermal stability of the glass can be described by (T_c-T_g) and T_c [3,4]. Structure representation has demonstrated that there are Te-In and Te-Te bonds but not Ge-Ge bonds, which results from the smaller binding energy of Ge-Ge [5] compared with Ge-Te, Te-In and Te-Te. The addition of a small amount of In (6 at %) to the Ge-Te system leads to the cross-linking of the chains to a small extent, hence increasing the chain length which leads to an increase in theT_g of the system. Further, addition of In leads to an increase in the number of chains which results in a decrease of the T_g of the system [6].

The Effective Coordination Number And Number Of Constraints

The glassy networks are influenced by mechanical constraints (N_{con}) associated with the atomic bonding, and an average effective coordination number $<r_{eff}>$ which is also related to N_{con}. In a covalently bonded glassy network two types of constraints [7], bond-bending N^{β} and bond stretching N^{α} need to be counted. Knowing the average number of constraints $N_{con}=N^{\alpha}+N^{\beta}$ and the average coordination number 'r' for different composition of $Ge_{20}Te_{80-x}In_x$(x=0, 6, 10 & 16) glassy system, the effective average coordination number $<r_{eff}>$ can be calculated [8] using the formula

$$<r_{eff}> = (2/5)(N_{con} + 3) \quad \ldots\ldots (1)$$

The values of N^{α}, N^{β}, N_{con} and $<r_{eff}>$ for the $Ge_{20}Te_{80-x}In_x$(x=0, 6, 10 & 16) glassy system is given in Table1. It can be seen that all parameters exhibit the same trend with increasing In content.

Optical Band Gap (E_g) Determination

The optical band gap (E_g) of amorphous semiconductors is usually determined from the part of the absorption edge, where the density of state variation is significant, i.e., where the following relation holds [10].

$$(\alpha h\nu)^{1/2} = A(h\nu - Eg) \quad \ldots\ldots (2)$$

where α is the absorption coefficient and ν is the frequency. The plots of $(\alpha h\nu)^{1/2}$ versus $h\nu$ are used to evaluate the optical band gaps(E_g) for the indirect transitions. The values of energy gap (E_g) obtained from the graphs (figure 2) are listed in Table 1. It is seen that E_g decreases with increasing In content. This trend indicates the increasing density of localized states within the band gap as suggested [11] by Davis and Mott.

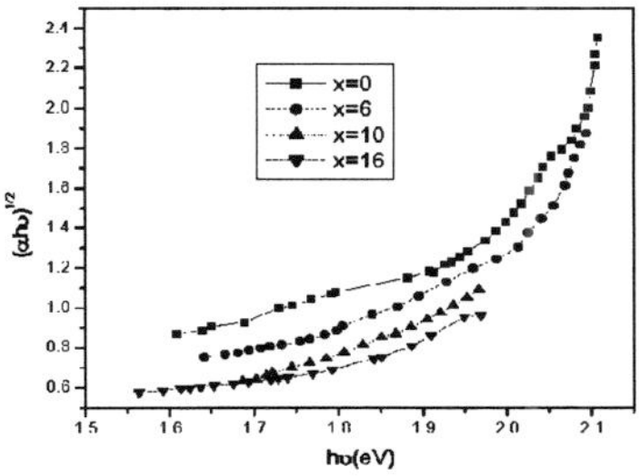

FIGURE2: Relation between $(\alpha h\nu)^{1/2}$and photon Energy $h\nu$ for $Ge_{20}Te_{80-x}In_x$(x=0, 6, 10&16) glass system

The Relationship Between The Average Heats Of Atomization, The Optical Gap And Coordination Number.

The average heat of atomization H_S is defined for the compounds $A_{\alpha}B_{\beta}C_{\gamma}$ as a direct measure of the cohesive energy and that of average bond strength, given by[11].

$$H_s = (\alpha H_s^A + \beta H_s^B + \gamma H_s^C)/(\alpha+\beta+\gamma) \quad \ldots\ldots (3)$$

The values of H_s obtained from equation 3 (the H_s values in units of kcal/gm.atom are 90 for Ge, 46 for Te and 58 for In) are listed in table 1. In order to correlate H_s with E_g in non crystalline solids, it is reasonable to use the average coordination number <r> instead of the isostructure of crystalline semiconductors. The relation between the energy gap and average heat of atomization was discussed by Aigrain and Balkanski [12]. According to their study a linear correlation exists for semiconductors of the diamond and zinc blende structure.

$$E_g = a(Hs-b) \quad \ldots\ldots\ldots\ldots \quad 4$$

where a and b are characteristic constants. By adding In to this glassy system, the average bond strength of the compound decreases, and hence E_g will decrease.

The calculated values of the parameters listed in Table1 for all the different compositions of $Ge_{20}Te_{80-x}In_x$ glassy system reveal that the addition of In leads to a decrease in the optical gap and an increase in the parameters like, average heat of atomization and average coordination number.

TABLE 1: Values of N^α, N^β, N_{con}, $<r_{eff}>$, H_s and E_g for the $Ge_{20}Te_{80-x}In_x$(x=0,6,10 & 16) glass system

$Ge_{20}Te_{80-x}In_x$	N^α	N^β	N_{con}	$<r_{eff}>$	H_s(eV)	E_g(eV)
X= 0	1.2	1.8	3	2.4	2.377	1.844
X= 6	1.26	2.04	3.3	2.52	2.417	1.71
X= 10	1.3	2.2	3.5	2.6	2.429	1.691
X=16	1.36	2.44	3.8	2.72	2.461	1.581

REFERENCES

1. D.Alder,Sci.Amer,**236** (1977) ,36
2. C.A.Hogarth and A.A.Hosseini, J.Mat. Sci 18(1983) 2697.
3. .Mahadevan.S,Giridhar.A and Singh.A.K. J.Non-Cryst.Solids 88 (1986)11.
4. N. B. Mahajan, N.S.Saxena, D.D.Paudyal, M.Husain, Phy.Stat.Sol (a) 178(2000)663.
5. J.Bicerano and S.R.Ovishinsky, J.Non.Cryst.Solids74 (1972) 75.
6. .R.M.Mehra, G.Kaur, P.C.Mathur, J.Mat.Sci.26 (1991)3433-3437.
7. S.A.Fayek,A.F.Maged,M.R.Balboul,Vacuum 53, 447(1999).
8. M.Fadel,Vacuum 48(1), 73(1997).
9. J.Tauc, R. Grigoroviei, A. Vancu, Phys. Status Solidi. 15 (1966) 627
10. E.A.Davis, N. F. Mott, Philos. Mag. 22 (1970) 903.
11. C.Benoit,P.Aigrain and M.Balkanski, "Selected Constants Relative to Semiconductors". (Pergamon Press, New York,1961).
12. L.Tichy and H. Ticha, J. Non-Cryst Solids 189 (1995)141.

Structural, transport, magnetic and thermal properties of $La_{0.7}Ca_{0.24}Sr_{0.06}MnO_3$ single crystal

Ashish Bodhaye[a], Sankar P. Sanyal[a] and Y. M. Mukovskii[b]

[a]*Department of Physics, Barkatullah University Bhopal 462026, India.*
[b] Moscow State Steel and Alloys Institute, Moscow, 117936, Russia

Abstract. $La_{0.7}Ca_{0.24}Sr_{0.06}MnO_3$ single crystals with high quality have been prepared by using floating zone technique. The structural, transport, magnetic and thermal properties of the hole doped $La_{0.7}Ca_{0.24}Sr_{0.06}MnO_3$ single crystal, have been studied using various experimental techniques. The resistivity data shows the metal to insulator transition (MI) occurs at T_{MI} = 281 K along the c-axis and T_{MI} = 270 K along the *ab*-plane. It is observed that the T_{MI} data is higher along the c-axis as compared to that in the *ab*-plane, thus indicating more favourable hoping of electrons is along the c-axis. This MI transition and large magnetoresistance peak 79 % is closely near to the ferromagnetic - paramagnetic transition T_c, which was confirmed from the magnetization data (at ~ 275 K), it was also confirmed from the ac susceptibility data (at ~ 275 K) and the specific heat data (at ~ 270 K), which in term is indicative of the onset of long range order. The entropy change associated with this transition is found to be 2.3 J/mol K. For bolometric applications point of view, the temperature coefficient of resistance (TCR) along ab-plane is found ~ 23 % K^{-1}.

Keywords: Single crystal; Manganites; Colossal Magnetoresistance; Metal-Insulator Transition
PACS: 75.47.Gk, 75.47.Lx, 71.30.+h, 91.25.F-, 65.40Ba

INTRODUCTION

The recent interest in the perovskite-based manganese oxides, $R_{1-x}A_xMnO_3$, where R is a trivalent lanthanides ion and A is a divalent alkaline earth ion, have been studied extensively for their remarkable colossal magnetoresistance (CMR) properties that have technological applications [1-3]. The fundamental mechanism of the CMR phenomenon is generally understood on the basis of the double exchange (DE) theory [4-6]. The CMR behavior occurs near the ferromagnetic (FM) transition temperature and this remarkable phenomenon is attributed to the magnetic coupling between Mn^{+3} and Mn^{+4} ions as well as to the strong electron-phonon coupling arising from Jahn-Teller splitting of Mn 3d levels [7]. In these materials, a lattice distortion, or a tilting of MnO_6 octahedra, depends on the averaged ionic radius of R and A cations. The (R, A) site composition controls an effective one-electron bandwidth W, because with a smaller averaged radius of the (R, A) site the transfer interaction of an e_g-orbital electron becomes smaller in a network of MnO_6 octahedra. In $La_{1-x}Sr_xMnO_3$ (x>0.3), a typical example with a relative large W [8], the e_g orbital carriers tend to be itinerant, and due to the strong ferromagnetic interaction (Hund's-rule coupling J_H) between local t_{2g} and itinerant e_g electrons, the ferromagnetic and metallic state is stabilized [9-10]. The ferromagnetic and metallic state has been well interpreted by the DE

CP1004, *Thermophysical Properties of Materials and Devices-NCTP '07*,
edited by P. Predeep, S. Prasanth, and A. S. Prasad

mechanism. Bulk single crystals are more preferable for the right investigations, but manganite crystals of the nominal composition demonstrate significantly different magnetic and electric characteristics depending on their mosaicity and/or point defect structure. In this paper we report on the structural, transport, magnetic and thermal properties of $La_{0.7}Ca_{0.24}Sr_{0.06}MnO_3$ single crystals, in which the W is increased by substitution of Ca with Sr, while the hole doping level (x=0.3) used to maximize T_C, is kept constant.

EXPERIMENTAL DETAILS

The single crystals of $La_{0.7}Ca_{0.24}Sr_{0.06}MnO_3$ were prepared by using Floating Zone technique [11]. X-ray diffraction data confirmed the quality of the crystal. The composition of the crystals was determined by energy dispersive X-ray analysis (EDAX) using INCA Oxford analyzer attached with scanning electron microscope. The electrical resistance and magnetoresistance (MR) of the crystal were measured as a function of temperature and magnetic field by standard four-probe technique (using a constant current) in a superconducting magnet with the maximum applied field 8 T and temperature range 4 K to 320 K [12]. The field direction was parallel to the current direction. The MR is defined as $\Delta\rho/\rho_0 = [\rho(H)-\rho(H=0)]/\rho(H=0)$. Magnetization was measured using Quantum Design PPMS-7. AC-Susceptibility measurement was performed using homemade automated ac susceptometer [13]. The specific heat measurements are carried out by the semi-adiabatic heat pulse method.

RESULTS AND DISCUSSION

The typical EDAX data on the above single crystal are reported in Fig. 1 (a). The composition obtained from the EDAX spectra were close to the expected calculated composition.

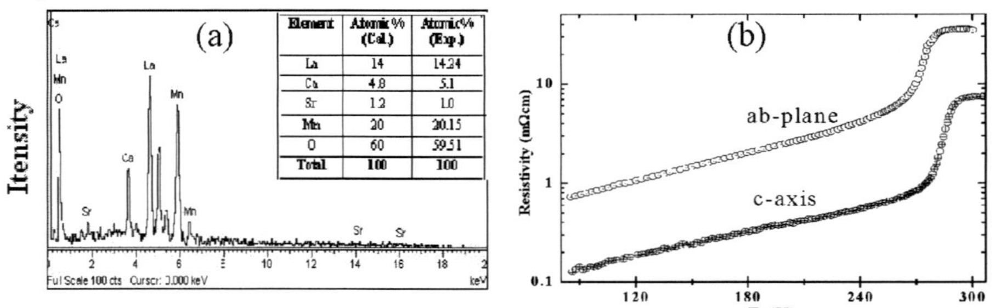

FIGURE 1. (a) EDAX spectrum recorded for $La_{0.7}Ca_{0.24}Sr_{0.06}MnO_3$ single crystal. (b) Resistivity as a Function of Temperature for $La_{0.7}Ca_{0.24}Sr_{0.06}mno_3$ Single Crystal along C-axis and ab-plane.

Typical plot of resistivity versus temperature for $La_{0.7}Ca_{0.24}Sr_{0.06}MnO_3$ single crystal along c-axis and *ab*-plane is shown in Fig.1 (b). Resistivity, along *ab*-plane shows a large increase in resistivity of about five times. Here the resistivity shows a peak at 270 K, which separates the high temperature paramagnetic insulating phase from the low temperature ferromagnetic metallic phase. The resistivity data along c-axis of the same crystal is also plotted in Fig.1 (b). It is noted that the MI transition temperature is higher and its magnitude is less than that of *ab*-plane. This is expected, as the Mn-O-Mn bond along c-axis is linear (i.e.180^0) while the same along *ab*-plane is non-linear. Thus the hoping of electron is favourable along the c-axis and not in the ab-plane.

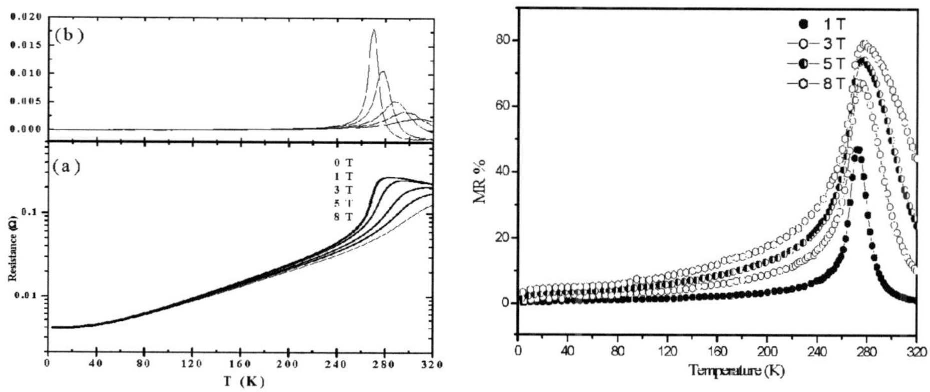

FIGURE 2. (a)Temperature profiles of (dR/dT) for $La_{0.7}Ca_{0.24}Sr_{0.06}MnO_3$ single crystal. (b) Temperature dependence of resistance at different magnetic fields (indicated in figure). (c) The temperature dependence magneto-resistance.

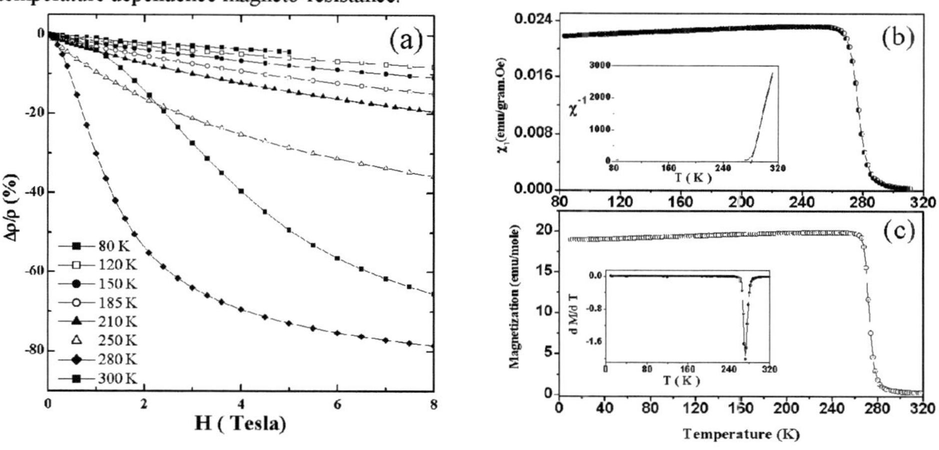

FIGURE 3. (a) $\Delta\rho/\rho_0$ versus H (magnetic field) at a number of temperatures from T<80 to 300 K. (b) Temperature dependence of the magnetization. The inset shows the temperature dependence of the differential magnetization (dM/dT). (c). Real Part of ac-susceptibility as a function of temperature. Inset shows the inverse of ac-susceptibility. The straight line corresponds to the Curie-Weiss behavior.

In Fig. 2(a), we show the temperature-variation of electrical resistance of same single crystal at different magnetic fields up to 8T and Fig. 2(b) shows the dR/dT as a function of temperature. The single crystal at zero field exhibits a sharp metal-insulator (MI) transition at around 281 K. Both the resistance and maximum MR 79 % (see Fig. 2(c)) peak found at T_c. This is precisely the temperature where the resistance and MR are largest. In fact, application of a magnetic field will give the field-induced ferromagnetic ordering and hence reduce the resistance mostly near the transition temperature. T_{MI} is manifested as a peak in the (dR/dT) curve or a decrease in resistance, and increases with the increase in magnetic fields. The temperature dependence of the MR% at various magnetic fields up to 8 Tesla are plotted in Fig. 4. The MR ratio has a sharp peak at a T_{MI} slightly R_{max} and is very small at low temperatures. Furthermore, it is observed that with increasing H the MR ratio increases and the peak moves closer to the T_{MI} R_{max} in zero field. The MR behaviour near T_{MI} in the manganites has been attributed to strong coupling between the conduction electrons and the local magnetic moment through the so-called DE mechanism [4].

The field-dependent resistivity of the $La_{0.7}Ca_{0.24}Sr_{0.06}MnO_3$ single crystal is displayed in Fig. 3(a). Below 80 K the negative MR is extremely small ($\Delta\rho/\rho_0$ = -0.04 % at 80 K in 5 Tesla). The data in Fig. 2(c) show the MR grows as temperature increases, reaching its largest saturation values near T_c. Despite the large MR near T_c in 8 Tesla, the low-field MR is very small ($\Delta\rho/\rho_0$ = -0.32 % at 1 tesla and -0.12 % at 0.5 Tesla).

The temperature versus magnetization data measured at 0.5Tesla of $La_{0.7}Ca_{0.24}Sr_{0.06}MnO_3$ single crystal is plotted in Fig. 3(b). The single crystal exhibits a sharp ferromagnetic transition with a T_C of 275 K. Inset shows that the magnitudes and shapes of the dM/dT curve are varied depending on the sample homogeneity. As can be seen clearly from the dM/dT curves, the single crystal has the sharpest transition. Fig.3(c) shows the temperature dependence of the real part of *ac*-susceptibility along ab-plane measured at an applied field of 3.87 Oe and an exciting frequency of 131.11 Hz. $La_{0.7}Ca_{0.24}Sr_{0.06}MnO_3$ undergoes a transition from paramagnetic to a ferromagnetic state at about T_c = 275 K. Inset shows the inverse of the magnetic susceptibility as a function of temperature. The Curie-Weiss behaviour is followed above T_c but below this temperature the behaviour clearly indicates the presence of short-range ferromagnetic correlations in good agreements with the resistivity and specific heat results.

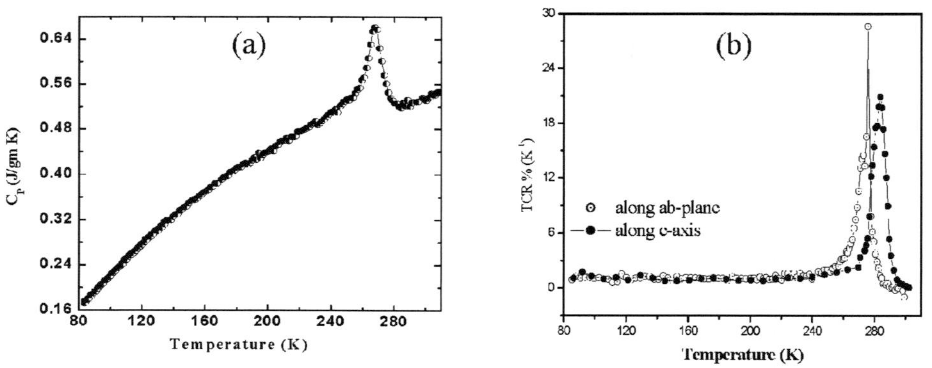

FIGURE 4. (a) Specific heat as a function of temperature. (b) Temperature Dependence of TCR Values for the $La_{0.7}Ca_{0.24}Sr_{0.06}mno_3$ Single Crystal.

The temperature dependence of specific heat is shown in Fig.4 (a). It shows a sharp peak around 268 K, which is close to the transition temperatures obtained from other measurements. The sharp peak in the heat capacity data indicates the onset of long range ordering. To estimate the entropy change associated with the transition we subtracted the background, which is obtained by fitting the measured heat capacity data excluding peak region with a polynomial. The magnetic heat capacity thus obtained showed an entropy change of 2.3 J/mol K.

Fig.4 (b) shows the TCR % as function of temperature. These were obtained by point-by-point differentiation of Fig. 1(a). Interestingly, fairly high TCR values are realized in these single crystals. Along ab-plane TCR is 28.6%K^{-1} and for along c-axis it is 20.9 % K^{-1}. Such large TCR values are a highly desirable goal in the context of the development of high responsively bolometer.

CONCLUSION

We have reported the structural, transport magnetic and thermal properties of single crystal of $La_{0.7}Ca_{0.24}Sr_{0.06}MnO_3$. All these measurement are consistent with each other to reveal the electronic and magnetic phase transition. The obtained crystals present a small mosaicity, a sharp metal-insulator transition at the curie temperature, where the magnetization and the resistance are very sensitive to the applied magnetic field, with the metallic behavior subsisting down to 4 K and a significance CMR of 79 % for H=8T.

ACKNOWLEDGMENTS

Authors are grateful to UGC-DAE-CSR, Indore for providing experimental facilities and partial financial support. The authors are also thankful to DST [SR/S5/NM-12/2005] and UGC [F.No.31-4/2005 (SR)] for financial support.

REFERENCES

1. S. Jin, T. H. Tiefel, M. McCormack, R. A. Fastnacht,and H. Chel, Science **264**, 413 (1994).
2. R. Von Helmolt, J. Wecker, B. Holzapfel, L. Schultz, K. Samwer Phys. Rev. Lett. **71,** 2331 (1993).
3. C. N. R. Rao, A. K. Cheetham, R. Mahesh Chem. Mater. **8,** 2421 (1996).
4. C. Zener, Phys. Rev. **82**, 403 (1951).
5. P. W. Anderson and H. Hasegawa, Phys. Rev. **100,** 675 (1955).
6. P. -G. de Gennes, Phys. Rev. **118,** 141 (1960).
7. S. E. Lofland, S. M. Bhagat, C. Kown, S. D. Tiagi, Y. M. Mukovskii, S. G. Karabashev, A. M. Balbashov, J. Appl. Phys. **81**, 5737 (1997).
8. H. Y. Hwang, S. W. Cheong, P. G. Radaelli, M. Marezio and Y. Tokura, Science **270**, 961 (1995).
9. Y. Tokura, A. Urushibara, Y. Moritomo, T. Arima , A. Asamitsu, G. Kido and N. Furukawa J. Phys. Soc. Jpn. **63**, 3931 (1994).
10. A. Urushibara, Y. Moritomo, T. Arima, A. Asamitsu, G. Kido and Y. Tokura Phys. Rev. B **51,** 14103 (1995).
11. D. Shulyatev, S. Karabashev, A. Arsenov, Ya. Mukovskii and S. Zverkov, J. of Crystal growth 237, 810 (2002).
12. Ashish Bodhaye, Sankar P. Sanyal and Y. M. Mukovskii, Solid State Physics (India), **51,** 695 (2006).
13. A. Bajpai and A. Banerjee, Rev. Sci. Instrum., **68,** 4075 (1997).

The Effect of Heat on Structural Characteristics and Water Absorption Behavior of Agave Fibers

Dip Saikia

Department of physics, Duliajan College, Duliajan-786 602, Assam, India

Abstract. The structural characteristics and water absorptions behavior agave fibers were investigated over a range of temperature by using XRD, IR, TG and gravimetric methods. Three distinct thermal processes were observed during heating the fiber in the temperature range 310 – 760 K in air, oxygen and nitrogen invariably. The cellulose structures of the fibers were unaffected on heating up to 450 K. The samples showed thermal decomposition processes beyond 500 K. Fibers displayed a two-stage diffusion behavior. The structural parameters and kinetic of water absorption of the fibers at specific temperatures were analyzed.

Keywords: Degree of crystallinity, Diffusion coefficient, Thermal stability.
PACS: 81.05.Lg.

INTRODUCTION

In order to use natural fibers as the best alternative materials for glass-fiber reinforced composites in structural applications, their structural characteristics and thermal effect must be well known so that one can predict the properties of the final product [1- 4]. Moisture absorption of natural fiber plastic composites is also a major concern in their outdoor applications [5-7]. So, attempts have been made to study the structural characteristics, thermal stabilities and water absorptions behaviors of agave fibers available in North East India at different temperatures by using XRD, IR, TG and ordinary gravimetric absorption methods.

EXPERIMENTAL

The fibers of agave collected from different localities of North East India were processed by the methods described elsewhere [8]. The plant fibers under study had cellulose content of 70-73 % and lignin content of 5-7 %. Average diameter and density of the fiber were 0.19 mm and 1.52 g/cm^3 respectively. The fibers were scanned at constant temperatures 310, 450 and 530 K in a reflectance mode ranging from 4^0 to 40^0 on a Philips X'Pert PRO automatic powder diffractometer with high temperature attachment by using Cu-K$_\alpha$ X-radiation (λ = 1.54 Å) operated at 40kV and 20mA. Crystallite size and interplanar spacing were determined by using Scherrer equation and Bragg equation

CP1004, *Thermophysical Properties of Materials and Devices-NCTP '07*,
edited by P. Predeep, S. Prasanth, and A. S. Prasad

respectively. Degree of crystallinity was computed by using Harmans and Weidinger equation. IR spectra of the fibers were recorded in air on a Perkin Elmer –1600 FTIR spectrometer in the frequency region 4000—500 cm^{-1}. The TG profiles were recorded at the heating rate 10K / min in air, oxygen and nitrogen in temperature range (310-760 K) on a Mettler-Toledo TGA/SDTA 851. In case of gravimetric moisture absorption method, each sample having weight of 0.1 g were washed thoroughly with distilled water and dried in an oven at 100^0C for 2hr. The dried samples were immersed fully in distilled water kept at different constant temperatures (300,310,320,330 and 340K). The immersed samples were removed at different interval of times and immediately weighted on Mettler AE 200 Analytical Balance (accuracy = ± 0.0001 g). The water content absorbed by the sample at time'tʼ was calculated by using Eq.1. In all the cases, mean ± standard deviation data from three repeated experiments were taken to ensure authenticity of the results. Following the analysis by Carter-Kibler[9] and Gurtin-Yatomi[10], the kinetic of water absorption of the fibers at specific temperatures had been analyzed on the basis of the Eq.2. The diffusion coefficient D for a given temperature was determined by using Eq.3. Applying Arrhenius equation to experimental data activation energy of diffusion E_D can be calculated by using Eq.4. By plotting the log D versus 1/T, E_D can be determined from the slope of the linear fitting.

$$M_t = \left[\frac{M_w - M_d}{M_d} \times 100\right]\% \tag{1}$$

$$M_t = M_m\left[1 - \frac{8}{\pi^2}\sum_{n=o}^{\alpha}\frac{1}{(2n+1)^2}\exp\left\{-\frac{D\cdot t}{h^2}\pi^2(2n+1)^2\right\}\right] \tag{2}$$

$$D = \pi\left(\frac{h}{4M_m}\right)^2 \times \left[\frac{M_t - M_i}{\sqrt{t}-\sqrt{t_i}}\right]^2 \tag{3}$$

$$D = D_0\exp\left(\frac{-E_D}{RT}\right) \tag{4}$$

In Eq.1, M_d is the mass of dried sample (in g) and M_w is the mass of wet sample after time t (in g). In Eq.2, M_m is the maximum water content absorbed at equilibrium, h is the initial diameter and n is the concentration of absorption per unit volume of the fibers. In Eq.3, M_i is the water content absorbed in initial time t_i In Eq.4, D_0 is the pre-exponential factor of the diffusion, R is the molar gas constant and T is the Absolute temperature.

RESULTS AND DISCUSSION

Figure1 (a) shows the X-ray powder diffractograms of agave fiber under different thermal conditions (unheated and heated at 370, 450 and 530 K) and their structural parameters are furnished in Table 1. Although the crystalline structure of fibers remained

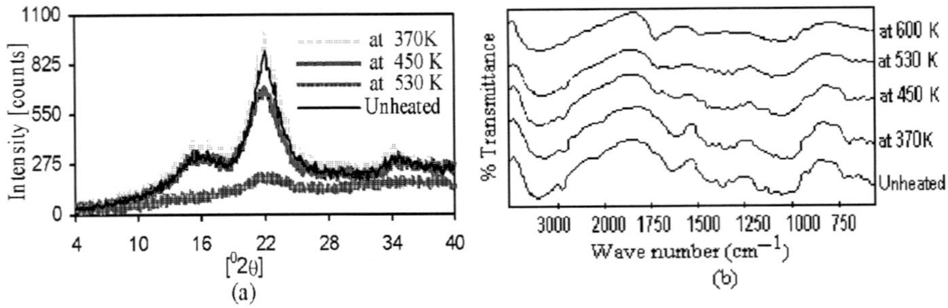

FIGURE 1. (a) XRD scans and (b) IR spectra of agave fibers under unheated and heated conditions

unaffected, slight change in their interplanar spacings and crystallite sizes were noticed at 370K, which might be due to the change in dipole interactions and the effect on weak bend interactions of the crystalline materials at low temperature heating [11]. The significant changes in crystallite size and degree of crystallinity were observed for the sample heated at 450 K, which might be occurred due to gradual degradation of the fiber. XRD of the fiber at 530K showed the transformation of the fiber crystalline structure to amorphous state as a result of their thermal decompositions.

IR spectra of agave fibers (Fig. 1(b)) under ambient and different thermal conditions (at 370, 450, 530 and 600K) are similar to those of cellulose fibers [12]. No significant changes were observed in the IR spectra of the sample heated at 370 and 450 K. The intensities of the peak owing to the OH in the regions 3600—3125 cm^{-1} were gradually decreased on heating and it was attributed to the loss of water molecules from the fibers during heating. A weak peak assigned as C=O stretching was found at 1730 cm ^{1} for the sample heated at 450 K, which might be occurred from degradation of the cellulose and their subsequent opening of a sugar unit to generate carboxyl functional group. The sample heated at 530K showed a rapid declination in the intensity of CH_2 deformation at 1430 cm^{-1} and this result attributed to the occurrence of decomposition processes in fibers. The intensity of the peaks occurred about 2940 cm^{-1} due to CH_2 asymmetric stretching was also found decreasing and there was a declination in the ring vibration at 1164 cm^{-1} on heating the sample at 530K. This declination of C—O bond might be attributed due to the cleavage of bridging C—O—C bond in the cellulose, followed by loss of crystallinlty of the fibers [13]. This was well agreed with our XRD study of the

TABLE 1. Interplanar Spacing, Crystallite Size and Degree of Crystallinity of agave fibers

Sample	Interplanar Spacing (in Å)			Crystallite Size (in Å)			Degree of Crystallinity (in %)
	Peak - I	Peak -II	Peak-III	Peak -I	Peak -II	Peak -III	
Unheated	5.427	4.015	2.583	32.46	31.56	33.46	51.23
At 370K	5.426	4.014	2.575	31.03	32.06	33.42	52.45
At 450K	5.424	4.007	2.560	27.68	28.37	28.72	40.91
At 530K	Unresolved low intensity peak						

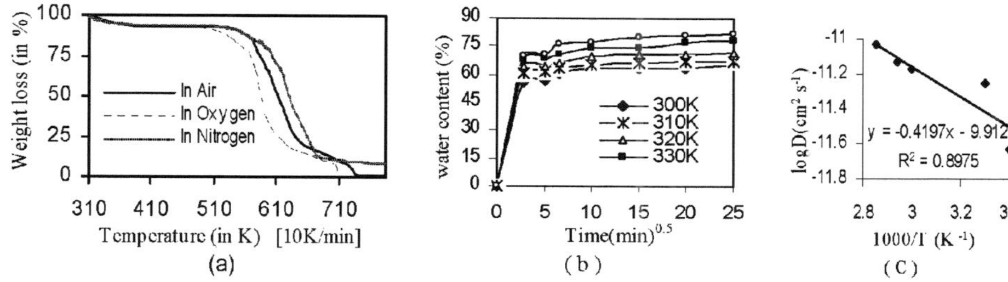

FIGURE 2. (a) TG profiles of agave fibers in air, oxygen and nitrogen; (b) The water content absorbed by agave fibres as a function of square root of time at different temperatures; (c) Liner fitting of logarithm of diffusion–coefficient versus reciprocal of absolute temperature

sample at 530 K. The C=O stretching absorption at 1730 cm^{-1} was increased significantly for the sample on heating at 530 K. This again supports earlier discussion on the formation of C=O functional group through thermal decomposition. Similar trend was also observed in case of IR of the sample heated at 600K.

TG of the agave (Fig. 2(a)) showed the occurrences of three distinct thermal processes in air, oxygen and nitrogen invariably in the temperature range 310 - 760K. The weight loss occurred during the first step of kinetic reaction may be attributed due to the dehydration process [14]. The fibers showed thermal stability up to approximately 500K. During second step, the difference of weight loss found in TG profiles might be attributed due to oxygenation of the fibers in presence of air and oxygen. In nitrogen such oxygenation reaction could not be taken place. In case of nitrogen there was more weight loss than that of oxygen. The XRD and IR data at this range of temperature confirmed the decomposition of the fibers .The third step of kinetic reaction is occurred only when the TG are performed in oxygen and air. Such kinetic reaction did not occur in inert atmosphere. The changes observed during this third stage of thermal decomposition were attributed due to the formation of carbon dioxide and carbon monoxide gas from the carbonized material formed during the third segment [15].

The general shapes of the water absorption curves for agave (Fig.2 (b)) are similar to those of cellulose and other natural fibers [16]. The corresponding results exhibited two stages absorption behavior of the fibers. The first stage occurred very rapidly having linear uptake to above 60% of sorption, which might be due to the capillary action of the fibers following Fick's law of diffusion [17]. The second stage of absorption began very

TABLE 2. Maximum water content absorbed (M_m), Diffusion Coefficient (D), Pre-exponential factor (D_0) and Activation energy of diffusion (E_D)of agave fiber at different temperatures (T)

T (K)	M_m (%)	D (cm^2 s^{-1}) x 10^{-5}	D_0 (cm^2/s)	E_D (kJ/mol)
300	62	0.52		
310	64	1.11		
320	67	1.28	0.001	11.781
330	76	1.32		
340	80	1.36		

slowly and proceeded up to a completely saturation. These non linear uptake behavior exhibited a non- Fickian diffusion ($0.031<n<0.249$) [18]. The diffusion coefficient was found to be increased with temperature. Arrehenius plots obtained from the dependence of the logarithm of diffusion–coefficient (D) on the reciprocal of constant absolute temperature (T) of water absorptions are shown in Fig.2(c). The corresponding calculated activation energy of diffusion is displayed in Table 2.

CONCLUSION

The cellulose structures of the agave fibers were unaffected on heating up to temperature 450 K. Fibers showed thermal stability up to temperature approximately 500 K. When heated from 500 to 760 K fibers showed two different closely related thermal decompositions which leaded to incomplete carbonization. The original structural features of the cellulose backbone were not completely lost even after heating at temperature 600 K. Fibers exhibited two stage absorption behaviour. The first stage was occurred very rapidly and that obeyed Fick's law of diffusion. The second stage of absorption represented a non- Fickian diffusion. Water diffusion in the fibers was a thermally activated process.

REFERENCES

1. E. A. Turi, M. Jaffe, "Fibres" in *Thermal characterization of polymeric materials*, second edition, Academic Press Ins., London, Chap.7. 709pp, 1981
2. I. Maries, B. Abderrahim, L. Umadevi, I. Laurent, C. Yves and T. Sabu, *Compos. Sci.Tech,*, 66,15, 2719-2725, 2006
3. J. B. Naik, S. Mishra, *Polym-Plastics Technol & Eng*, 44, 4, 687-693, 2005
4. D. .J. Vaughan, Fiberglass reinforcement (Hand book of Composites), 2nd edition, Chamman & Hall, London, pp131-154, 1998
5. O. L. S. Alsina, L. H. de Carvalho, F. G. Ramos Filho and J. R. M. d'Almeida, *Polym-Plastics Technol & Eng*, 46,5,515-520, 2007
6. I. C. Watt and M. Kabir, Tex Research J., Vol. 45, No. 1, 42-48,1975
7. C. Shen and G. S. Springer, "Moisture absorption and desorption of composite materials" in *Environment effects on composite materials*, Technomic publishing co., Lancaster, PA, 1981
8. D. Saikia, "Studies on Thermo-Physical Properties of Some Textile Fibres (Plant) Available in the North East India", Ph.D. thesis Gauhati University, Guwahati, India, pp31-42, 2003
9. H. G. Carter and K.G. Kibler, *J. composite Mater*, 12,118-131, 1978
10. M. E. Gurtin and C. Yatomi, *J. composite Mater*, 13,126-130,1979
11. L. Howarld, Needles and W. Cynthia, *J Appl Polym Sci*, 47, 249-262, 1991
12. H. G.Higgins, *J Polym Sci*, 28(118), 645-648, 1953
13. D. E. Goldberg, Fundamental of chemistry, WCB/McGraw Hill, New York, pp427, 1998
14. M. N. Bora. D. Saikia, R. Saikia and C.L. Talukdar, *High Temp- High Pres,* Vol. 29, 683-688, 1997
15. M. N. Bora, G. C. Baruah and C.L. Talukdar, *Thrmochimica Acta*, 218, 425-434, 1993
16. D. Saikia and M. N. Bora, *Indian J Pure & Appl Phys*, New Delhi, Vol. 41, No. 6, 484-487, 2003
17. L. L. March, R. Lasky, D.P. Seraphim and G.S. Springer, IBM J. Res. Develop, 28, 6, 655-661, 1984
18. J. Fan, X. Cheng and Y. S. Chen, Experimental Thermal and Fluid Sc., 27,723-729, 2002

Effect of Post-deposition Annealing on the Physical Properties of DC Magnetron Sputtered Molybdenum Oxide Films

V. Nirupama, M.Chandra Sekhar, P.Sreedhara Reddy, O.M. Hussain and S. Uthanna

Department of Physics, Sri Venkateswara University, Tirupati - 517502, India

Abstract: Molybdenum oxide films were deposited on glass and single crystal silicon substrates held at room temperature by sputtering of molybdenum target in an oxygen partial pressure of $2x10^{-4}$ mbar. The MoO_3 films were annealed in oxygen atmosphere at various temperatures in the range 473 - 673 K. The effect of annealing temperature on the structure, chemical binding configuration, and optical properties of the MoO_3 films was studied. The X-ray diffraction studies revealed that the as-deposited MoO_3 films were amorphous. It crystallizes into a layered orthorhombic phase by the post-deposition annealing at a temperature of 673 K. The Fourier transform infrared spectrum of the MoO_3 films annealed at 473 K exhibits the absorption bands at 560, 840 and 995 cm^{-1} related to the active stretching modes of MoO_6 octahedra. When the films annealed at 673 K the intensity of the absorption band observed at 560 cm^{-1} diminished and the band of 995 cm^{-1} split into two bands of 995 and 1010 cm^{-1} which are related to the zig - zag rows of layered structure of MoO_3. The optical band gap of the as-deposited and annealed films was 3.07 and 3.20 eV respectively. The refractive index increased with increase of annealing temperature.

Keywords: Molybdenum oxide, Dc magnetron sputtering, Structure, Optical properties
PACS: 74.25 Gz, 81.15 Cd, 81.40 Ef and 81.70 Jb.

INTRODUCTION

Molybdenum oxide (MoO_3) has an upsurge of interest in gas sensors, electrochromic and thermochromic devices [1-3]. Several studies have been carried out to investigate the structural and optical properties of molybdenum oxide films. However annealing the films at different temperatures can change the colour of the film and state of cystallinity. Various deposition methods such as electron beam evaporation [1], thermal evaporation [2], dc reactive magnetron sputtering [3,4]and spray pyrolysis [5] have been employed for preparation of MoO_3 films. Among these deposition methods, sputter deposition is one of the best methods to form uniform film over large area substrates. The influence of oxygen partial pressure on the structural and optical properties of dc magnetron sputtered MoO_3 films was reported earlier [6]. However, the as-deposited films appear light blue with low optical transmittance in visible range. The aim of the present work is to investigate the effect of post-deposition annealing on the structural and optical properties of MoO_3 films formed by dc reactive magnetron sputtering.

CP1004, *Thermophysical Properties of Materials and Devices-NCTP '07*,
edited by P. Predeep, S. Prasanth, and A. S. Prasad

EXPERIMENTATION

Thin films of molybdenum oxide were deposited on glass and single crystal silicon substrates held at room temperature (303 K) by dc reactive magnetron sputtering of metallic molybdenum target in mixture of oxygen and argon atmosphere. The vacuum in the sputtering chamber achieved by diffusion and rotary pump combination was $5x10^{-6}$ mbar. In the present study, the oxygen partial pressure maintained was $2x10^{-4}$ mbar and the sputtering pressure of $5x10^{-2}$ mbar. The flow rates of both argon and oxygen gases were individually controlled by Tylan mass flow controllers. After deposition, the films were annealed in oxygen atmosphere at different temperatures in the range 473 - 673 K for an hour. The deposited films were characterized by studying the crystallographic structure, chemical binding configuration and optical properties. The structural studies of the films were carried out using Siefert X-ray diffractometer (model 3003 TT) with Cu k_α radiation source in the 2θ range 10 - 50°. Nicolet (Model 5700 FTIR) Fourier transform infrared spectrophotometer was employed for recording the optical transmittance of MoO_3 films deposited on crystalline silicon substrates. The optical transmittance of as-deposited and annealed films was recorded using Hitachi (U-3400) UV-Vis-NIR double beam spectrophotometer in the wavelength range 300 - 1200 nm.

RESULTS AND DISCUSSION

The X-ray diffraction profiles of as-deposited and annealed MoO_3 films are shown in Fig.1. The as-deposited film shows amorphous nature as no diffraction peak has been observed. The physical colour of as-deposited film was light blue. This colour is related to the presence of oxygen vacancies. In the case of the films formed at room temperature the mobility of adatoms on the substrate surface is low and the probability

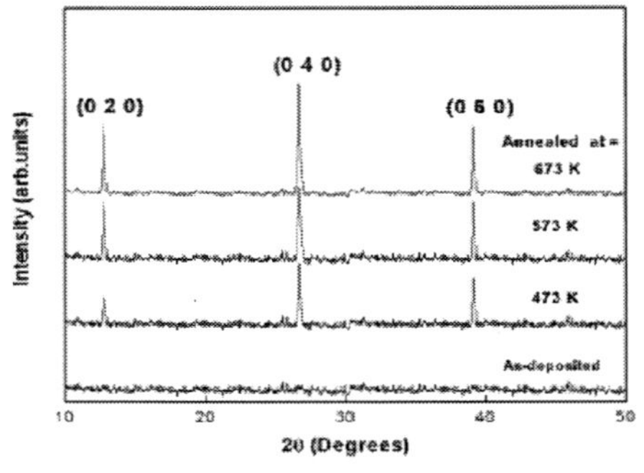

FIGURE 1. XRD profiles of as-deposited and annealed MoO_3 films

of interaction between the molybdenum and oxygen will be low hence of amorphous. For improving the film structure and optical properties, post-deposition annealing was carried out in the temperature range of 473 - 673 K. The colour of the films changes from light blue to transparent state by annealing the films in oxygen atmosphere at higher temperatures due to oxidation. The films annealed at higher temperatures exhibited the diffraction peaks indicated the polycrystalline nature. The X-ray diffraction spectra of the annealed MoO_3 films exhibits (020), (040) and (060) orientations representing the orthorhombic α- phase. The evaluated lattice parameter from XRD profile of the film annealed at a temperature of 673 K was a = 0.396 nm, b = 1.386 nm and c = 0.369 nm which are in good agreement with the JCPDS data of orthorhombic MoO_3 phase. The amorphous films have many defects in structure and leads to non-stoichiometry. On annealing, the oxygen reacts with the unoxidised molybdenum then leads to nearly stoichiometric MoO_3 films. Apart from this, adatom mobility is also increased with annealed temperature hence the films were polycrystalline with improved crystallinity. Mohamed and Venkataraj [3] also observed the similar behavior in as-deposited dc magnetron sputtered MoO_3 films formed at an oxygen partial pressure of $4x10^{-3}$ mbar were amorphous whereas those annealed at 623 K were polycrystalline and at higher temperatures the intensity of the β-phase decreased and completely transformed to α- MoO_3. Sain et al. [7] deposited amorphous MoO_3 films at room temperature by thermal evaporation technique were found to crystallize into a stable orthorhombic phase on annealing in air at 653 K. The crystallite size of films calculated from the X-ray diffraction peaks using the Scherrer's relation increased from 25 to 40 nm with the increase of annealing temperature from 473 to 673 K.

The films deposited on silicon substrates were used for FTIR measurements. The IR transmittance spectra of MoO_3 films annealed at different temperatures are shown in Fig. 2. The infrared spectra of films annealed at 473 K showed the absorption bands at

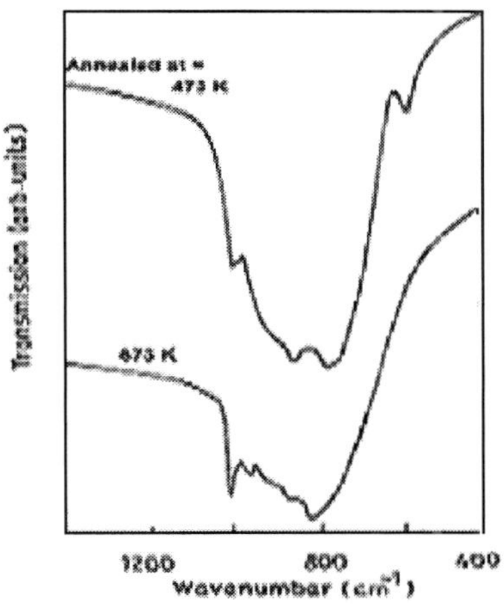

FIGURE 2. FTIR spectra of MoO_3 films annealed at different temperatures

560, 810, 840 and 995 cm^{-1}. This spectrum can be analyzed on the basis of the fact that the structure of MoO_3 formed from MoO_6 octahedra for which the stretching and bending vibrational modes occur in the wavenumber range 800 – 1000 cm^{-1} that corresponds to the stretching modes of O-Mo terminates, which is in correlation with the normal mode calculations from MoO_6 octahedra [8]. The absorption peaks observed were related to the infrared active stretching modes due to alternating bond lengths in MoO_6 octahedra. The shoulder observed at 995 cm^{-1} corresponds to the vibrations along the zig - zag rows. These results are in good agreement with the reported α-orthorhombic MoO_3 crystals [9]. When the films annealed at 673 K the intensity of the absorption band observed at 560 cm^{-1} diminished and the band at 995 cm^{-1} split into two bands of 995 and 1010 cm^{-1} which are related to the zig-zag rows of layered structure of MoO_3. It revealed that the films annealed at 673 K leads to the growth of layered structure of α- MoO_3.

Fig.3 shows the optical transmittance spectra of as-deposited and annealed MoO_3 films. The sub-stoichiometry MoO_3 films contain a number of oxygen vacancies, which are positively charged structural defects capable of capturing one or two electrons. Electrons may be promoted into these positively charged defects. The oxygen vacancies occupied by these electrons acts as donor centers, which is responsible for broad band absorption at around 800 nm. It clearly shows that the annealing process enhances the transmittance of the films. It is seen from the figure that the optical transmittance increased with the increase of annealing temperature. This increase can be attributed to the oxidation of the unreacted metallic Mo atoms in the films [3].

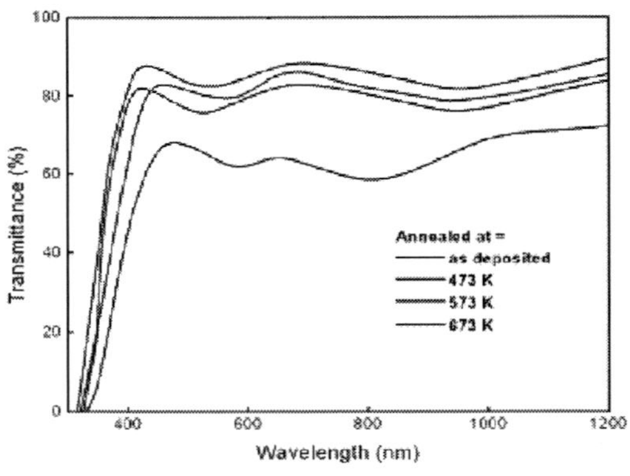

FIGURE 3. Optical transmittance spectra of films as deposited and annealed MoO_3

The optical absorption edge also shifted towards lower wavelength side. The optical absorption coefficient (α) was evaluated from the transmittance data where reflection losses were taken into consideration. By extrapolating the linear portion of the $(\alpha h\nu)^2$ as a function of photon energy ($h\nu$) curve to the absorption coefficient equal to zero the optical band gap was obtained. The optical band gap of the as-deposited film is 3.07 eV. The optical band gap of the oxygen annealed films increases from 3.13 to 3.20 eV with the increase of temperature from 473 to 673 K. The increase in the optical band gap with the annealing temperature was due to partial filling of oxygen vacancies [1].The band gap values are in consistent with the value of 3.16 and 3.27 eV obtained in thermal evaporated films by Sain and Reddy [2] for as-deposited and annealed molybdenum oxide films respectively. The refractive index of the films was evaluated from the optical transmittance interference data using Swanepoel envelope technique. At a fixed wavelength of 500 nm, the refractive index of the as-deposited film was 2.03. It increased to 2.08 when annealed in oxygen at a temperature of 673 K. The increase in the refractive index with the increase of annealing temperature is due to the improvement in the crystallinity and packing density of the films.

CONCLUSIONS

Molybdenum oxide films were prepared by dc reactive magnetron sputtering in an argon and oxygen atmosphere on glass and single crystal silicon substrates held at room temperature. The effect of oxygen annealing on the structural and optical properties of MoO_3 films has been studied. The as- deposited shows amorphous nature, while those annealed in oxygen at 473 K indicate the onset of polycrystalline and crystallinity of the films improved further increase in the annealing temperature to 673 K. The Fourier transformed infrared optical transmittance of the MoO_3 films annealed at 473 K shows the optical absorption bands. When the films annealed at 673 K the intensity of the absorption band at 560 cm^{-1} diminished and the band observed at 995 cm^{-1} split into two bands of 995 cm^{-1} and 1010 cm^{-1} related to the zig-zag rows of layered structure of MoO_3. The optical band gap of the as-deposited film was 3.07 eV and it increased to 3.20 eV with increase in the annealing temperature to 673 K.

REFERENCES

1. R. Sivakumar, R. Gopalakrishnan, M. Jayachandran and C. Sanjeeviraja, *Current Appl. Phys.* **7**, 51-59 (2007).
2. T.S. Sain, and G.B. Reddy, *Solar Energy Mater. Solar Cells* **82**, 375-386 (2004).
3. S.H. Mohamed and S. Venkataraj, *Vacuum* **81**, 636-643 (2007).
4. J. Okumu, F. Koerfer, C. Salinga, T.P. Pedersen, and M. Wuttig, *Thin Solid Films* **515**,1327-1333 (2006).
5. A. Bouzidi, N. Benramdane, H.T. Derraz, C. Mathieu, B. Khelifa, R. Desfeus, *Mater. Sci.Eng,* **B 97**, 5-8 (2003).
6. V. Nirupama, P. Sreedhara Reddy. O.M. Hussain and S. Uthanna, *Ionics* DOI 10.100/S11581-007-0132-9 (2007).
7. T.S. Sain, G.B. Reddy and M. Shivaprasad, *Jap. J. Appl. Phys.* **43**, 6248-6254 (2004).
8. N. Sotani, K. Eda, M. Sadamatu and S. Takagi, *Bull. Chem. Soc. Japan* **62**, 903-910 (1989).
9. G.A. Nazri and C. Julien, *Solid State Ionics* **53**,376-384 (1992).

Influence of Annealing Temperature on Electron Beam Evaporated $LiMn_2O_4$ Thin Films

V. Kalai Vani[a] and O.M Hussain[a]*

[a]*Thin film laboratory, Department of physics, Sri Venkateswara University, Tirupati-517 502, India

Abstract. Thin films of $LiMn_2O_4$ were prepared by electron beam evaporation technique from a lithium rich target at a substrate temperature of 473 K in an oxygen partial pressure of 4-5 x 10^{-4} mbar. The as deposited films were found to be X-ray amorphous. The films were annealed in the temperature range 500-1000 K for 4 hours in a controlled oxygen environment. The initial crystallization was observed for the $LiMn_2O_4$ films annealed at temperature greater than 673 K. The films exhibited predominantly (111) orientation representing cubic spinel structure. The grain size increased with the increase of annealing temperature. The AFM and FTIR studies were also performed to understand the microstructure of films for effective utilization as cathode material.

Keywords: Thin films, Electron beam evaporation technique, Surface morphology
PACS:: 68.37.Ps, 81.05.Gc, 61.05.cp

INTRODUCTION

All solid state thin film rechargeable microbatteries have drawn considerable attention due to their wide potential application as an active back up power source in micro and nanoelectronic minimized devices viz MEMS, smart cards, remote sensors etc [1-2]. The realization of such microbatteries originates from the identification of novel thin film cathode materials with high energy density, high specific capacity and structural stability towards lithium insertion. Lithiated transition metal oxides have received considerable attention in recent years as high voltage cathode material in order to reduce weight for portable electronics and to increase the demand for lighter and thinner batteries for new generation of all solid state rechargeable lithium batteries. Among various lithium based cathode materials, $LiMn_2O_4$ is one of the most important cathode material due to its high energy density, high cell voltage, low cost and low toxicity [3-5]. The growth of these cathode materials in thin film form with preferred orientation is of great interest as a result of their possible use in the fabrication of all solid state rechargeable microbatteries.

Thin films of $LiMn_2O_4$ were prepared by several physical and chemical vapour deposition techniques [6-10]. The performance of these films as a active cathode layer are mainly dependent on the microstructural properties which depend on the deposition technique and process parameters using during the growth. In the present investigation, electron beam evaporation technique has been chosen for the preparation of $LiMn_2O_4$ thin films in which the contamination probabilities is least, decomposition of the starting material can be minimized and high deposition rates can be achieved [11].

CP1004, *Thermophysical Properties of Materials and Devices-NCTP '07*,
edited by P. Predeep, S. Prasanth, and A. S. Prasad

EXPERIMENTAL

The $LiMn_2O_4$ thin films were prepared by electron beam evaporation technique using Balzers BA510E high vacuum coating unit. Pure powders was made into pellets of 20 mm diameter and 3 mm thick and calcinated at 1073 K for about 12 hours. These pellets were kept in a water cooled copper crucible. In the evaporation system, electrons were extracted from a tungsten filament (DC heated cathode) through the electric field, pass the anode and deflected through an angle of 270° by magnetic field. The surface of the pellet was scanned with the resultant electron beam, which has an accelerated voltage of 6 kV and a power density of 1.5 kW/cm^2. The distance between source and substrate was 15 cm. The deposition was done onto chemically, ultrasonically and glow discharge cleaned (100) silicon wafers.

The X-ray diffraction patterns of the films was recorded in the scanning angle range 10-60° using a Siefert X-ray diffractometer with Cu K_α radiation (λ =1.546 Å). The surface morphological characteristics of the films have been studied by atomic force microscopy (Digital instruments, dimensions:3100 series). The vibrational studies of $LiMn_2O_4$ thin films was studied by Fourier transform infrared spectroscopy (Nicolet IR-5700).

RESULTS AND DISCUSSION

Thin films of $LiMn_2O_4$ were prepared by electron beam evaporation technique. The target for the deposition was prepared by mixing an addition of 15% Li_2O to the pure $LiMn_2O_4$ powder to avoid lithium deficiency in the films [12,13]. An oxygen partial pressure of 4-5 x 10^{-4} mbar is maintained in the vacuum chamber in order to avoid oxygen deficiency in the $LiMn_2O_4$ thin films. The films were annealed at various temperatures in a controlled oxygen environment for 4 hours. The microstructural properties of the films were studied in relation to the annealing temperature.

The X-ray diffraction patterns of $LiMn_2O_4$ thin film at various temperatures on silicon substrate as shown in Fig 1(a). The broad and diffused X-ray diffraction pattern indicates that the films are X-ray amorphous. The as deposited films are annealed at various temperatures in the range 500-1000 K in a controlled oxygen environment for 4 hours. The onset of crystallization is observed for the film annealed at temperature greater than 673 K. As the annealing temperature increased, the crystallinity of the film increased. The film annealed at higher temperatures exhibited characteristics peaks corresponding to the (111), (400) and (222) in which (111) orientation was predominant with cubic spinel structure of Fd3m symmetry. The full width at half maximum value of the (111) diffraction peak was utilized to assess the crystallinity of the film [14]. The average grain size of the films annealed at 773 K and 973 K is around 115 nm and 220 nm as shown in Fig 1(b). As the annealing temperature increases, the rearrangement of lithium, manganese and oxygen atoms takes place leads to form polycrystalline $LiMn_2O_4$ thin films with larger size. The evaluated lattice parameter for the films

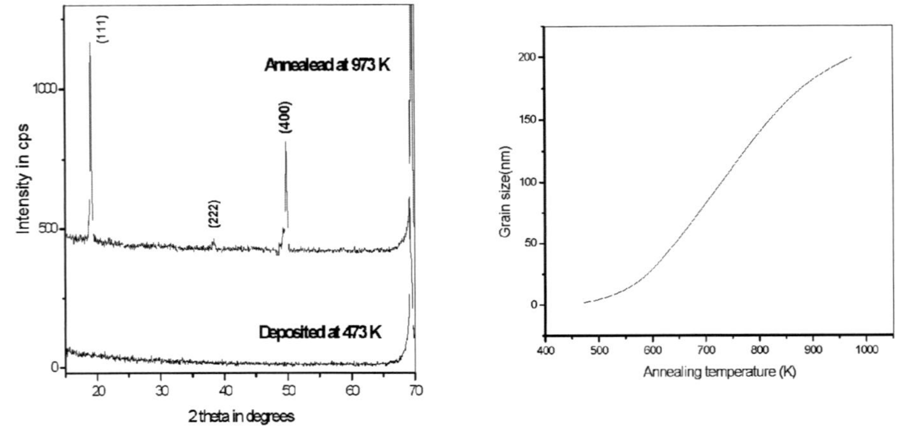

FIGURE 1. (a) X-ray diffraction pattern of $LiMn_2O_4$ thin films at various annealing temperatures (b) Variation of grain size of $LiMn_2O_4$ thin films at various annealing temperatures

lattice parameter for the films annealed at 773 K and 973 K were 8.18 Å and 8.24 Å. The lattice parameter of the bulk $LiMn_2O_4$ cubic spinel structure of Fd3m symmetry is 8.241 Å in which the oxygen ions are placed on a face-centred cubic array. The lithium ions occupy the tetrahedral 8a sites of the oxide network whereas the Mn is placed in the octahedral 16c sites [14-15]. This indicates that the films annealed at 973 K shows better crystallinity with predominant (111) orientation representing cubic structure of Fd3m symmetry with a lattice parameter nearly equal to bulk value. As the annealing temperature increased greater than 973 K, the crystallinity of the film is decreased and intensity of (111) peak was reduced representing the impurity phases in the $LiMn_2O_4$ thin films.

The surface morphological images of $LiMn_2O_4$ thin films at 773 K and 973 K are shown in Fig 2. The films exhibited fairly uniform grain sizes on smooth surface. The crystallization in the films starts at a annealing temperature of 673 K. The average grain sizes of annealed films at a temperature of 773 K and 973 K are 110 nm and 210 nm [17]. As the annealing temperature gradually increased in the films, the islands combines, nucleation occurs and coalesced with each other form a continuous film with larger grain size. The film annealed at higher temperature has a coalescence speed than that annealed at lower temperature which leads to the formation of grain size of the former is greater than latter. At an annealing temperature of 973 K, the film is continuous with larger grain size.

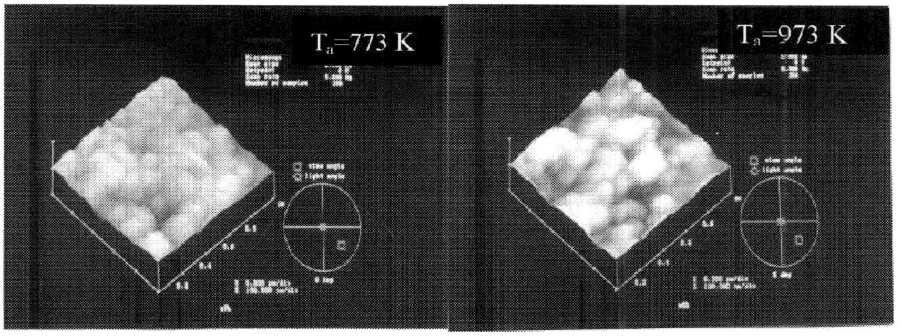

FIGURE 2. AFM images of $LiMn_2O_4$ thin films at various annealing temperatures

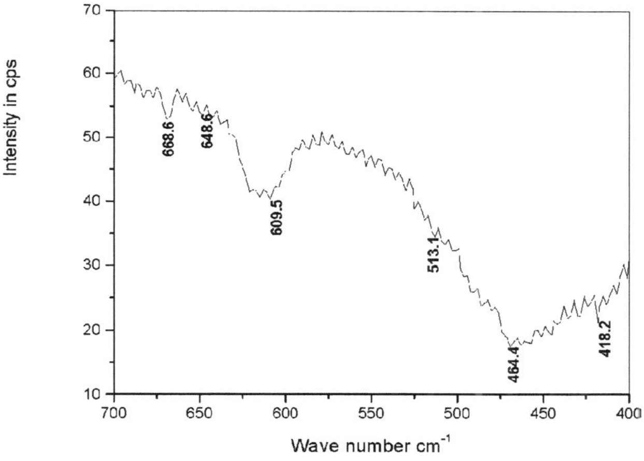

FIGURE 3. FTIR spectrum of $LiMn_2O_4$ thin film annealed at 973 K

Fig 3. shows the FTIR spectrum of annealed film at temperature of 973 K.The lithium manganese oxide $LiMn_2O_4$ crystallizes with a cubic spinel structure. sites. The spinel structure is characterized by structural groups MnO_6 octehedra and LiO_4 tetrahedra [18]. The FTIR spectrum of 773 K exhibited high frequency bands of the tranmittance spectrum of $LiMn_2O_4$ observed at approximately 675 cm^{-1} and 519 cm^{-1} and low frequency band at 424 cm^{-1}. As the annealing temperature increased at 973 K, high frequency bands of the spectrum of $LiMn_2O_4$ is shifted at approximately 648 cm^{-1} and 513 cm^{-1} are attributed to the asymmetric stretching modes of MnO_6 group and low frequency band at 418 cm^{-1} belongs to mode of LiO_4 group. FTIR line intensities at 973 K remained in good agreement with the Fd3m space group.

CONCLUSIONS

$LiMn_2O_4$ thin films were prepared by electron beam evaporation technique from a lithium rich target. The films annealed at a temperature of 973 K exhibited predominantly (111) orientation with a cubic spinel structure of Fd3m symmetry whose lattice parameter equal to its bulk value of $LiMn_2O_4$ thin film. The grain size of the films increased with the increase of annealing temperature as observed from XRD and AFM data. FTIR line intensities at an annealing temperature of 973 K of $LiMn_2O_4$ thin film remained in good agreement with the Fd3m space group. In conclusion, the electron beam evaporated $LiMn_2O_4$ thin films with cubic spinel structure were highly applicable in the fabrication of solid state microbatteries.

REFERENCES

1. Minko Balkanski, *Physics world*, **3**, 29-33 (1990).
2. J.M. Tarascon and M. Armand, *Nature*, **414**, 359-367 (2001).
3. S. H. Ye, J. Y. Lu, X.P. Gao, F. Wu and D. Y. song, *Electrochimica Acta*, **49**, 1623-1628 (2004).
4. Pawel piszora, Wojciech paszkowicz, Carsten Baehtz and Emilia Wolska, *J. of Alloys and Compounds*, **382,** 119-122 (2004).
5. C.M. Julien, *Mater.Sci.Eng.* R: Reports, **40,** 47-102 (2003).
6. Ki-Lyoung Lee, Ju-Yound, Seung-Won Lee, Hee-soo Moon and Jong-Wan Park, *J. of Power Sources*, **130**, 241-246 (2004).
7. Izumi Yamada, Takeshi Abe, Yasutoshi Iriyama and Zempachi Ogumi, *Electrochemistry Communications*, **5,** 502-505 (2003).
8. P. Piszora, *J. of Alloys and Compounds*, **382,** 112-118 (2004).
9. Young Ho Rho and Kiyoshi kanamura, *J. of Electroanalytical Chemistry*, **559,** 69-75 (2003).
10. L. Shui, G.S. Jiang, S. Xie and C.H. Chen, *Electrochimica Acta*, **49**, 2209-2213 (2004).
11. F. K. Shokoohi, J.M. Tarascon, B.J. Wilkens, D. Guyamard and C.C. Chang, *J. Electrochem. Soc.*, **139**, 1845-1849 (1992).
12. Shu-Juan Bao, Wen-Jia Zhon, Yan-Yu Liaing, Ben-Lin He and Hu-Lin Li, *Materials Chemistry and Physics*, **95,** 188-192 (2006).
13. V. Kalai Vani, K. Hari Krishna and O.M. Hussain, Workshop on Emerging trends in Power Source Systems for Aerospace and Related Applications (PSSARA-2006).
14. Xian Ming Wu, Xin Hai Li, Zhuo Wang, Zhuo Bing Xiao, Jianben Liu and Wen Bin Yan, *Materials Chemistry and Physics*, **83**, 78-81, (2004).
15. D. Singh, W.S. Kim, V. Cracium, H. Hofmann and R. K. Singh, *Applied Surface Science*, **197-198,** 516-521 (2002).
16. H. Gadjov, M. Gorava, V. Kotzeva, G. Avdeev, S. Uzunova and D. Kovacheva, *J. of Power Sources*, **134,** 110-117 (2004).
17. C.M. Julien and M. Massot, *Materials Science and Engineering B*, **97,** 217-230 (2003).

Effect of Annealing on Thermal & Optical Properties of Polypyrrole

Rashmi Saxena, Manasvi Dixit, Kananbala Sharma, Narendra S. Saxena and Thaneshwar P. Sharma

Semi-conductor & Polymer Science Laboratory, 5, Vigyan Bhawan, Deptt. of Physics, University of Rajasthan, Jaipur- 302004, India.
Corresponding Author: rashmi_saxena29@yahoo.co.in, n_s_saxena@rediffmail.com

Abstract. Pure polypyrrole sample (S_1) was synthesized by chemical oxidation method using NaOH as reducing agent in aqueous HCl medium. The polypyrrole pellet sample (S_2) was then annealed at 200 ^{0}C for 4 hrs. The amorphous nature of both annealed and as- prepared polypyrrole samples was confirmed by XRD. FTIR spectra of both samples were taken, which indicate the significant change in annealed sample (S_2) compared to as prepared sample. Temperature dependence of effective thermal conductivity of both samples (S_1, S_2) was studied by Transient plane source (TPS) technique. The effective thermal conductivity (λ_e) obtained for S_1& S_2 exhibits a variation with temperature and a peak was observed for the two samples at 150^{0}C &120^{0}C with a value 0.17 W/mK & 0.18 W/mK respectively. The shift of thermal conductivity peak of annealed sample towards the lower temperature side is explained on the basis of removal of voids and defects on annealing. The absorption spectra of these samples were recorded by USB-2000 spectrophotometer at room temperature in the wavelength range 300-800nm. From the analysis of absorption spectra, optical band gap of S_1& S_2 were determined. It was found that the values of optical band gap for sample S_1& S_2 are 2.39 eV&2.24 eV respectively.

Key words. Polypyrrole, annealed, effective thermal conductivity, band gap.
PACS: 66.70.HK

INTRODUCTION

In the past few years conducting polymers have become one of the most extensively studied materials over the world due to their unique properties which have led to some commercial applications such as manufacturing of printed circuit boards, corrosion protection, electrochromic displays electrolytic capacitors, rechargeable batteries and conductive coating for textile sensors, light emitting diodes electromagnetic shielding and so on [1-4]. Also the ease of processing makes these materials attractive. Conducting polymers such as polythiophene, polyaniline, polypyrrole are becoming increasingly important for their technological applications due to their electrical, optical, thermal properties and their high air, chemical, and electrical stability at ambient conditions [5-6].

CP1004, *Thermophysical Properties of Materials and Devices-NCTP '07*,
edited by P. Predeep, S. Prasanth, and A. S. Prasad

During oxidative polymerization, pyrrole typically polymerizes by linkage at α-position, along with the loss of a proton at each of these positions shown in Fig.1 [7].

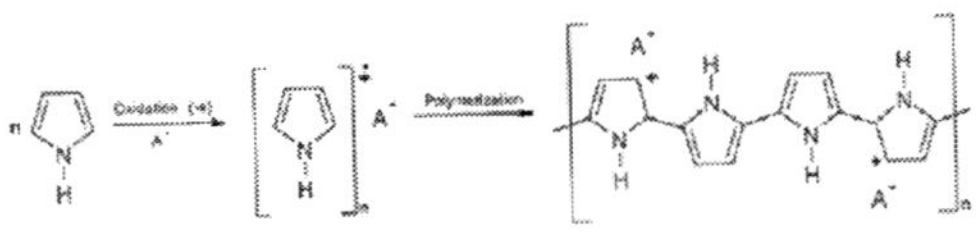

FIGURE 1.Oxidation of polypyrrole

A good deal of work has been done on polypyrrole to determine its various properties [8-10]. Though many studies have reported on the optical and thermal properties of polypyrrole in doped form, but very little information is available on the effect of annealing on thermal &optical properties of the polypyrrole. So the main aim of the present study is to determine the thermal & optical properties of as prepared and annealed polypyrrole.

MATERIAL PREPARATION

The polypyrrole was prepared by chemical oxidation process [11]. Sample S_2 was annealed at 200^0 C at 4 hrs. For thermal measurements sample in the form of pellets (diameter 12 mm and thickness of approx. 2m) were prepared by compressing the fine powder of polypyrrole and annealed polypyrrole under a load of 6 tons. For optical measurements polypyrrole powder of both the samples were dissolved in DMSO (Dimethyl sulfooxide) solution and stirred for 6 hrs at room temperature to obtain homogeneous solution & then absorption spectra were taken for the band gap determination.

The IR spectra of these samples were recorded on a model 84005 Shimadzu photometer in KBr medium at room temperature. The FTIR spectra for these samples are presented in Fig.3 where as the attributions of the observed bands are indicated in Table1. The spectra exhibit similar feature for the main polymer chain and the observed difference is attributed to the presence of side groups attached to the aromatic rings.

TABLE 1. Structure information though FTIR

Band Assignment	As- prepare sample (cm^{-1})	Annealed sample (cm^{-1})
N-H Ster.	3850-2350	3850-2350
Pyrrole ring Vibration	1700	1700
C=C Str.	1500-1650	1650-1500
Presence of PPY	750-780	750-780
N-H bending	900-650	900-650

X-ray diffraction studies were carried out on these samples using model 1840 X-ray diffractometer. X-Ray diffraction pattern of as-prepared sample and annealed samples show no sharp peak in the Fig.2, which confirms the amorphous nature of the material.

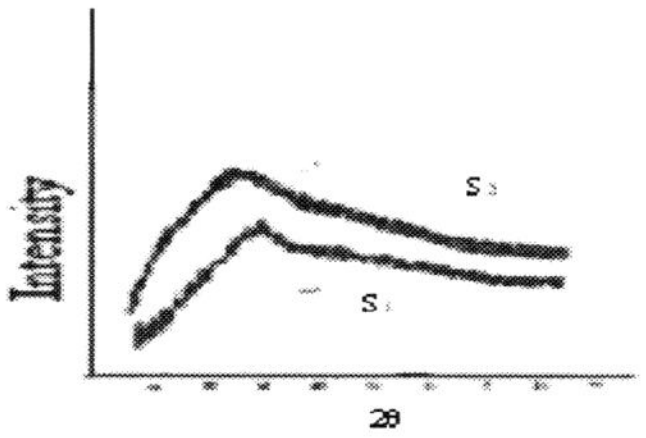

FIGURE.2. XRD of both samples

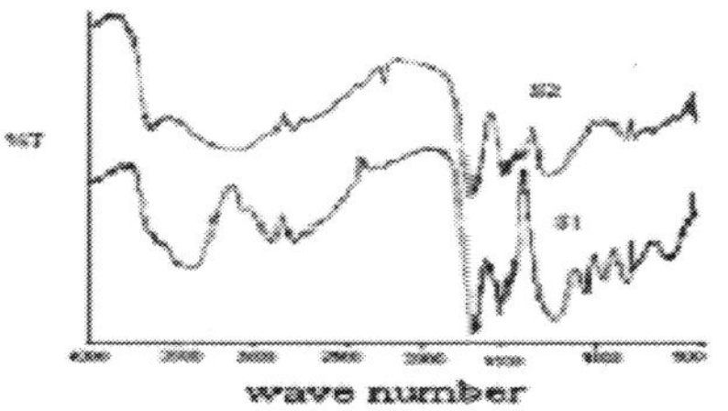

FIGURE.3 IR Spectra of both samples

THERMAL AND OPTICAL PROPERTIES

Effective thermal conductivity (λ_e) of these two samples was measured by transient plane source (TPS) method, which is an extension of the transient hot strip (THS) method introduced by Gustafsson [12]. Detailed theory of TPS is given elsewhere [13].

The optical absorption of samples (S_1&S_2) was recorded at room temperature in the wavelength range from 450 to 800nm. Using Ocean optics USB2000 spectrophometer. According to Tauc relation [14] the absorption coefficient for direct band gap material is given by

$$\alpha\, h\nu = A\, (h\nu - E_g)^{1/2} \quad (1)$$

where hν is photon energy, E_g the energy band gap and A is constant which is different for different transitions. A plot of $(\alpha\, h\nu)^2$ vs. photon energy (hν), when extrapolated to zero absorption provides the value of energy band gap as depicted in Fig. 4.

RESULT AND DISCUSSION

The thermal and optical properties of polypyrrole depend on the electronic structure and on the chemical nature of repeated unit. Annealing process results in changes in the electrical, optical, thermal& structural properties of the polymer [15]. The effective thermal conductivity & optical bang gap values of both the samples are given in Table 2 and Fig. 5 shows the plot of thermal conductivity vs temperature of both the samples S_1&S_2. From Fig. 5 it has been seen that effective thermal conductivity of as-prepared polypyrrole sample increases with increasing temperature initially and becomes maximum around 150^0C for sample S_1 When temperature is increased beyond 150^0C, value of effective thermal conductivity starts decreasing, also sample S_2 (annealed) shows same trend but the peak values of effective thermal conductivity are shifted from150^0C to 120^0C.

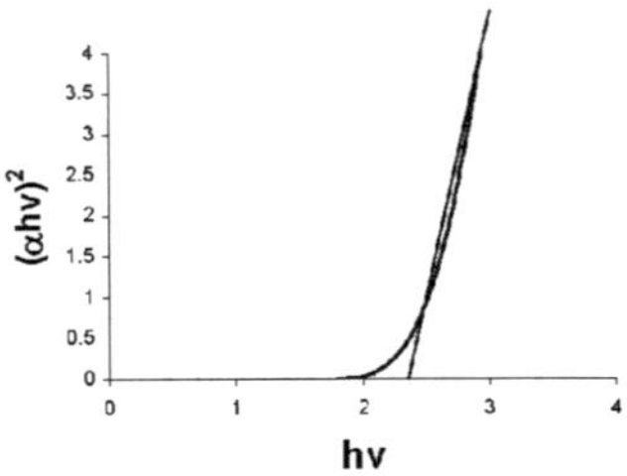

FIGURE 4. Optical band gap of sample S_2

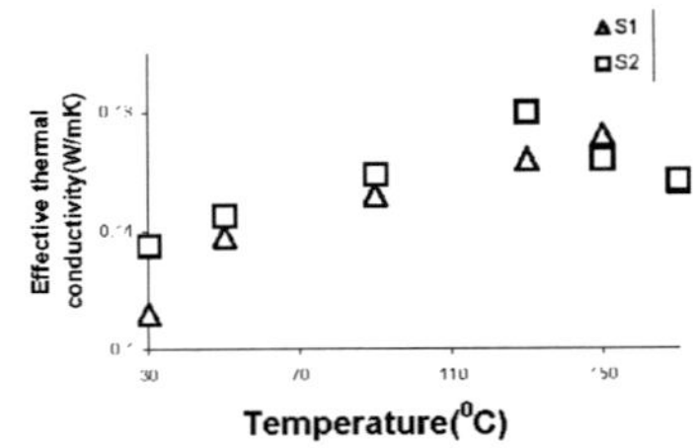

FIGURE 5. Variation of effective thermal conductivity with temperature (S_1- as prepared sample) (S_2-annealed sample)

TABLE 2 Values of thermal conductivity &optical band gap of both the samples

Samples	Optical band gap (eV)	Thermal Conductivity (W/mK)
S_1	2.39	0.11
S_2	2.24	0.13

In XRD Fig.2 hump is observed at low angle scattering of both samples. In annealed sample (S_2) hump is shifted to low angle, indicating that particle size of this sample increases as compared to the as prepared sample (S_1). The IR spectrum of annealed sample shows a drastic change in the range of 3550-2950cm^{-1}, 1400-1050cm^{-1} & 600-500cm^{-1}. This range indicate the moisture present in the sample and Sample S_1 shows the peak around these ranges which is assignable to the OH bending vibration of H_2O and it is absent in the annealed sample i.e. S_2, which indicates removal of moisture. The result in annealed sample indicates that the heat–treatment changes the structure of the annealed sample. [16].

The rise in thermal conductivity is due to expansion of the material increasing the area of contact between the particles. which further increases of phonon propagation in the medium and as a result the thermal transport is enhanced. Above150^{0}C, substantial changes occur in the structure due to breaking of the chains in material inhibiting further increase in heat transfer & also loss of coordinated H_2O occur hence reducing the thermal conductivity.

However, in the case of annealed sample (S_2), the maxima in the conductivity occur at a lower temperature. In this case the sample is already dehydrated and there are no trapped water molecules. The heat which was earlier case (S_1) used up both in expansion of particle and in dehydrating, it is now used up only in of the material, effectively causing an expansion of the material hence the conductivity maxima is obtain at lower temperature 120^{0}C. This is the only observed variation in thermal conductivity. The trend of variation remains identical. The almost constant value of thermal conductivity obtained in the as prepared & annealed sample shows that there in no degradation in the sample & no loss of dopant occurs, which confirm by FTIR data [17]. So, it can be said that the annealing process plays important role in the thermal properties. This is important according to the duration of the processing of the polymer product applications.

Similarly in optical study it is found that the band gap values of sample S_1 is 2.39eV & sample S_2 2.24eV.This lower value of band gap in sample S_2 indicate change in structure & particle size as is evident from XRD.This is suggestive of the fact that annealing process brings slightly change in the chemical structure, so finally the annealing process results in an improvement of the sample structure while retaining the essential features and increasing of charge carriers in the annealed sample [18].

CONCLUSION

Structural characterization suggests the amorphous nature of the samples and annealing effect traced by FTIR analysis. The thermal and optical properties of these samples indicate the effect of annealing and suggest that the annealing process brings about an improvement of the sample structure and there is no loss of dopant all the temperatures of study, which is important for processability of the material.

ACKNOWLEDGEMENT

One of the authors (Rashmi Saxena) is thankful to Mr. Neeraj Jain, Ms. Vinodini Shaktawat, Mr.Dinesh Patidar and Ms. Deepika who helped in various ways during the course of this work.

REFERENCES

1. Colin Pratt, Application of conducting polymers (1996).
2. M.M. Chehimi, E.Abdelijalie, *Synth. Met.***145,** 15-22 (2004).
3. G Smitha, Kundya, R Rajagapalam, Jo Iroh *Journal of Applied polymer science* **83** 1970-1977(2002)
4. JR Fried, "Polymer Science and Technology", New Delhi: Prentice Hall India 461-465. (1999)
5. K.Anuar, S.Murali, A.Fariz, H.N.M.Mahmud, Ekramul, *Mater.Sci.***10**, 255-258 (2004).
6. A.Kassim, Z.B.Basar, H.N.M.E.Mahmud, *Chem.Sci.***114,** 155-162 (2002).
7. M.P.Stevens, "Polymer chemistry", 3rd ^{Ed}, Oxford University press, Newyork 412. (1999)
8. R.Turcu, L.V.Giurgia, R.Ordcan, R.Grecu, M.Bric *Synth.Metal* **19,** 287-288 (2001).
9. B.A.Lunn, J.Unsworth, N.G.Booth, P.C.Innis, *Journal of Materials Science*, **28,**5092-5098 (1993)
10. Novato Tsutsumi, Satoshi Ishida, Tsuyoshi Kiyotsukuri, *Journal of polymer Science part B*, **32,** 1899-1906 (2003)
11. V.Shaktawat, N.Jain, R.Saxena, N.S.Saxena, K.Sharma&T.P.Sharma *Polymer Bulletin*, **57** 535-543 (2006)
12. S.E.Gustafsson. *Rev Sci Instrum*, **62** 797-799. (1991)
13. G.P. Joshi, N.S.Saxena T.P.Sharma, V.Dixit, S.C.K.Mishra, *Journal of physics and chemistry of solids* **64** 2391-2396 (2003)
14. J.Tauc, "Amorphous & Liquid Semiconductors" (J.TaucEdi) New York: Plenum 159-162 (1974)
15. Yasuokudoh.et al *Synthetic Metals* **79** 17-22 (1996).
16. E.M.Scherc, A.G.MacDiarmid et al *Synth Met.* **41** 735-738 (1991)
17. R.Gurdap, S.Chatwal&K.Shan, Anand, "Spectroscopy (Atomic Molecular) Book" Fifth AD.2.55-2.60
18. Aliel-Shekeil et al *Journal of Macro Molecular Science* **41** 649-667 (2004).

First Principles Calculations of Electronic and Thermal Properties of AlRE (RE=La, Ce and Pr) Compounds

Vipul Srivastava[a,*] M. Aynyas [b], M. Rajagopalan[c] and S. P. Sanyal[a]

[a] *Department of Physics, Barkatullah University, Bhopal-462026, India.*
[b] *Department of Physics, S. V. College, Bairagarh, Bhopal-462030, India.*
[c] *Department of Physics, Anna University, Chennai-600025, India*

Abstract. Electronic properties of non-magnetic cubic B_2-type AlRE (RE= La, Ce and Pr) compounds have been derived from self-consistent tight binding linear muffin tin orbital method at ambient pressure. These compounds show metallic behaviour under ambient conditions. While thermal properties like Debye temperature and Grüneisen constant are calculated at T=0K within the Debye-Grüneisen model and compared with the others theoretical results. We have also performed a pressure induced variation of Debye temperature. We have found a decrease in Debye temperature around 40 kbar in all the AlRE compounds.

Keywords: Electronic structure; Thermal properties; Debye temperature; Inter-metallic compounds
PACS: 65.40.-b; 71.15.Ap; 71.20.Lp

INTRODUCTION

The AlRE (RE= La, Ce and Pr) compounds crystallize in cubic cesium chloride structure (B_2-type, *Pm3m*, Space Group, *221*). The B_2-type AlRE intermetallics are very important compounds with respect to their thermal and electronic properties [1]. These compounds are having significant importance to the novel material design and further scientific and technical investigations. Some experimental information of structure of AlRE equiatomic compounds can be found in literature [2]. The mixing of Aluminum with rare earth has made a new class of highly ordered and ductile compounds. The f-electrons present in RE specie, play a crucial role in exploring structural and electronic properties, therefore it becomes important to understand the electronic properties of this class of compounds. It is revealed from the literature, that the role of *f*-electrons in RE-ion with Al on the structural, electronic and elastic properties at normal and as well as at high pressure, has not been extensively studied and it is yet to be explored in details. In the present paper, ground state electronic and thermal properties for B_2-type AlRE are studied by combined the first principles tight binding linear muffin tin orbital (TB-LMTO) method and Debye-Grüneisen model. The calculated lattice parameters, bulk modulus, Debye temperature, Grüneisen constants are good agreement with the others reported values [3].

CP1004, *Thermophysical Properties of Materials and Devices-NCTP '07*,
edited by P. Predeep, S. Prasanth, and A. S. Prasad

METHODOLOGY

TB-LMTO Method

The total energy and structural stabilities have been studied by using TB-LMTO method with atomic sphere approximation (ASA) [4]. von-Barth and Hedin parameterization scheme has been used for exchange correlation potential within the local density approximation (LDA). The AlRE compounds crystallize in the B_2-type structure, and positioned at Al: (0, 0, 0) and RE: (0.5, 0.5, 0.5). The Wigner-Seitz sphere is chosen in such a way that the boundary potential is a minimum and charge flow between the atoms is in accordance with the electro-negativity criteria. The total energy is computed by changing the volume from 1.0 V_0 to 0.65 V_0, where V_0 is the experimental equilibrium cell volume. The calculated total energy is fitted to Birch equation of state [5] to obtain the pressure, volume, bulk modulus etc.

Debye-Grüneisen Model

To calculate important thermal properties of a vibrating Debye lattice we have used Debye-Grüneisen (DG) model [6]. Debye temperature Θ_D previously measured by the following expression

$$(\Theta_D)=67.48\sqrt{(rB/M)} \tag{1}$$

where r in a.u., B is bulk modulus in kbar and M is atomic weight. To evaluate $(\Theta_D)_0$ using above expression, at $r = r_0$ (at equilibrium) and putting calculated (or experimental) value of bulk modulus yields much higher value of experimental Debye temperature. To overcome from this situation a scaling factor had been introduced in the modified expression of Θ_D. In this modified expression, we put theoretical $r = r_0$ and bulk modulus derived from the above first principles calculation at r_0.
With above considerations the Debye temperature $(\Theta_D)_0$ can be calculated by

$$(\Theta_D)_0=41.63\sqrt{(r_0B/M)} \tag{2}$$

and Grüneisen constant γ is calculated by

$$\gamma=\partial \ln\Theta_D/\partial \ln V \tag{3}$$

where V is the volume

RESULTS AND DISCUSSION

The electronic band structure calculations are performed to estimate the total energy of AlRE compounds by using first principles TB-LMTO method. By calculating the total energy at different volumes we obtained E(V) equation of states, which is depicted in Fig. 1 for the case AlLa only.

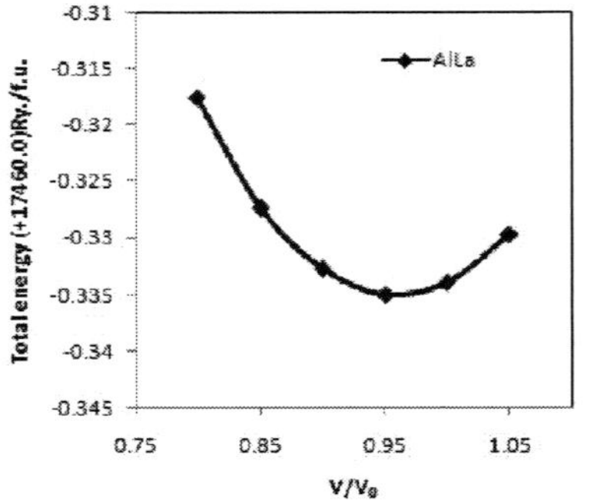

FIGURE 1. Total energy variation with V/V_0 for AlLa

The minimum of the E(V) curve defines the equilibrium volume V_0 (or equilibrium separation r_0) which is found 50.488, 45.69 and 44.099 $Å^3$ and corresponding lattice parameters 3.696, 3.575, 3.533 Å for all AlRE compounds, respectively and are tabulated in Table-1. These calculated values of lattice parameters are in good agreement with the others calculated values of 3.804, 3.795 and 3.759 Å for AlRE compounds, respectively. Though experimentally [1] reported values are 3.791, 3,85 and 3.82 Å for AlRE compounds, respectively, which are little higher as compare to our results. This may be due to usage of local density approximation (LDA) in the present calculation. Regarding the LDA contraction, it is often found that the LDA leads to some overbinding that yields lattice parameters that are somewhat smaller when compared with experiment. However, full potential calculations with LDA may yield a smaller equilibrium volume, which leads to better overall agreement with experiment.

TABLE 1 Calculated properties for AlRE under ambient conditions.

Solid	a_0 (Å)	B_0 (GPa)	r_0 (Å)	$(\Theta_D)_0$ (K)	γ_0
AlLa	3.696 , 3.791[a]	56.3, 50.91[a]	1.820	219, 231[a]	1.099, 0.901[a]
AlCe	3.575, 3.795[a]	69.3, 53.07[a]	1.760	201, 223[a]	1.032, 1.026[a]
AlPr	3.533, 3.759[a]	65.5, 55.11[a]	1.739	219, 227[a]	1.0692 1.127[a]

[a]Ref.[1]

We have presented band structure along the high symmetry directions for all the AlRE compounds in Fig. 2. The lowest energy bands are due to Al-'s' states. In Fig. 2 for AlLa, La-'d' states can be seen above the Fermi level, which hybridize with Al-'p' states near the Fermi level. For AlCe, Ce-'f' states (fat band just above E_f) can be seen near the Fermi level, which hybridize with 'Al-p' states and show metallic behaviour, which can be clearly seen in density of states (DOS) plot. Similar analysis can be done for AlPr. From the present calculation of band structure plot one can understand the increased metallic character as we go from -La to -Pr. In Fig. 2 we have also plotted total density of states (DOS) for AlLa, AlCe and AlPr under ambient conditions. The hybridization of Al-'p' and RE-'d' can be seen near E_f in Fig. 2 for AlLa. While f-states of f-electron systems can be seen as a sharp peak in Figs. 2 for AlCe and AlPr.

From the present calculations we have found that number of states increases if we go from –La to –Pr.

After knowing the metallic character of AlRE from first principles calculation, it is important to understand thermal properties of AlRE. Thermal properties are calculated by using DG model [6]. This method has been gained wide use for its simplicity. In this method, the Debye temperature can be obtained from the calculated bulk modulus and Grüneisen constant can be obtained from anharmonicity of the binding curve. The bulk modulus (B_0), Wigner-Seitz radii (r_0), Debye temperature $(\Theta_D)_0$ and Grüneisen constants (γ_0) are calculated for all AlRE compounds and compared with the others theoretical calculations [1] in Table-1. Our results show good agreement.

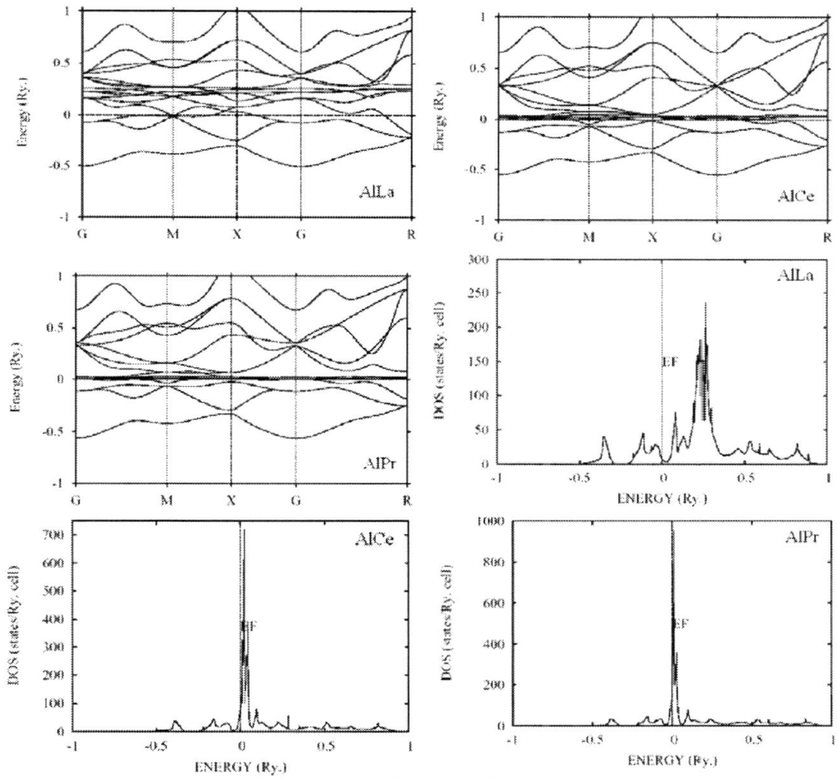

FIGURE 2. Band structure and density of states for AlRE compounds

We have also performed variation of Θ_D with respect to pressure for all the AlRE compounds and presented in Fig. 3. It can be seen from the Fig. 3 that a decrease in AlLa, AlCe, AlPr, is found at different pressures within the pressure range of 90 kbar. We have noticed this decrease in Θ_D at pressure of 40.3, 40.1 and 40 kbar for AlLa, AlCe and AlPr, respectively. Such interesting feature in three compounds indicates some structural phase transformation, which is yet to be confirmed. On the other hand, we have also calculated DOS under pressure for three AlRE intermetallics. In AlLa increase in DOS is noticed in between the compression value of V/V_0 = 0.90 to 0.80 (Fig. 4). In case of AlCe such increase in DOS is noticed twice, firstly for compression V/V_0 = 0.90 to 0.80 and secondly for V/V_0 = 0.75 to 0.65. A decrease in DOS is noticed as we compress AlPr but increase in DOS is also noticed at very low value of V/V_0 = 0.70 to 0.65. Such interesting feature could be another evidence of structural phase transformations in these intermatallics.

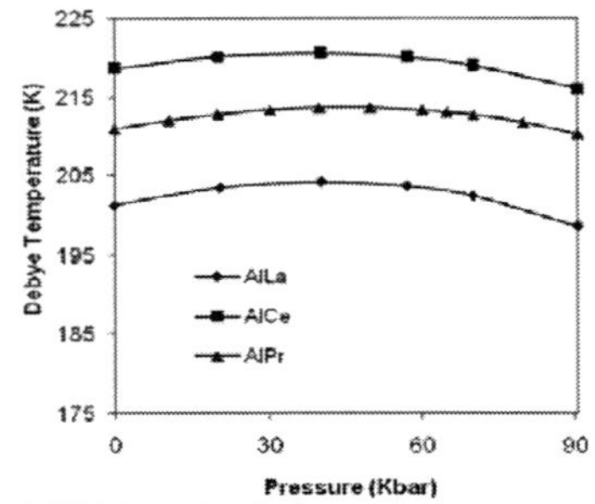

FIGURE 3. Variation of Debye temperature with Pressure for AlRE compounds

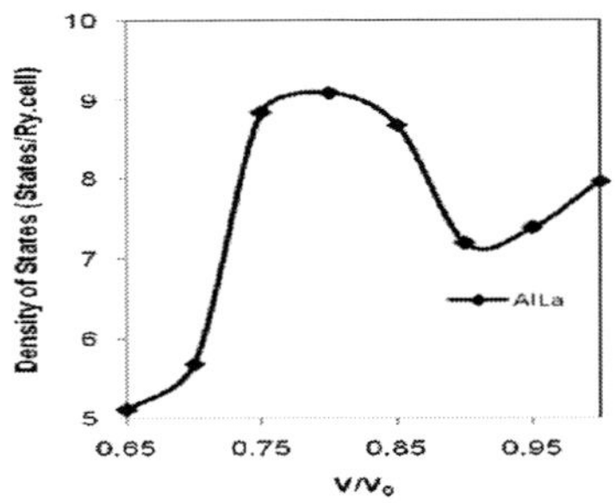

FIGURE 4. Variation of Density of states with V/V_0 for AlLa compound

CONCLUSION

We have made a theoretical study of electronic and thermal properties of some AlRE compounds. It is found from electronic study that AlRE compounds are metallic and metallicity increases from –La to –Pr. Debye-Grüneisen model is used to calculate Debye temperature and Grüneisen constant. We have noticed one interesting property in the variation of Debye temperature with pressure. At certain pressure Θ_D decreases for AlRE. On the other hand, under pressure density of states increases, which is another indication of some structural phase transformation, which needed some experimental and theoretical evidences.

ACKNOWLEDGMENTS

VS is thankful to DST, New Delhi for the award of Young Scientist under the Fast Track scheme and for financial support.

REFERENCES

1. X. Tao, Y. Ouyang, H. Liu, F. Zeng, Y. Feng and Z. Jin, *Physica B* (2007), (in Press) doi:10.1016/j.physb. 2007.05.037 and reference therein.
2. P. Villars, L. D. Calvert, Pearson's *Handbook of Crystallographic Data for Intermetallic Phases*, vol. 1-4, ASM International, Materials Park (OH, USA),1991.
3. X. Tao, Y. Ouyang, H. Liu, F. Zeng, Y. Feng and Z. Jin, *Computational Material Science* **40**, 226 (2007)
4. O. K. Andersen, *Phys. Rev.* ***B*12** 3060 (1975).
5. F. Birch, *J. Geophys. Rev.* **83** 1257 (1978).
6. P. Debye, Ann. D. *Physik* **39** 786 (1912).

Thermo physical Properties of Multiferroic Rare Earth Manganite $GdMnO_3$

Renu Choithrani and N. K. Gaur

Department of Physics, Barkatullah University, Bhopal – 462026. India.
Corresponding author: renuchoithrani@gmail.com*

Abstract. We have investigated the thermophysical properties of multiferroic rare earth manganite $GdMnO_3$ in the temperature range 15K $\leq T \leq$ 300K. We have applied interatomic potential to study the Specific heat (C) as a function of temperature. The calculated Specific heat values are closer to the available experimental data. At room temperature, the orthorhombic $GdMnO_3$ phase is indicative of a strong Jahn-Teller distortion. In addition, we have reported the cohesive energy (ϕ), molecular force constant (f), compressibility (β), Restrahalen frequency (ν_0), Debye temperature (θ_D) and Grüneisen parameter (γ) at temperature $15K \leq T \leq 300K$.

Keywords: Rare-earth manganites; Manganates; Multiferroic materials; Specific heat, Rare-earth ion; Thermophysical properties.
PACS: 74.25.Bt; 75.47.Gk; 75.47.Lx

INTRODUCTION

The Multiferroic manganites have recently attracted considerable attention because of their magnetic and electric ordering coexistence [1-2]. Among such materials, rare earth manganite $GdMnO_3$ with orthorhombically distorted perovskite structure are prototypical and show unusual interplay of magnetic and electric properties [3]. The material shows *colossal magnetoresisitance* effect (CMR) [4-10] which leads to the vast technological applications such as magnetic sensors, reading head devices and magnetic refrigeration [11]. The large magnetoelectric effects in the rare-earth manganites $RMnO_3$ has reopened the field of the so-called *multiferroic materials*. Multiferroic means, that several ferro-type orders, such as ferromagnetism, ferroelectricity or ferroelasticity coexist. Among such manganites, $GdMnO_3$ is one of the multiferroic rare-earth manganite where magnetic and ferroelectric order coexist [12]. $GdMnO_3$ have a very rich phase diagram depending on the doping concentration, temperature and pressure, being either an anti-ferromagnetic insulator, ferromagnetic metal or charge-ordered insulators. As electron-phonon interaction is one of the most relevant contributions in determining the conduction mechanism in these materials [13]. The Strong coupling between the electron and the lattice brings about by a small change in the chemical composition, like the ratio between trivalent and divalent ions at the A site or the average ionic radius of the ions on the A site, a large changes in the

CP1004, *Thermophysical Properties of Materials and Devices-NCTP '07*,
edited by P. Predeep, S. Prasanth, and A. S. Prasad

physical properties [14]. Thus, the thermophysical properties of this material may be taken as a starting point to a consistent understanding of the more complex physical properties of the compound.

The crystal structure of $GdMnO_3$ (space group pnma) has an orthorhombic structure and four formula units per unit cell with four 3d electrons: three t_{2g} and one e_g electrons. Since three t_{2g} electrons form an orbitally closed shell, many physical properties are believed to be determined by its e_g electron. The occurrence of the spin and orbital ordered state has been understood in terms of the cooperative Jahn-Teller Distortion [15-16].

In the process of development of new and novel materials an important input is the fundamental understanding of the thermophysical properties in terms of the microscopic structure. Therefore, we have presented a detailed study of the thermophysical properties which include cohesive energy (ϕ), molecular force constant (f), compressibility (β), Restrahalen frequency (ν_0), Debye temperature (θ_D), specific heat (C) and Grüneisen parameter (γ) of $GdMnO_3$ using Rigid Ion Model (RIM).

The interaction potential is given in the next section followed by the Results and Discussion.

ESSENTIAL FORMULATION

We have formulated the Rigid Ion Model (RIM) by including the effects of the long-range Coulomb attraction, Van der Waal's interaction due to dipole-dipole interaction [17-19] and the short-range Hafemeister Flygare type overlap repulsion. The total interaction potential (ϕ) is given by,

$$\phi = -\frac{e^2}{2}\sum_{kk'} Z_k Z_{k'} r_{kk'}^{-1} - b\sum_{kk'} \beta_{kk'} \exp\left\{\frac{r_k + r_{k'} - r_{kk'}}{\rho}\right\} - \sum_{kk'} c_{kk'} r_{kk'}^{-6} \tag{1}$$

Here, k (k′) denotes the positive (negative) ions and sum is taken over all the ions (kk′). $\beta_{kk'}$ is the Pauling coefficients given by

$$\beta_{kk'} = 1 + (z_k / n_k) + (z_{k'} / n_{k'}) \tag{2}$$

with z_k (z_k') and n_k (n_k') are the valence and the number of electrons in the outermost orbits and $r_{kk'}$ and r_{kk}(= $r_{kk'}$) are the first and second neighbour separations respectively. In the above expression (1), the first term represents the long-range Coulomb, the second represents the short-range Hafemeister Flygare interaction operative upto second nearest neighbour atom and the last term is Van der Waal's interaction due to dipole-dipole interaction.

The hardness and range parameters are determined from the equilibrium condition:

$$[d\phi(r) / dr]_{r=r_0} = 0 \tag{3}$$

and the bulk modulus:

$$B = (9Kr_0)^{-1}[d^2\phi(r) / dr^2]_{r=r_0} \tag{4}$$

Where, r is the nearest neighbour (nn) seperation and r_o is the equilibrium nn separation. K as the crystal structure constant

The specific heat is given as

$$C = 9R\left(\frac{T}{\theta_D}\right)^3 \int_0^{\frac{\theta_D}{T}} \frac{e^x x^4}{e^x - 1} dx \tag{5}$$

Which is Debye T^3 approximation.

The thermophysical properties of $GdMnO_3$ are computed using the expressions given in our earlier papers [19-21].

RESULTS AND DISCUSSION

Using the input data from ref. [1, 22-25] and the van der Waals Coefficient calculated from the Slater Kirkwood Vibrational (SKV) Method [26] on the lines of our earlier papers [17-21], we have calculated the model parameters b and ρ using equation (3) and (4) for the temperature range $15K \leq T \leq 300K$ and listed them in Table-1.

TABLE 1. Model parameters of $GdMnO_3$ at $15\ K \leq T \leq 300K$

T (K)	Mn-O		Gd-O	
	$b_1(10^{-12}$ erg)	ρ_1 (Å)	$b_2(10^{-12}$ erg)	ρ_2 (Å)
15	1.883	0.6328	2.157	0.8397
40	1.880	0.6323	2.155	0.8390
65	1.877	0.6318	2.153	0.8383
90	1.874	0.6312	2.151	0.8377
115	1.871	0.6307	2.149	0.8370
140	1.867	0.6302	2.146	0.8363
165	1.864	0.6296	2.144	0.8357
190	1.861	0.6291	2.142	0.8350
215	1.858	0.6286	2.140	0.8343
240	1.855	0.6280	2.138	0.8337
265	1.852	0.6274	2.136	0.8330
290	1.848	0.6270	2.134	0.8324
300	1.847	0.6267	2.133	0.8321

Using the values of model parameters, we have computed the thermophysical properties such as cohesive energy (ϕ), molecular force constant (f), compressibility (β), Restrahalen frequency (ν_0), Debye temperature (θ_D) and Gruneisen parameter (γ) of $GdMnO_3$ and listed them in Table II.

TABLE 2. Cohesive and Thermal properties of $GdMnO_3$						
T (K)	ϕ (eV)	f (10^4dyn/cm)	β(10^{-12}dyn^{-1}.cm^2)	ν_0 (THz)	Θ_D (K)	γ
15	-146.71	12.97	3.887	9.49	455.58	2.67
40	-146.74	12.99	3.882	9.50	455.96	2.68
65	-146.76	13.01	3.878	9.50	456.37	2.68
90	-146.78	13.04	3.874	9.51	456.76	2.69
115	-146.81	13.06	3.866	9.52	457.16	2.69
140	-146.83	13.08	3.861	9.53	457.55	2.70
165	-146.86	13.10	3.857	9.54	457.94	2.70
190	-146.88	13.13	3.853	9.55	458.33	2.71
215	-146.90	13.15	3.849	9.55	458.72	2.71
240	-146.93	13.17	3.845	9.56	459.11	2.72
265	-146.95	13.19	3.841	9.57	459.50	2.72
290	-146.98	13.21	3.832	9.58	459.69	2.73
300	-146.99	13.22	3.821	9.58	460.05 (474K)^{exp}	2.73

exp. [22-23]

It can be seen from the Table II that the cohesive energy is almost constant in the temperature range 15K $\leq$T$\leq$ 300K. The negative values of the cohesive energy show the stability of the compound. We have computed the cohesive energy (ϕ) for $GdMnO_3$ to test the validity of our interaction potential and found that the calculated cohesive energy for $GdMnO_3$ is closer to the magnitude of the parent compound $LaMnO_3$ [27] obtained by De Souza et al. We have also computed the molecular force constant (f), compressibility (β) and Restrahalen frequency (ν_0) as a function of temperature but our comment on the reliability are restricted until the report of the experimental data on them. The calculated values of Debye Temperature (θ_D) are in good agreement with their experimental data [22-23] available only at room temperature.

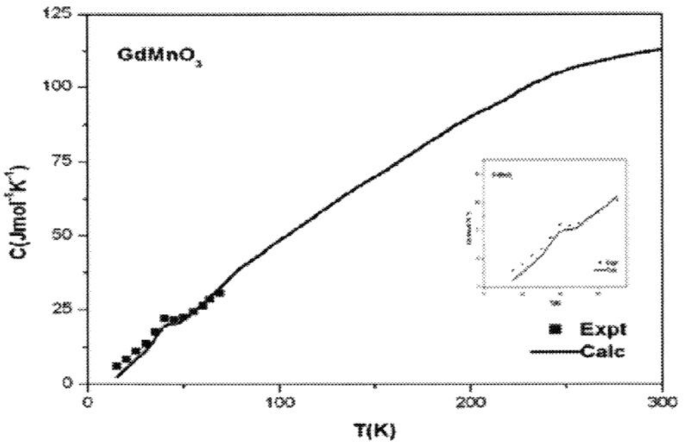

FIGURE 1. Specific heat of $GdMnO_3$ where solid line (—) and the squares (■) show present and the experimental [22-23] results respectively.

The higher values of Debye temperature indicate the higher phonon frequencies of the $GdMnO_3$. The calculated value of Debye temperature at $15K \leq T \leq 300K$ are 455K-460K, which seems quite reasonable as the value is within the range of Debye temperature often found in perovskite manganites (300K-550K). The values of Gruneisen parameter (γ) as a function of temperature $15K \leq T \leq 300$ K lies in between 2 and 3, which are the similar results as reported by Dia et al [28]. The calculated values of Cv for $GdMnO_3$ in the temperature range $15K \leq T \leq 300K$ are displayed in Fig. 1. Our results on Specific heat are in good agreement with the measured data [24-25] available only at low temperature i.e. below 80K (Inset Fig.1).

It is observed that there is a magnetic phase transition at 40K. Below 16-20K, spin canting of the manganese moments occurs, polarizing the Gd 4f spins and, for further decreasing temperatures, a weak ferromagnetic moment evolves. Finally, at 6.5 K due to the interaction of the 4f spins, long range order of the Gd moment evolves, yielding a canting of the Gd spins with the FM (ferromagnetic) component antiparallel to the FM moment of the canted manganese spins [23]. The anomalies between the present and the experimental Specific heat in the low temperature range i.e. below 40K (Fig. 1) might be due to the exclusion of ferromagnetic spin wave contribution in our interaction potential. Finally, it may be concluded that the thermodynamic calculations of structurally more complex manganites have largely involved an atomistic approach involving semiempirical interatomic potential.

ACKNOWLEDGMENTS

The authors are thankful to the University Grants Commission (UGC), New Delhi for providing the financial support.

REFERENCES

1. T. Arima, T. Goto, Y. Yamasaki, S. Miyasaka, K. Ishii, M. Tsubota, T. Inami, Y. Murakami , Y. Tokura , *Phys. Rev. B* **72**, 100102 (R) (2005).
2. K. Noda, S. Nakamura, J. Nagayama, H. Kuwahara, *J. Appl. Phys.* **97**, 10C103 (2005).
3. A. A. Mukhin, V. D. Travkin, A. S. Prokhorov, A. M. Balbashov, J. Hemberger, A. Loidl , *J. Magnetism and Mag. Mat.* **272-276**, 96-97 (2004).
4. T. Katsufujj, H. Takagi, *Phys. Rev. B* **64,** 054415 (2001).
5. J. Nogues, Vassil Skumryev, J. S. Munoz, B. Martinez, J. Fontcuberta, L. Pinsard, A. Revcolev-Schi, *Phys. Rev. B* **64**, 024434 (2001).
6. T. Mizokawa, D. I. Khomskii, G. A. Sawatzky, *Phys. Rev. B* **63**, 024403 (2000).
7. O. Chmaissem, B. Dabrowski, S. Kolesnik, J. Mais, J. D. Jorgensen, S. Short, Phys. Rev. B **67**, 094431 (2003).
8. A. Munoz, M. T. Casais, J. A. Alonso, M. J. Diaz Fernandez, *Inorg. Chem.* **40**, 1020-1028 (2001).
9. R. Wrang, J. Gui, C. Dong, *Phys. Stat. Solidi b* **229**, 1145 (2002).
10. Takashi Hotta, Mohammad Moraghebi, Adrian Feiguin, Adriana Moreo, Seiji Yunoki, Elbio Dagotto, *Phys. Rev. Lett.* **90**, 247203 (2003).
11. A. E. Nikiforov, S. E. Popov, S. Y. Shashkin, *Physica B* **276**, 772-773 (2000).
12. J. Baier, D. Meier, K. Berggold, J. Hemberger, A. Balbashov, J. A. Mydosh, T. Lorenz, *Phys. Rev. B* **73**, 100402 (R) (2006).
13. K. Noda, S. Nakamura, J. Nagayama, H. Kuwahara, *J. Appl. Phys.* **97**, 10C103 (2005).

14. A. Szewezyk, M. Gutowska, B. Dabrowski, *Phys. Rev. B* **72**, 224429 (2005).
15. E. E. Rodriguez, Th. Proffen, Llobet, Ryne, *Phys Rev B* **71**, 104430 (2005).
16. J. A. Ionso, M. J. Martinez-Lope, M. T. Casais, M. T. Fernandez-Diaz, *Inorg, Chem.* **39**, 917 (2000).
17. E G Rini, Mala N Rao , N. K. Gaur , S. L. Chaplot, R. K. Singh, *Phys. Rev. B* **75**, 214301 (2007).
18. N. Kaur, R. Mohan, N. K. Gaur, R. K, Singh, *Physica C* **451**, 24-30 (2007).
19. N. Kaur, N. K. Gaur, R. K, Singh, *J. Phys. Chem. Solids*, (2007) (in Press).
20. Renu Choithrani, N. K. Gaur, *J. Magnetism and Mag. Mat.*, (2007) (in press).
21. N. K. Gaur, Renu Choithrani, Archana Srivastava, *J. Sol. Stat. Commun.*, (2007) (in press).
22. T. Kimura, S. Ishihara, H. Shintani, T. Arima, K. T. Takahashi, K. Ishizaka, Y. Tokura , *Phys Rev B* **68**, 060403 (R) (2003).
23. J.Hemberger, S. Lobina, H.-A. Krug von Nidda, N. Tristan, V. Yu. Ivanov, A. A. Mukhin, A. M. Balbashov, A. Loidl, *Phys Rev B* **70**, 024414 (2004).
24. J. Baier, D. Meier, K. Berggold, J. Hemberger, A. Balbashov, J. A. Mydosh, T. Lorenz, *Phys Rev B* **73**, 100402 (R) (2006).
25. J. Laverdier, S. Jandl, A. A. Mukhin, V. Yu. Ivanov, V. G. Ivanov, M. N. Iliev, *Phys Rev B* **73**, 214301 (2006).
26. J. C. Slater, K. G. Kirkwood, *Phys. Rev. Lett.* **37**, 682 (1931).
27. Roger a. deSouza, M. Saiful Islam, Ellen Ivers – Tiffe, *J. Mater. Chem.* **9**, 1621 (1999).
28. P. Dai, Zhang Jiandi, H. A. Mook, S. H. Liou, P. A. Dowben, E. W. Plummer, *Phys Rev B* **54**, R3694 (1996).

Study of Extended Rigidity Percolation in Se–Te–Sn Glasses Using Dsc

Sushama .D[a], Achamma George[b], and P. Predeep[c]
Condensed Matter Physics Laboratory, Department of Physics
Sree Narayana College, Kollam-691001, Kerala, India 691 001
a.M.S.M.College,Kayamkulam, Kerala, India
b.Achamma George, St.Stephen's College, Pathanapuram,Kerala,India
c. P.Predeep, National Institute of Technology, Calicut, Kerala
ppredeep@gmail.com

Abstract. The effect of addition of Sn on amorphous Se_{80} Te_{20} system has been studied using UV VIS NIR Spectrum. The samples are prepared by conventional melt quenching method. It has been found that the addition of Sn reduces the crystalline nature of Se_{80} Te_{20} chalcogenide alloy and amorphous samples are obtained for Se_{80-x} Te_{20} Sn_x (x= 0,2,4,6,8,10,15). The metal influences both the disordered network structure of the glass matrix and the transport properties in accordance with its chemical properties. The glass transition temperature and the crystallization temperatures show detectable changes at composition x = 4, 6 and 10. From the observed DSC results, it is seen that the compositions x = 4 and x = 6 denote the onset and completion of rigidity percolation and x = 10 corresponds to the chemical threshold of the system. The mean bond energy calculated using the covalent bond approach of Tichy and Ticha shows the same anomalies at x=4, 6 and 10
Keywords. Chalcogenide glasses, Rigidity percolation, bond energy
PACS: 61.43.-j, 61.43 Dq

INTRODUCTION

Because of their flexibility and chemical durability, chalcogenide glasses are attractive candidates for broad transmission range, stability against crystallization and far- infrared transmitting fibers [1]. Se based glasses are widely used for various applications in many fields as optical recording media, xerography, and electrographic applications such as photoreceptors in photocopying and laser printing, infrared spectroscopy, laser fibre technique etc [2-4]. Amorphous Se-Te alloys have greater hardness, higher crystallization temperature, higher photosensitivity and smaller ageing effects than pure Se [5]. The additions of third element (Sn) expands the glass forming area and also create compositional and configurational disorder in the system. The lattice perfection and the energy band gap of the material play a major role [6] in the preparation of devices for a particular wavelength, which can be modified by the addition of dopants. In the present work systematic investigation of the thermal stability and variation of glass transition temperature for Se_{80-x} Te_{20} Sn_x (x=0, 2, 4, 6, 8, 10, 15) system has been made. Calorimetric studies are made under non-isothermal conditions. In systems containing group IV-V-VI and V-III-VI elements, rigidity percolation threshold and chemical threshold can occur [7-9] at $<r>$ values other than 2.4 and 2.67. Extended rigidity percolation leads to an inter mediate phase existing between the floppy and the rigid

CP1004, *Thermophysical Properties of Materials and Devices-NCTP '07*,
edited by P. Predeep, S. Prasanth, and A. S. Prasad

phases [10]. In the present work, DSC measurements have been carried out to, explore the possibility of extended topological thresholds in Se_{80-x} Te_{20} Sn_x glasses.

EXPERIMENTAL

Bulk samples of the $Se_{80-x}Te_{20}Sn_x$ (x=0,2,4,8,10,15) are prepared by the melt quenching technique. Appropriate amounts of 5N purity elements are sealed in quarts ampoules and placed in the rotating furnace at a temperature of 700 ^{0}C for 24 hours for homogeneous mixing of the elements. The melt is then rapidly quenched in ice-cold water. DSC measurements are made for all compositions of the sample at a heating rate of 10^0C/min in order to scan the samples through their glass transition temperature (T_g), crystallization temperature(T_c) and melting temperature (T_m).

RESULTS AND DISCUSSION

DSC curves for Se_{80-x} Te_{20} Sn_x (x=0, 2, 4, 6, 8, 10, 15) are shown in figure (1). Figure (2)shows the variations of T_g values (given in Table 1) of the samples as a function of average coordination number (<r>).

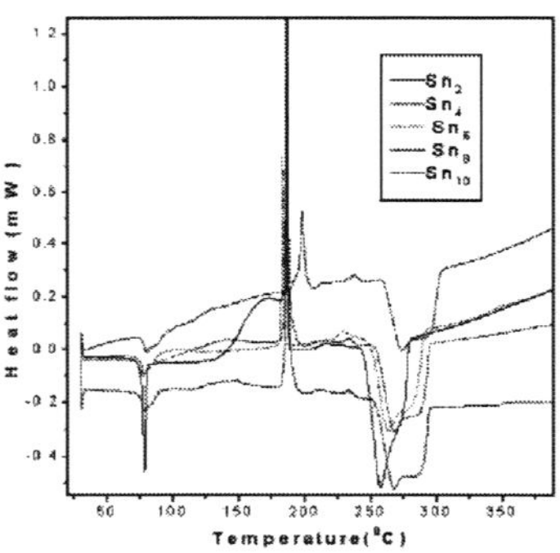

FIGURE 1. DSC thermograms of Se_{80}-x Te_{20} Sn_x glasses at the heating rate of 10^0C/min

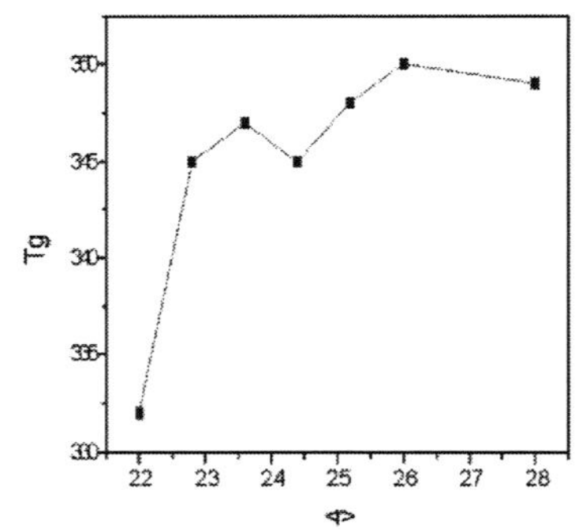

FIGURE 2. Variation of T_g with <r>

In chalcogenide glasses, the composition dependence of T_g and T_c are determined by factors such as coordination of additive element, rigidity percolation, chemical ordering etc. Here for the calculation of <r> the coordination number of Se is taken as 2, and that of hexagonal Te is 3 [11]. Sn is considered as [11,12] a six fold coordinated element with metallic- like bonding. Addition of Sn to the Se-Te system cross-links the already existing Se-Te chains [13] thus causing an increase in the chain length and hence a more rigid structure of the material is obtained. Some anomalies are present in figure (2). It can be seen that T_g increases up to an average coordination number <r> = 2.36 (x = 4) then decreases up to <r> = 2.44 (x = 6) and again increases. When it reaches at the point corresponding to <r> = 2.6 (x = 10) again it shows a decreasing trend.

Considerable attention has been given to understand the influence of two topological thresholds, namely the Rigidity Percolation Threshold (RPT) and Chemical Threshold (CT) respectively on the properties of chalcogenide glasses [14-16]. Rigidity percolation deals with network connectivity and at the rigidity percolation threshold equilibrium is established in a chalcogenide network glass, between the degrees of freedom per atom and the number of constraints acting on it [17-20]. For a glassy network with purely covalent bonding, the rigidity percolation threshold occurs [17-20] at a mean coordination number < r_c> = 2.4. The Chemical Threshold (CT) is usually taken [21] to be the critical composition at which chemical ordering is maximized in a chalcogenide network glasses, with the bonding being fully heteropolar. The Chemical Threshold (CT) is expected to occur [22] at a mean coordination number <r> = 2.67. Anomalies in various properties of chacogenide glasses have been observed at the Rigidity Percolation Threshold (RPT) and Chemical Threshold (CT)[23-28].

TABLE 1. Variation of T_g, T_c & (T_c-T_g) with <r>

<r>	T_g	T_c	T_c-T_g
2.2	332	343	11
2.28	345	411	66
2.36	347	455	108
2.44	345	452	107
2.52	348	453	105
2.6	350	463	113
2.8	349	383	34

In recent studies by Boolchand et al [29-31] it has been proposed that the rigidity percolation transition occurs over a range of mean coordination numbers leading to three distinct phases namely; floppy, intermediate and rigid. The occurrence of extended rigidity percolation has been observed in glasses such as Si-Se [36], As-Se [31] and Ge-Se [32]. Based on our present results, we propose that the floppy –rigid transition in $Se_{80-x}Te_{20}Sn_x$ glasses occurs over a range of average coordination numbers with the existence of an intermediate phase. Here the composition x = 4 (<r> =2.36) corresponds to the onset and x = 6 (<r> = 2.44) to the completion of rigidity percolation. The composition x = 10 (<r> = 2.6) corresponds to chemical threshold of the system. A higher coordination of metallic atoms in a chalcogenide network demands partially ionic bonding, [33] which explains the shifting in the RPT. This threshold behavior at <r> = 2.60 seems to support Tanaka's ideas [34] of formation and development of layered structures in network glasses. As per our studies Ge-Sb-Se, Ge-As-Te and Si-As-Te glasses show anomalous behavior near <r> = 2.60. The observed threshold in IV-V-VI glasses at <r> = 2.60 instead of the predicted value of <r>=2.67 can be attributed to a variety of reasons. Phillips and Tanaka have considered the interaction between atoms to be purely covalent while arriving at the balance conditions. Such assumption may [34] be valid to Ge-Se and Ge-As-Se glasses, but not for systems containing heavier elements like Te, Sb, Pb, Sn etc. The presence of such heavier elements leads to partial covalent bonding. This can affect the balance conditions. There are also some other shortcomings in the simple theory such as the assumption of the 8-n rule and the omission of the effect of Vander Waals' and dihedral angle forces. There are reports [35] that the coordination of elements like Te changes with composition and the simple assumption of the coordination number of the elements on the basis of the 8-n rule in the balance condition is not strictly valid for such cases.

The difference between the crystallization temperature and the glass transition temperature gives [36] the tendency of the system towards crystallization. Greater the difference smaller the tendency towards crystallization of a glassy amorphous system. The present result shows that $\Delta T = T_c - T_g$ is highest for $Se_{70}\,Te_{20}\,Sn_{10}$. When Sn is added to the Se-Te system formation of $SnSe_4$ tetrahedral promotes glass formation and improves the glass properties. Sn is less electronegative than Se. This is the reason [1] for the strong glass-forming tendency of Se-Te-Sn system

Application Of Covalent Bond Approach In Se-Te-Sn System

The covalent bond approach of Tichy and Ticha [37,38] to the glass transition temperature may be considered as a first approximation in the case of chalcogenide glasses. The glass transition temperature is considered to be proportional to the mean bond energy <E>, which depends on factors like mean coordination number, degree of cross linking, bond energy and the nature of bonds. Taking account of all these factors, they obtained a good correlation between Tg and <E> in the form,

$$T_g = 311[\langle E \rangle - 0.9] \qquad (1)$$

which satisfies [38] the Arrehenius relation for viscosity .

TABLE.2. Variation of <E> & H_s with average coordination number <r>

<r>	<E>(eV)
2.2	1.967
2.28	2.009
2.36	2.015
2.44	2.009
2.52	2.018
2.6	2.025
2.8	2.022

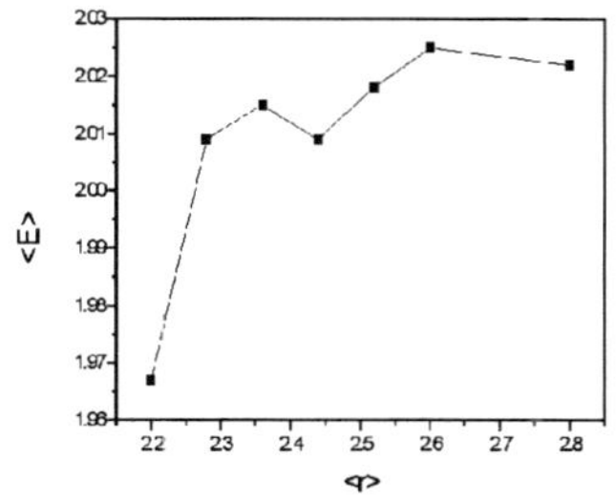

FIGURE 3 Variation of <E> with <r>

Applying this model in our problem, we have evaluated mean bond energies for various compositions of Se-Te-Sn system, and it can be seen that T_g is proportional to mean bond energy, <E>. The value <E> also shows a disorder corresponding to RPT and CT. The values of <E>thus evaluated are also listed in Table 2. Figure (3) shows the variation of <E> with average coordination number. The major limitation of this model is that, it does not account for the molecular interactions, which play a vital role in relaxation process in glass transition region.

REFERENCES

1. Zhiyu Wang, Chongjiang Tu, Yuanmei Li, Quanqing Chen J.Non. Cryst. Solids **191** (1995) 132-137
2. M.Horie,T.Ohno, N.Nobukuni, K.Kioyo,T.HahizumeTech.Digest, ODS2001 MCI (2001) **37**.
3. T. Akiyama, M. Uno, H. kituara, K. Narumi, K. Nishiuchi, N. Yamada, Jpn. J. Appl. Phys.**40** (2001) 1598.
4. T. Ohta, J. Opto-electron. Adv. Mater. **3** (2001) 609.
5. S.O.Kasap,T.Wagner,V.Aiyah, O. Krylouk, A.Bekirov, L.Tichy ,J. Master.Sci.**34** (1999) 3779
6. E.Marquez, T. Wagner, J.M.Gonzalez – Leal, A.M. Bernal- Olive, R. Prieto- Aleon, R. Jimenez – Garay, P.J.S.Ewen, J.Non.Cryst. Solids **274** (2000) 62
7. A. Giridhar and Sudha Mahadevan, J.Non- Cryst. Solids **134** (1991) 94
8. Sudha Mahadevan and A. Giridhar, J.Non- Cryst. Solids **143** (1992) 52
9. Sudha Mahadevan and A. Giridhar, J.Non- Cryst. Solids **152** (1993) 42
10. P.Zaheerudeen Saheb, B.H.Sharmila, S.Asokan, K. Appaji Gowda . Solid state Communications **129** (2004) 765-768
11. Liang Zhenhua J.Non-Cryst. Solids **127** (1991) 298-305
12. MIN Szukwei, Yang Hanmei and ZHANG Xiaowei, J.Non-Cryst. Solids **112** (1989) 204-206.
13. G.Kaur, T.Komatsu, R.Thangaraj.J.Mater.Scie.**35** (2000) 903-906.
14. R.Aravinda Narayanan, S.Asokan, A.kumar, Phys. Rev.**B54**, 4413 (1996)
15. R.Aravinda Narayanan, S.Asokan, A.kumar, Phys. Rev.**B63**, 092203-1 (2001)
16. J.T.Devaraju, B.H.sharmila,S.Asokan, K.V.Acharya, Philos. Mag. B (2001)
17. J.C.Phillips, J.Non-Cryst. Solids **34**,153 (1979)
18. J.C.Phillips, M.F.Thorpe, Solid State Commun. **53**, 699 (1985)
19. P.Boolchand, M.F.Thorpe, Phys.Rev. B **50**, 10366 (1996)

20. A.K.Varshneya, A.N.Sreeram, D.R.Swiler, Phys. Chem. Glasses **34**, 179 (1993)
21. G.Lucovsky, T.M.Hayes in Amorphous Semiconductors Ed. M.H.Brodsky, Springer-Berlin, **197**a, P.215
22. R.Aravinda Narayanan and A.kumar, "Unified approach to the constraint counting theory of glasses," Phys. Rev. B **60**, pp. 11859-11862,1999
23. S.Asokan, M.V.N.Prasad, G.Parthasarthy and E.S.R.Gopal ,"Mechanical and chemical threshold in IV-VI chalcogenide glasses," Phys.Rev.Lett.**62**,pp.808-810,(1989).
24. S.Mahadevan, A. Giridhar and A.K. Singh,"Volumetric effect of topological in chalcogenide glass system" J.Non-Cryst. Solids, **169**,pp133-142 (1993)
25. A.K.Varshneya, A.N. Sreeram and D.R.Swiler,"A review of the average coordination number concept in multicomponent chalcogenide glass system," Phys.Chem. Glasses, **34**,pp.179-192,(1993)
26. U.Senapati and A.K. Varshneya,"configurational arrangements in chalcogenide glasses; a new perspective on Phillips constraint theory,"J.Non-Cryst.Solids, **185**,pp.289-296, (1995).
27. B.Uebbing and A.J.Sievers,"Role of network topology on the vibrational lifetime of an H_2O molecule in the Ge-As-Se glass series," Phys.Rev.Lett.,**76**,pp.932-935,(1996).
28. X.Feng,W.J.Bresser and P.Boolchand,"Direct evidence for stiffness threshold in chalcogenide glasses."Phys.Rev.Lett.,**78**,pp.4422-4425,(1997).
29. D.Selvanathan, W.J. Bresser, P.Boolchand, Phys. Rev. B 61(2000) 15061.
30. P. Boolchand, J.Optoelect. Adv. Mater.**3** (2001) 703.
31. D.G.Georgiev, P. Boolchand, M. Micoulaut, Phys. Rev. B **62** (2000) R9228.
32. X.Feng,W.J.Bresser, P. Boolchand.Phys. Rev.Lett. **78** (1997) 4422
33. Devaraju J T, Sharmila B H, Acharya K V, Asokan S and Gopal E S R. Proc.SPIE Vol **4412**
34. E.S.R.Gopal, A.Srinivasan,S.Asokan.Indian Journal of pure and Appl.Phys.Vol.**31** April 1993
35. Liu J Z and Taylor P C, Phys. Rev, **B41** (1990) 3163
36. K.Ramesh, S.Asokan, K.S.Sangunni and E.S.R.Gopal The Europian Physical Journal B **6** 207-212 (1998)
37. .L.Tichy, H.Ticha, J.Non-Cryst.Solids **189,** 141 (1995).
38. Tichy , H.Ticha, Mater.Lett.**21,** 313 (1994) .

Phase Transformation and Kinetics of Crystallization of $Ge_{0.5}Sn_{0.5}Se_{2.5}$ Glass

Deepika[a*], Vibhav K. Saraswat[a,b], Praveen K. Jain[a], Narendra S.Saxena[a*], Kananbala Sharma[a], Thaneshwar P. Sharma[a] and Sandeep K. Dhawan[c]

[a]*Semi-conductor & Polymer Science Laboratory, 5-6, Vigyan Bhawan, University of Rajasthan, Jaipur-302004, INDIA*
bDepartment of Physics, Banasthali University, Banasthali-304022, INDIA
[c]Polymeric & Soft Materials Section, National Physical Laboratory, New Delhi-110012, INDIA
*Corresponding E-mail: n_s_saxena@rediffmail.com, deepika.spsl@gmail.com

Abstract. In the present work, Differential Scanning Calorimetry (DSC) has been performed at different heating rates under non-isothermal conditions to study the crystallization kinetics of the $Ge_{0.5}Sn_{0.5}Se_{2.5}$ chalcogenide glass. Four different methods of analysis (Kissinger equation, Ozawa equation, Matusita equation and Augis and Bennett method) of the non-isothermal DSC data and various kinetic parameters i.e the activation energy of glass transition (E_t), the activation energy of crystallization (E_c), order parameter, rate constant, frequency factor have been determined. The rate constant K (indicator of the rate of to crystalline phase transformation) has been evaluated using the method of Augis and Bannett. Further, the rate of crystallization and dimensionality of growth has also been investigated for the above system.

Keywords: Chalcogenide glass, Crystallization Kinetics, Activation energy, Rate Constant.
PACS: 64.70.dg

INTRODUCTION

Great attention has been given to chalcogenide glasses over the past several years due to their wide range of applications. Chalcogenide glasses in Ge-Se system are interesting materials due to their large range of transparency and good mechanical properties such as hardness, adhesion, low internal resistance and water resistance [1]. The addition of third element in tetrahedral structure of Ge-Se glass converts the glass into an interesting material and new promising properties of the material are expected [1]. With this point of view Ge-Se system with Sn additive has been studied. The aim of the present paper is to throw some light on the phase transformation and the crystallization kinetics of $Ge_{0.5}Sn_{0.5}Se_{2.5}$ glassy system. The study of kinetics is always connected with the concept of activation energy. The values of this energy are associated with nucleation and growth mechanism that dominate the divtrification of most of the glassy solids [2,3]. The activation energy of both the phase transformations(glass transition and crystallization) has been obtained using theoretical models [4-5] such as Kissinger's equation, Ozawa equation, Matusita and Sakka theory and Augis-Bennett approximation. Besides, rate of crystallization and dimensionality of growth of the above system have also been investigated.

CP1004, *Thermophysical Properties of Materials and Devices-NCTP '07*,
edited by P. Predeep, S. Prasanth, and A. S. Prasad

EXPERIMENTAL

Glassy alloy of $Ge_{0.5}Sn_{0.5}Se_{2.5}$ was prepared by melt-quenching technique. High purity (99.999%) of all constituent materials in appropriate atomic weight percentage proportions were weighed into a quartz ampoule. The content of the ampoule (2g) was sealed in a vacuum of 10^{-5} Torr and then heated in furnace where temperature was raised at a rate of 2-3°C per minute up to 900 °C and kept around that temperature for 15 hours. The ampoule was frequently rocked to ensure the homogeneity of the sample. The molten sample was rapidly quenched in ice-cooled water to get glassy state. The ingot of so produced glassy sample was taken out of the ampoule by breaking the ampoule and then grinded gently in mortar and pestle to obtain powder form. The amorphous nature of the sample was ascertained through XRD of the sample shown in fig.1.The absence of any sharp peak confirms the amorphous nature of the sample. The thermal behavior of the sample has been investigated using Mettler-Toledo DSC 822^e.

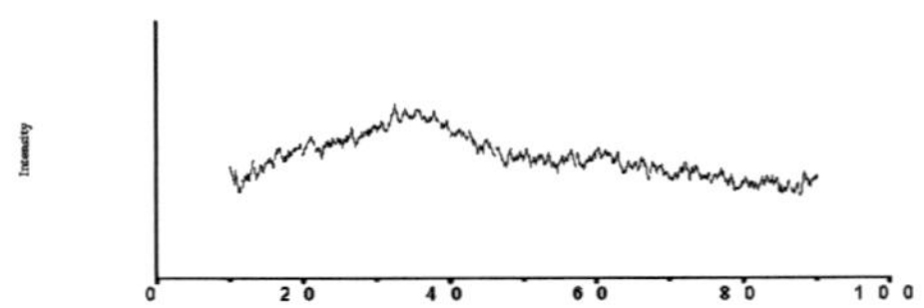

FIGURE 1. XRD pattern of $Ge_{0.5}Sn_{0.5}Se_{2.5}$ glass

RESULTS AND DISCUSSION

Differential Scanning Calorimetry (DSC) has been employed at five different heating rates (5,10,15,20 and 25 °C) under the non-isothermal conditions. A typical DSC thermogram for $Ge_{0.5}Sn_{0.5}Se_{2.5}$ glass at heating rate of 10 K/min is shown in fig.2.Two characteristic phenomena are evident in the DSC trace. The first one corresponds to the glass transition region (endothermic reaction) and the other to the crystallization region (exothermic reaction). All the samples show a single glass transition temperature and a single crystallization peak indicating the sample as homogeneous. The variation of glass transition temperature (T_g) with heating rate is shown in fig.3.Theoritically, T_g is defined as the temperature at which the relaxation time τ becomes equal to the experimental time of observation time $\tau_{obs.}$ At the same time T_g varies inversely as the relaxation time [6]. With increasing heating rate, τ_{obs} decreases hence the glass transition temperature increases.

Glass Transition

An amount of energy called the glass transition activation energy E_t, must be given to groups of atoms in the glassy region so that a jump from a metastable state to a more stable state takes place. The evaluation of E_t has been widely [7-8] carried out using Kissinger's equation [9] inspite of the fact that this equation was used for the evaluation of the crystallization activation energy. The equation may be written as –

$$\ln(T^2_{gp} / \alpha) = E_t / RT_{gp} + Const. \tag{1}$$

where, T_{gp} is the peak temperature in the endothermic reaction in glass transition region. Fig.4 shows the plot of ln (T^2_{gp} / α) as a function of $1/T_{gp}$.The slope of the resulting straight line gives E_t.

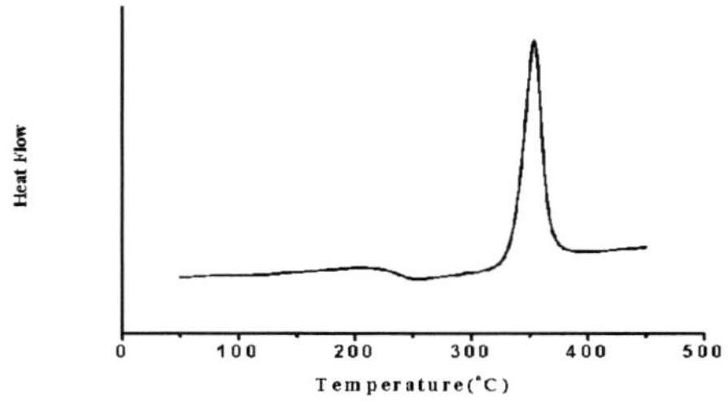

FIGURE 2. DSC thermogram of $Ge_{0.5}Sn_{0.5}Se_{2.5}$ glass at heating rate of 10K/min.

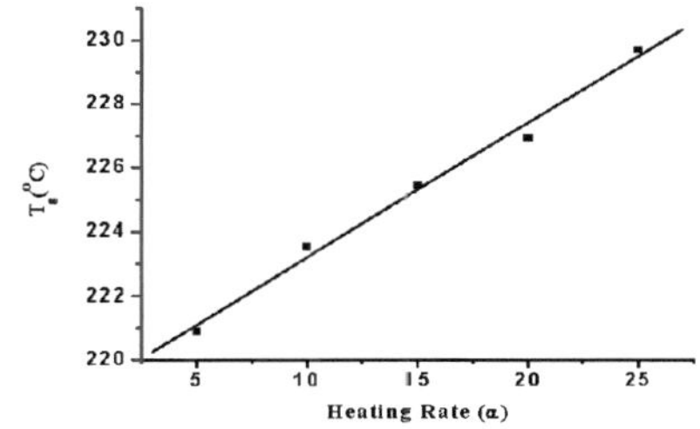

FIGURE 3. Variation of glass transition temperature T_g with heating rate α in $Ge_{0.5}Sn_{0.5}Se_{2.5}$ glass

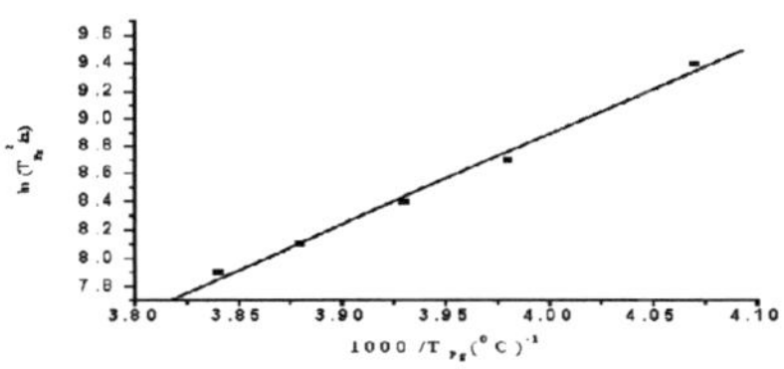

FIGURE 4. Plot of ln (T^2_{gp} / α) against $1000/T_{gp}$ of $Ge_{0.5}Sn_{0.5}Se_{2.5}$ glass(for the glass transition activation energy)

Crystallization

The crystallization process under non-isothermal conditions has been analysed using three different models. Besides, regular features, dimensionality of growth and avrami exponent have also been computed.

According to Matusita and Sakka [10] theory, for non-isothermal crystallization-

$$\ln[-\ln(1-x)] = -n \ln \alpha - 1.052\, mE_c / RT + Const. \tag{2}$$

Where x is the fraction of volume x crystallized at any temperature T, m and n are are constants relate to the crystallization mechanism. E_c is the activation energy for crystallization and R is a constant. For the evaluation of E_c, the slopes of the plot of ln [-ln(1-x)] against 1/T at different heating rates are computed. The curves are shown in fig.5 (a). The curve is a straight line for most of the temperature range but shows a break at higher temperatures, which is attributed to the saturation of nucleation sites in the final stages of crystallization [11] or to the restriction of crystal growth by the small size of the particles [12].

The value of Avrami exponent 'n'is obtained from the slope of ln [-ln(1-x)] versus ln α curves (fig.5(b)).The value of n has been evaluated from the slopes of the straight line fit of these relations. It is clear from the figure that n is temperature independent and hence an average value can be calculated. The average value of n is 3.5, which gives m=2 (2.5$\approx$2) suggesting bulk nucleation with two-dimensional growth. The value of m is taken to be n-1 since no prior heat treatment was given to the samples before the thermal analysis runs were taken.

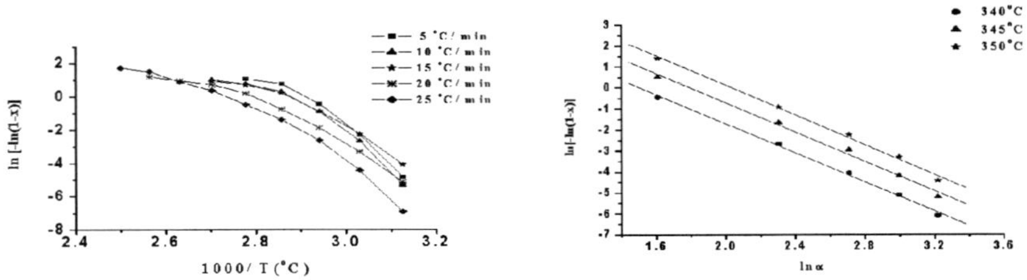

(a) (b)

FIGURE 5. (a)Plot of ln [-ln (1-x)] against 1000/T at different heating rates. (b) Plot of ln [-ln (1-x)] against ln α of $Ge_{0.5}Sn_{0.5}Se_{2.5}$ glass at different temperatures

The activation energy E_c of crystallization can also be conveniently calculated from the variation of the onset crystallization temperature T_c with the heating rate through the use of Ozawa's equation [13].

$$\ln \alpha = - E_c / R\, T_c + \text{Const.} \quad (3)$$

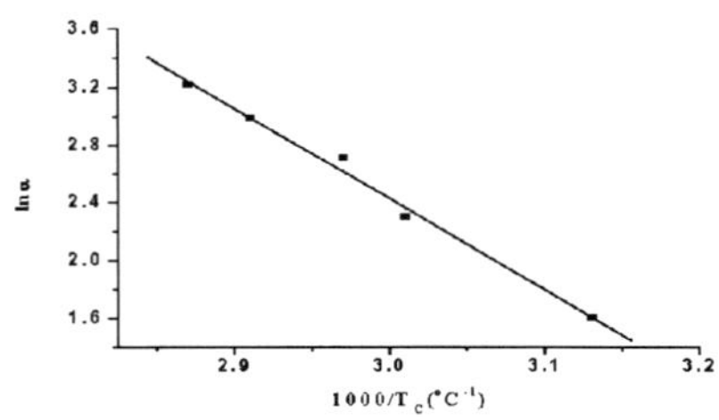

FIGURE 6. Plot of ln α against $1000/T_C$ of $Ge_{0.5}Sn_{0.5}Se_{2.5}$ glass.

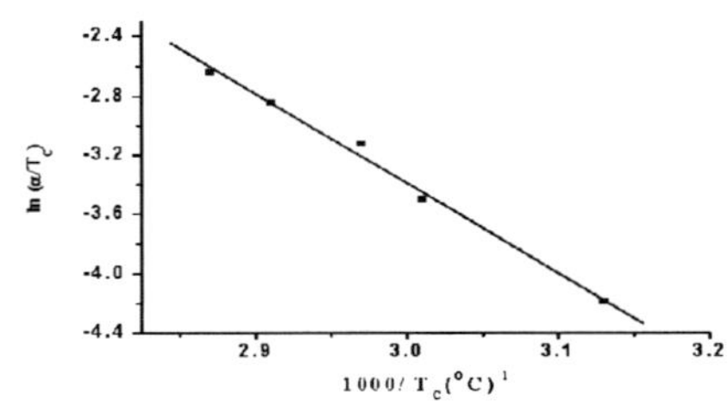

FIGURE 7. Plot of Ln (α/T_c) against 1000/ T_c ($^oC^{-1}$) of $Ge_{0.5}Sn_{0.5}Se_{2.5}$ glass

Figure 6 shows the plot of ln α against 1 / T_c. It is found to be linear for the entire heating range. From the slope of this plot, the activation energy E_c of crystallization is obtained.

The activation energy E_c of crystallization can also be determined by an approximation method developed by Augis and Bennett [14]. The relation used by them is of the form:

$$\ln (\alpha / T_c) = - \quad E_c / R\, T_c + \ln K_o \quad (4)$$

The plot of ln (α /T_c) against 1/T_c (shown in fig.7) gives activation energy E_c of crystallization. This method has an extra advantage that the intercept of ln (α/T_c) against 1/T_c gives the value of pre-exponential factor K_o(frequency factor) of Arrhenius equation.

$$K= K_o \exp (E_c / RT) \tag{5}$$

The value of Frequency factor (K_o) and Rate constant as obtained from Eq. (4) and Eq.5 is 2.70x10^6 $(s)^{-1}$ and 1.30 respectively.

The glass transition activation energy E_t calculated from equation (1) and comparison of E_c values obtained from equation (2-4) are shown in table 1. The values of E_c are in good agreement with each other.

TABLE 1. The calculated values of the glass transition activation energy E_t , the activation energy of crystallization E_c using different models

Alloy	E_t (kJ/mol)	E_C (kJ/mol)		
	ln (T^2_{gp} / α) Vs 1000/T_{gp} ($^oC^{-1}$)	ln[-ln(1-x)] Vs 1000/T($^oC^{-1}$)	ln α Vs 1000/T_c ($^oC^{-1}$)	ln (α/T_c) Vs 1000/ T_c ($^oC^{-1}$)
$Ge_{0.5}Sn_{0.5}Se_{2.5}$	54.20	58.83	52.21	50.38

CONCLUSIONS

It has been found that E_c values are in good agreement with each other. Thus one can use any of the three methods to calculate the activation energy of crystallization. The Avrami exponent n is temperature dependent with value ranging from 2.66-3.93, which indicates bulk nucleation with two-dimensional growth.

ACKNOWLEDGMENTS

Authors are thankful to Ms. Vinodini Shaktawat, Ms. Manasvi Dixit, Mrs. Rashmi Saxena and Mr. Kuldeep S Rathore for their help in various ways.

REFERENCES

1. N.Mehta, S.K. Agrahari and A. Kumar, *Materials Letters* **61** (3), 837-841 (2007).
2. S. S. fouad, S. A Fayek and M. H. Ali, *Vacuum* **49** (1), 25-30 (1998).
3. P. Predeep, N. S. Saxena, M.P. Saxena and A. Kumar, *Physica Scripta*, **54** 207-209 (1996).
4. Rohit Jain, N. S. Saxena, Deepika Bhandari and S. K. Sharma, *Mater. Sci. Engg. A* **298**, 51-55 (2001).
5. G. Kaur, T. Komatsu and R. Thangaraj, *J. Mater. Sci.* **35**, 903-906 (2000).
6. S.A. Fayek, *Vacuum* **72**, 11-20 (2004).
7. J. Vazquez, P.L. L-Alemany, P. Villares, R. Jimenez-Garay, Mater. Chem. Phys. **57**, 162-168 (1998).
8. N. Mehta, A. Kumar, *Materials Chemistry and Physics* **96**, 73-78 (2006).
9. H.E. Kissinger, *J. Res. Nat. Bur. Standards* **57**, 217-227 (1956).
10. K. Matusita, T. Komatsu, R. Yokota, *J. Mater. Sci.* **19**, 291-300 (1986).
11. M.M.A Imran, N. S. Saxena, D. Bhandari and M. Hussain, *phys. Stat. Sol (a)* **181**, 357-368 (2000)
12. *G. Kaur, T. Komatsu, J. Mater. Sci.* ***36**, 4531-4533 (2001).*
13. T. Ozawa, *Polymer* **12**, 150-158 (1971).
14. J.A. Augis, J.E. Bennett, *J. Them. Anal.* **13**, 283-292 (1978).

Thermal Properties of Natural Rubber Composites with Organic and Inorganic Fillers

Najidha.S[a], P.Predeep[a*] and N S. Saxena[b],

[a]Condensed Matter Physics Lab., Sree Narayana College, Kollam, Kerala, India 691 001.
[b] Semiconductor and Polymer Science Lab, Department of Physics, University of Rajasthan, Jaipur

Abstract. The thermal transport properties of Natural Rubber/Polyaniline and Natural Rubber /Polyaniline/Carbon black composites have been investigated by Transient Plane Source (TPS) Technique at room temperature. The samples of different weight percentage (typically 20,30,40,50 and 60%) of fillers have been taken. The composites were prepared by dry mill mixing in a roll-mill and vulcanized in a hot press. It has been found that the effective thermal conductivity and effective thermal diffusivity of the both the composites increase as the fraction of filler increases.

Keywords: Polyaniline, Natural Rubber, carbon black, vulcanization, thermal transport properties.s
PACS: 73.61.Ph, 78.66.Qn

INTRODUCTION

Blending is a technique known in polymer technology that takes advantage of the processibility of polymers to produce new solid materials or composites with specific structural and physical properties, distinct from the ones of their components. Conductive vulcanized elastomers represent an important segment of the market for applications in rubber industry. These materials are used for shielding the sources of electromagnetic radiation interference to protect electronic equipment in modern vehicles. Presently, they are produced by mixing conductive carbon black in the rubber compound prior to vulcanization [1]. However, the percolation threshold for the conductivity is high and also the relative high filler content needed for high electrical conductivity may adversely affect the materials processability and properties. Therefore, there is an increasing interest in processable polymeric materials, using conventional techniques, whose electronic conductivity can be tailored for a given application, having attractive mechanical properties. A different approach in obtaining conducting composite systems with low percolation threshold involves mixing solutions of a matrix polymer such as nitrile rubber, Ethylene propylene rubber (EPDM), Styrene butadiene rubber, etc. and an intrinsically conducting polymer such as polyaniline, polypyrrole as fillers. Doped conducting polymers present a much higher conductivity; unfortunately their, mechanical

CP1004, *Thermophysical Properties of Materials and Devices-NCTP '07*, edited by P. Predeep, S. Prasanth, and A. S. Prasad

properties do not permit their use in substitution to vulcanized elastomers [2]. However, the processing problems and the cost effectiveness for obtaining high-quality homogeneous polymers are still challenging issues for many potential applications. Because of this, there is a growing demand for dry mill mixing method compared to the solution mixing. The roll mill mixing provides more uniformity and hence convenient method compared to the solution-processable composite in which the use of organic solvents arise problem.

Besides, the preparation of hybrid organic-inorganic composite materials are gaining more attention due to the promising technological application of these materials in field of non linear optics, coatings and in semipermeable membranes. Among the conductive fillers, carbon black is perceived as one of the best candidates because of its high electrical conductivity, environmental stability and low cost. However fillers like carbon powder have various limitations when used to provide conductivity to polymer matrices. Because of this there is an increasing demand for the combination of this inorganic filler with organic fillers such as polyaniline and various matrixes so that the hybrid composite can be tuned for various technical applications. Apart from known chemical and mechanical characteristics of polymeric composite systems their thermal characteristics are of significance for a number of technical applications. Heat transport properties of elastomers are attracting more attention throughout the rubber industry. In view of this, the thermal transport properties of elastomeric composites using different fillers have been studied by TPS technique

EXPERIMENTAL

Natural Rubber (ISNR-5) was supplied by the Rubber Research Institute of India. Aniline (Merck) was thrice distilled under reduced pressure and stored under nitrogen at 4^0 C. Ammonium per sulfate (APS) (Nice Chemical) as oxidant and hydrochloric acid (36.46%) (Merck) as dopant were used as received. Carbon black (commercial grade) Methanol and acetone (Merck) were used without further purification. Polyaniline was synthesized in powdered form by the conventional chemical method of oxidative polymerization of aniline with ammonium peroxy disulphate (APS) in an aqueous acidic medium [3].The composites were prepared by mechanical mixing in a roll-mill at room temperature according to the formulations indicated in table 1. NR was masticated and then blended with polyaniline and carbon black. In the case of NR/Polyaniline /Carbon black composite, the total filler content is fixed to 40phr so as to avoid polymer dilution effect. The samples were then vulcanized at 150^0C and 10 tonnes of pressure for 4 min using a Hot press Model. Vulcanized Circular samples of 13mm diameter are used for the experiment.

TPS technique is used to perform measurements of thermal conductivity of all the samples at room temperature. The measurements were performed with a 'resistive element'. It is made of 10μm thick nickel foil with an insulating layer made of 50μm-thick Kapton, on each side of the metal pattern. Each TPS element has a resistance of about 3.6Ω at room temperature and a temperature coefficient of resistance of around $4.6x10^{-3}$ K^{-1}.The technique is based on a three-dimensional heat flow inside the sample, which can be regarded as an infinite medium by limiting the total time of the transient recording. The experiment is performed by recording the voltage variations over the TPS element while its temperature is slightly raised by a constant electrical current pulse (width 15s and height 0.0729Amp). The transient potential difference across the terminals is recorded by a digital multimeter and the current through the TPS sensor with a digital power supply. Multiple readings are taken to ensure the accuracy of the results.

TABLE 1. Formulation of mixes used for crosslinking

Formulation	**Composition (g)**
Natural Rubber	100
ZnO	5
Stearic acid	2.5
Sulfur	2.5
HS	1
CBS	1
TMT	0.12
Polyaniline + Carbon black	40

RESULT AND DISCUSSION

The experimental results of effective thermal conductivity of Natural Rubber/ Polyaniline composites and Natural Rubber/Polyaniline/Carbon black composites have been plotted as a function of volume percentage of respective fillers in figure 1 & figure 2.

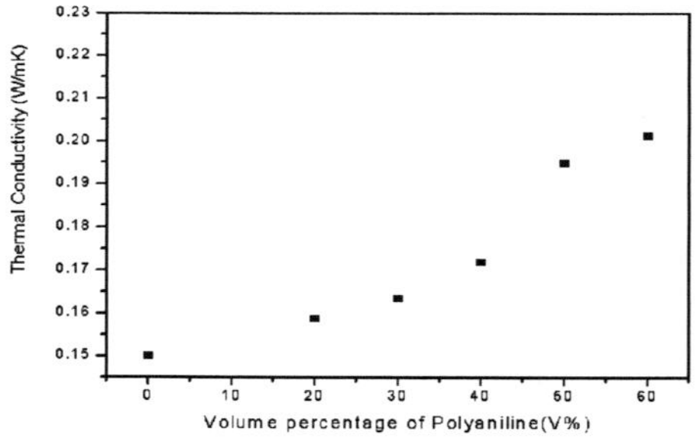

FIGURE 1. Thermal conductivity of composite vs. volume fraction of polyaniline

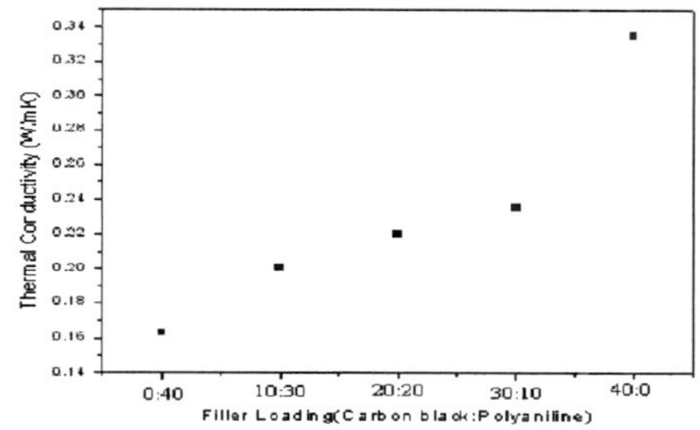

FIGURE 2. Thermal conductivity of hybride composite vs. volume fraction of filler

From figures it is clear that as the volume percentage of fillers increases the thermal conductivity of both the composite increases. The effective thermal conductivity of the composite increases due to the replacement of the matrix by the more conductive polyaniline. As the filler content increases, more and more filler particles are available for conductive chain formation in the NR matrix and hence the thermal conductivity of the composite increases. Besides, as the volume percentage of carbon black increases, the thermal conductivity of the hybrid composite increases much more than that of pure composite. Thus it can be seen that the inorganic filler, carbon black can be exerted higher thermal conductivity than the conjugated polymer, polyaniline in the NR matrix though the electrical conductivities of the two fillers are almost similar. This may be due to the fact that the addition of polyaniline decreases the crosslinking density due to the presence of acid in the polyaniline and hence the thermal conductivity increase is only due to the conductive chains of the polyaniline filler itself. That is addition of polyaniline lowered the crosslinking degree, but produced a reinforcing effect in the elastomer. But in the case of NR/carbon black/polyaniline composite the addition of black increases both crosslinking density and reinforcing effect simultaneously due to the formation of a continuous net of graphite layer which causes a greater increase in thermal conductivity than that of polyaniline filled NR. The variation of the thermal diffusivity with filler loading is shown in figure 3 & 4 .The trend of the thermal diffusivity is the same as that of the thermal conductivity since it is directly proportional to thermal conductivity.

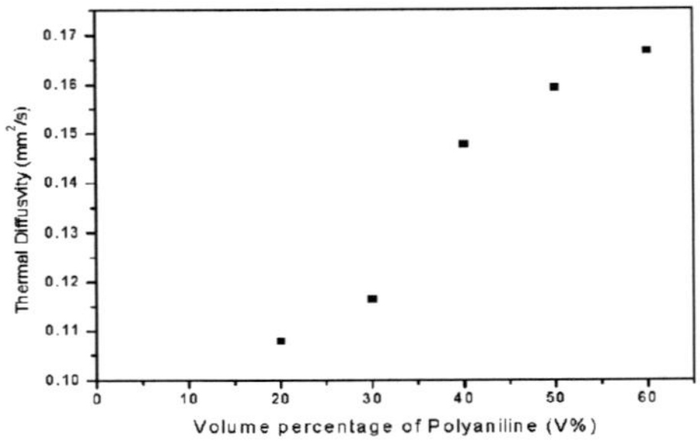

FIGURE 3. Thermal diffusivity of composite vs.volume fraction of polyaniline

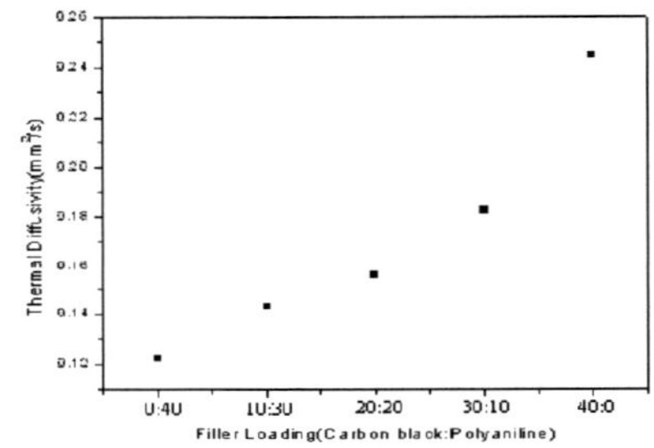

FIGURE 4. Thermal diffusivity of hybride composite vs. volume fraction of filler

REFERENCES

1. 1 M.R.Vallim,M.I.Felisberti,M.A.De Paoli. J.Appl.Polm.Sci. Vol.**75**(2000)677-684.
2. J.Laska,A.Pron,M.Zagorska,S.Lapkowski,S.Lanfrant.Synth.Met.**B** 7(1995)177.
3. A.G.Mac Diramid, J.C.Chiang, A.F.Richter, N.L. D.Somasiri and A. J. Epstein., Conducting Polymers. ed. L. Alcacer and D. Reidel Publishing Company.,(1987) 105.

Thermal Analysis of Crystallinity Variations in Semicrystalline Polymer Composites

S.Bhadrakumari[a] and P.Predeep[b]

[a]Department of Physics, M.S.M. College, Kayamkulam, Kerala.

[b]*Department of Science and Humanities, NIT, Kozhikode, Kerala.*

Abstract. Composite samples of $YBa_2Cu_3O_{7-x}$ and linear low density polyethylene (LLDPE) prepared by mixing both materials, melting the polymer followed by hot pressing, display improving flexibility and crystallinity with increasing polymer content. The crystallization behavior is studied using differential scanning calorimetry (DSC). Morphological changes like crystallite size, degree of crystallinity etc. are found to increase with increasing volume fraction of the polymer. The composition dependence of the overall rate of crystallization is explained in terms of the processes of nucleation and crystal growth. It is found that the addition of superconductor decreases the rate of nucleation of polymer and hence a reduction in the crystallinity.

Keywords: superconductor, polymer, composites, semicrystalline polymers.

PACS: 61.25.hk; 64.70.dg; 64.70.dj

INTRODUCTION

The discovery of high T_c oxide superconductors in 1986 has since generated an unprecedented challenge in both fundamental research and the applications of these materials. Unfortunately, the use of these ceramics is limited by their inherent brittleness and chemical instability. Incorporation of the brittle ceramics in to composites with polymer [1] is a good solution. In case of semicrystalline polymers the morphology is influenced by the crystallization behavior of the polymer [2]. The morphological features such as degree of crystallinity, domain shape and size are affected by relative areas of crystallization of polymer. Knowledge of the crystallization behavior of the thermoplastic composites is therefore necessary for effective manipulation of properties and for specifying the molding conditions.

In this study, composites containing superconducting $YBa_2Cu_3O_{7-x}$ (YBCO) are formed using Linear Low Density Polyethylene (LLDPE) as the matrix material. This polymer has been chosen mainly because it is readily worked at room temperature since it possesses low glass transition temperature. YBCO has been chosen as the filler material because they exhibit critical temperatures near 90K, [3] well above the boiling point of liquid nitrogen (77K).

CP1004, *Thermophysical Properties of Materials and Devices-NCTP '07*,
edited by P. Predeep, S. Prasanth, and A. S. Prasad

EXPERIMENTAL DETAILS

The $YBa_2Cu_3O_{7-x}$ powder used in this study is prepared by the usual solid state sintering reaction. The superconductor (YBCO) / polymer (LLDPE) composites are prepared by melt-blending of the components at various temperatures around the melting point of LLDPE and at varying mixing rates. The thermal characterization data are obtained at a cooling rate of 10°C/minute.

RESULTS AND CONCLUSIONS

The DSC thermograms of YBCO-LLDPE composites in the regions of crystallization, recorded during the cooling scan are depicted in figure 1. The DSC scan of 80% LLDPE sample shows a wide and sharp crystallization peak. It can be seen that the sharpness of the peak decreases with increasing percentage of superconductor. This indicates that the overall amount of crystallinity is decreasing in the composites as the superconductor content increases.

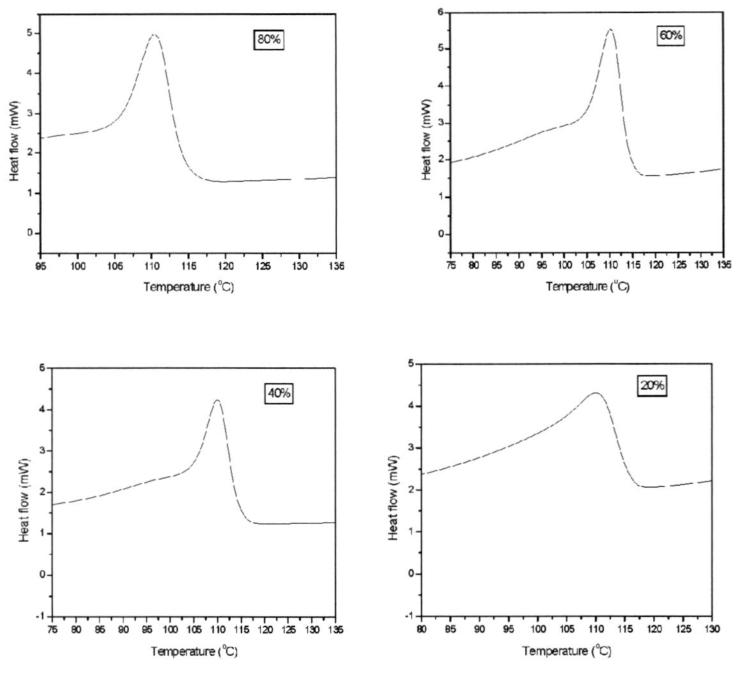

FIGURE 1. DSC scans of 80%, 60%, 40%, and 20% LLDPE composite in cooling mode.

The parameters significant in the crystallization behavior such as the temperature of onset of crystallization (T_{co}), which is the temperature where the thermogram initially departs from the baseline on the high temperature side of the exotherm, the peak temperature of the crystallization exotherm (T_c), and the heat of crystallization (ΔH_c) are shown in table1.

TABLE 1. Thermal behavior (crystallization) of 80%, 60%, 40%, and 20% LLDPE composites

Sample Composition	Onset	Peak °C	H_c x10^3 °C J/Kg
Y-123/LLDPE, 80%	114.24	110.79	294.81
Y-123/LLDPE, 60%	114.19	110.69	283.13
Y-123/LLDPE, 40%	114.16	110.95	272.26
Y-123/LLDPE, 20%	113.15	110.20	230.06

From this table it is clear that the heat of crystallization, ΔH_c decreases with increasing percentage of superconductor. The results presented here show that having increased surface area for crystal nucleation, ie, adding Y-123 particles is not enough to enhance crystallinity. Addition of increasing concentration of superconductor to LLDPE resulted in a decrease in T_{co} of LLDPE. A decrease in T_{co} in the composites clearly indicates that the inclusion of a superconducting phase in LLDPE results in delayed nucleation. The difference in T_{co} and T_c values of LLDPE in the composites were also decreases with increasing percentage of superconductor. Such a decrease can be attributed to an increase in the rate of crystallization [4]. Beck and Ledbetter [4] suggested that the small the difference between the onset and peak temperature (T_{co}-T_c), the faster the overall crystallization rate. The onset temperature indicates the beginning of the crystallization process, while the maximum of the exothermic peak indicates [5] the occurrence of the spherulite impingement. The free expansion of spherulites occurs between the onset and peak temperatures. The increase in the peak crystallization temperature of LLDPE in the lower superconductor content composites suggested that the crystal growth of LLDPE in the composite would takes place at a higher temperature leading to larger crystallite size, improved crystal perfection and narrower crystallite size distribution.

The composites may be regarded as consisting of two distinct crystalline regions(Y-123 crystallites and LLDPE crystallites). The process of recrystallization of the polymer upon cooling from the melt could be influenced by the presence of Y-123 particles. Since the development of crystallinity in a polymer involves two stages; nucleation and growth [6], some of the Y-123 particles may be able to act as nucleation centers that enable polymer chains to grow and be ordered on their surfaces. Limited crystallization could be expected during such growth if the spacing between nuclei (Y-123 particles) in the bulk is small. Therefore, it is reasonable to suggest that the percentage of crystallinity in the composite samples decreases with increasing volume of the superconductor.

REFERENCES

1. A.K.Sarkar, L.O.Salyer and B.Kumar , *J.Am. Ceram. Soc.*, **72**,1247 (1989).
2. V.M.Nadkarni and J.P.Jog, *Plastics Engg.* **40**, (8),37 (1984).
3. S.Bhadrakumari and P.Predeep, *Supercond. Sci. Technol.*, vol. **19**, 808-812 (2006).
4. H. N. Beck and H. D. Ledbetter, *J. Appl. Polym. Sci.*, **9**, 2131 (1965).
5. J.D. Muzzy, D.G. Bright and G.H. Hoyos, *Polym. Eng. Sci.*, **18**, 437 (1978).
6. A.Sharples, A.D.Jenkins (Ed.), *Polymer Science* Ch **.4**, p.285 North- Holland, Amsterdam (1972).

NANO MATERIALS

Synthesis of Micro / Nano Crystalline Ceramics

Philip Mathew,[a] Sheelakumari Isac [b], Rosalin Abraham ,[c] Jayakumari Isac[a *]

[a] *Centre for Condensed Matter, Department of Physics, CMS College, Kottayam, Kerala, India*
[a*] *Department of Physics, CMS College, Kottayam, Kerala, India*
[b] *Department of Chemistry, UC College, Aluva, Kerala, India*
[c] *Department of Physics, St. Dominics College, Kanjirappally, Kottayam, Kerala, India.*

Abstract. As a matter of fact almost every industrial production line, office and home is dependant on ceramic materials. Newly designed devices incorporate ceramic materials because of their useful chemical, electrical, mechanical, thermal and structural properties. The effectiveness of a large system depends critically on its ceramic components. That is ceramics are important, first because they comprise a large basic industry and second because their properties are essential for many applications. Ceramics are more stable than metals both in chemical and thermal environments. In this work the authors describes a method for the preparation of PBT ceramics material from lead oxide (PbO), barium carbonate and titanium oxide based on pre-calcinations of oxides. The results were analyzed by X-ray Diffraction (XRD), Infrared Spectroscopy (IR) and TGA. XRD and SEM studies revealed that its particle size is in nanometer range. Dielectric studies were conducted at a frequency range of 100 KHz to 13MHz. Mechanical properties were calculated.

Keywords: lead barium Titanate, ceramics, XRD, IR, dielectric studies.
PACS: 78.67Bf, 61.10Nz
Corresponding author: *jayacms@gmail.com

INTRODUCTION

Analysis of ceramics shows them to be a mixture of crystalline phases and glasses, each of many different compositions, usually combined with porosity, in a wide variety of proportions and arrangements. Properties such as heat conduction, optical properties, diffusion, mechanical deformation, cleavage and dielectric and magnetic properties are influenced by the point of view of not only lattices and ideal structures but also the randomness or ordering of atoms and lattices defects such as vacant sites, interstitial atom, and solid solutions. Historically ceramic materials predate metals. Ceramic materials have better dielectric properties than metals. Metals as a class have comparable tensile and compressive strength whereas ceramics are stronger in compression than in tension [1]. Ferro electric ceramics have a wide variety of applications, ranging from ultrasonic devices to pressure sensors. The basis for the high permittivity of barium titanates (εr = 2000 to 10,000) lies in its ferroelectric character, which is shared by many titanates, niobates and tantalates of similar crystal structure.

Piezoelectric materials are made up of an active ceramic phase. The properties of the ceramics depend on the connectivity of the phases, volume percent of ceramic, and the spatial distribution. The concept of connectivity developed by Newnham et al. [2]

CP1004, *Thermophysical Properties of Materials and Devices-NCTP '07*,
edited by P. Predeep, S. Prasanth, and A. S. Prasad

describes the arrangement of the component phases. It is critical in determining the electromechanical properties.

In this work the authors describes a method for the preparation of **lead barium Titanate** (PBT) ceramics and it was characterized to show good quality, homogeneity and the desired stoichiometry of the sample prepared. Dielectric studies were conducted at a frequency range of 100 KHz to 13MHz.

EXPERIMENTAL

This work describes a method for the preparation of super conducting ceramics materials based on pre-calcinations of oxides. In order to show the viability of the proposed method, super conducting ceramics were prepared in special furnace. The results were analyzed by X-ray Diffraction (XRD), Infrared Spectroscopy (IR) and TGA.

Synthesis Of Pbt Ceramics

The final electromechanical properties of ferroelectric ceramic components greatly depend upon the processing conditions of the ceramic. Each step of processing has to be carefully monitored and controlled to get the best product. A flowchart for a typical ferroelectric oxide ceramic manufacturing process is shown in Fig.1.

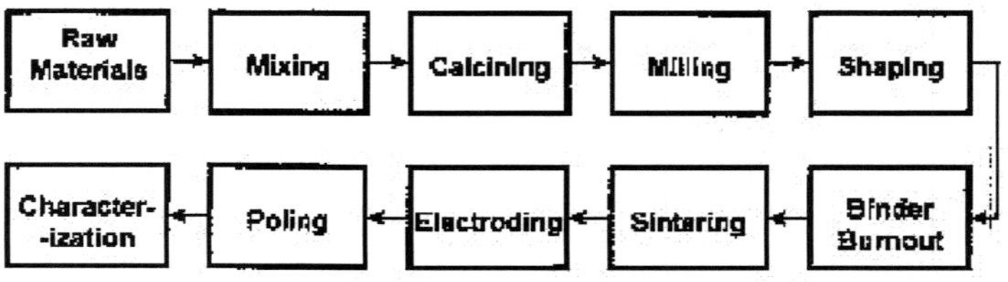

FIGURE 1: Flowchart for the processing of ferroelectric ceramics.

During the calcinations step the solid phase reaction takes place between the constituents giving the ferroelectric phase. For the calcination of PBT, the starting raw materials PbO, TiO_2 and $BaCO_3$ are mixed to obtain the perovskite phase. The higher the calcining temperature, the higher the homogeneity and density of the final ceramic product. Proper calcinations at the right temperature are necessary to obtain the best electrical and mechanical properties [3, 4].

RESULTS AND DISCUSSIONS

X-ray and SEM Analysis

Powder X-Ray diffraction (Fig.2) data (Table 1) were taken. The morphology and microstructure of the composites were analyzed by means of high resolution scanning electron microscopy using a JEOL JSM 840- microscope.

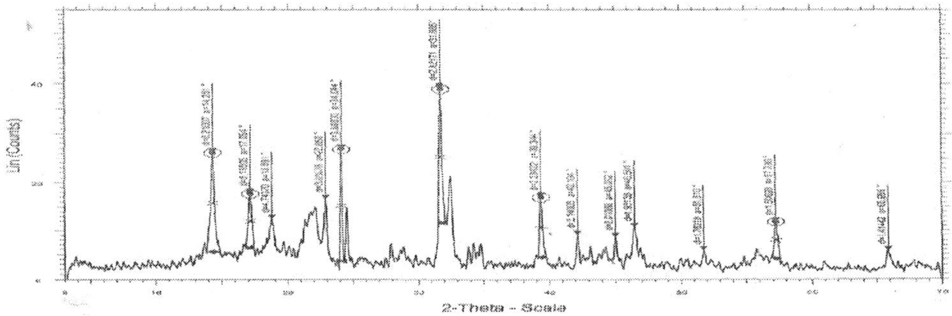

FIGURE 2: X-ray diffraction of PBT

The analytical method [5] for calculating lattice constants is used to calculate a, b, c for the crystal PBT where it was found that

a = b = 4.062 (A°) , c = 4.047 (A°), $\alpha = \beta = \gamma = 90^0$

A0 = 4.062000 C0 = 4.047000

Maximum Miller indices = 3 , Wavelength = 1.54180000(A°)

These calculated values of 'd' matching with the JCPDS file values, so that it can be concluded that these ceramics are crystalline in nature having tetragonal system.

The grain size of crystals was calculated using a well known Scherrer's Formula[5], which is given in table 2. $L = K\lambda / \beta \cos\theta$,Where L is the grain size of crystallite, λ (=1.54059) (A°). SEM also agrees with this value.

TABLE 1.XRD data of crystalline Ceramics of PBT

sl.no.	h	k	l	d (A°)	2 θ
1	0	0	1	4.04700	21.96239
2	1	1	0	2.87227	31.13747
3	1	0	1	2.86695	31.19670
4	1	1	1	2.34230	38.43087
5	2	0	0	2.03100	44.61369
6	0	0	2	2.02350	44.78800
7	2	1	0	1.81658	50.22140
8	2	0	1	1.81523	50.26127
9	1	0	2	1.81121	50.38075
10	2	1	1	1.65728	55.44116

TABLE 2. Particle size of PBT calculated

2θ	θ	β	L in nm
23.04	11.52	0.375	21.62
24.45	12.225	0.521	15.61
23.88	1.442	0.363	22.60
39.64	19.82	0.416	20.31
42.27	21.135	0.336	25.35
57.41	28.707	0.406	22.31
31.91	15.955	0.357	22.95

Infrared Spectroscopy

The Infrared absorption spectrum (Fig. 3.) of PBT was recorded on a thermo-Nicolet Avatar 370 spectrometer with a resolution of 0.9 cm-1 in KBr matrix.Vibrations in the finger print area 582.13, 692.97, 857.63(cm-1) characterizes the PBT ceramics powder.

T G / DTA Analysis

In thermo gravimetric analysis, the mass of the sample is recorded (Fig.4) continuously as a function of temperature as it is heated or cooled in a suitable environment, at a controlled rate [6,7]. Taking the initial weight as standard, the course of decomposition is analyzed from proportionate weight loss at each stage. The material decomposing 94.61 % at about 462.58^{0}C and the residue obtained is 1.828%.

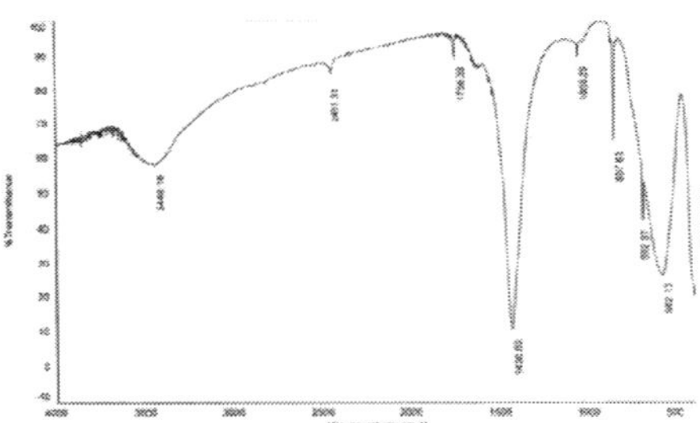

FIGURE 3: FTIR of PBT

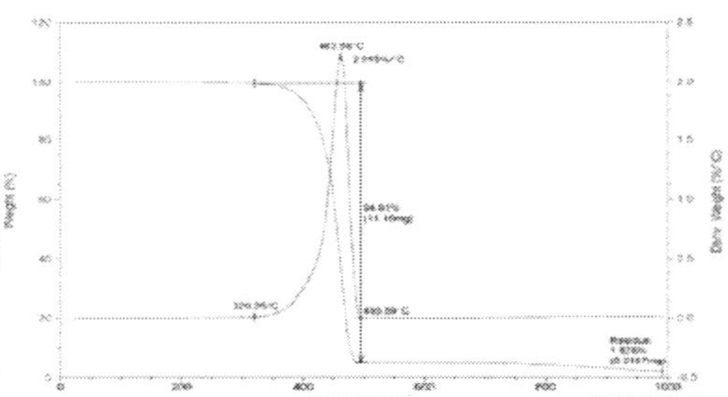

FIGURE 4: TG /DTA curve of PBT

Impedence Analysis

Dielectric measurements were performed with Hewlett-Packard-Japan HP 4192A impedance analyzer at a temperature of 28^{0}C at a frequency range of 100 KHz to 13 MHz. Well polished pellets of samples were given silver metallization for the measurement. The dielectric constants of the composites were calculated from the sample geometry and their capacitance.

At room temperature, the dielectric constant of PBaT is found to decrease with frequency (not shown) and all other temperatures the dielectric constant of PBT slightly decreases with frequency as in Fig.5. This indicates that at higher frequencies the mobility of polar groups in polymer chains is too slow to contribute to the dielectric constant. Fig. 6 depicts loss tangent of the PBT ceramics.

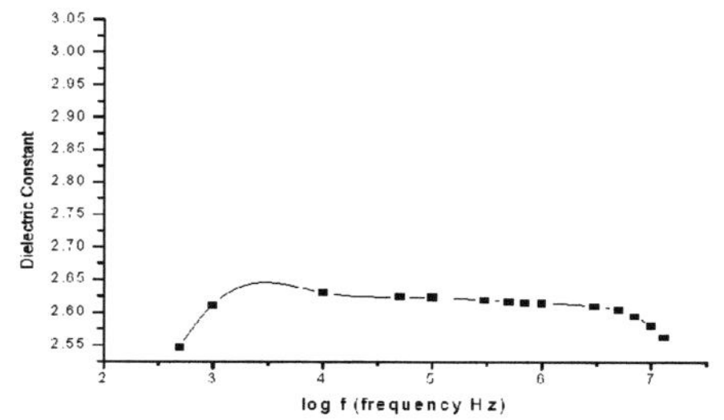

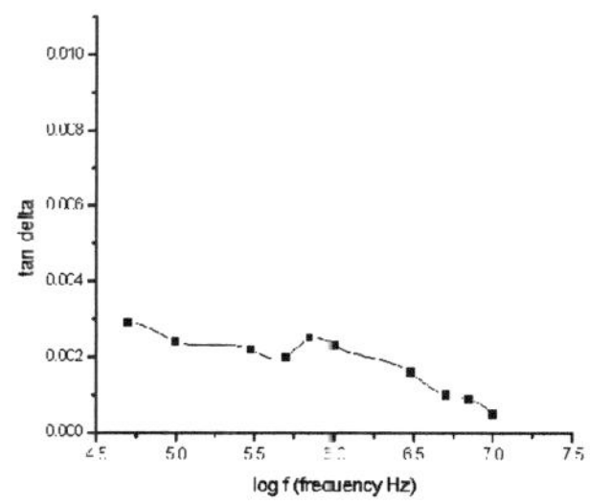

FIGURE 5. Frequency vss dielectric constant of PBT of

FIGURE 6. Dielectric losses (tan δ) of PBT

The electrical conductivity of PBT ceramics was determined at various frequencies. Logf (logarithm of frequency) versus electrical conductivity is plotted in Fig.7 which reveals that there is slight difference in the value of conductivity.

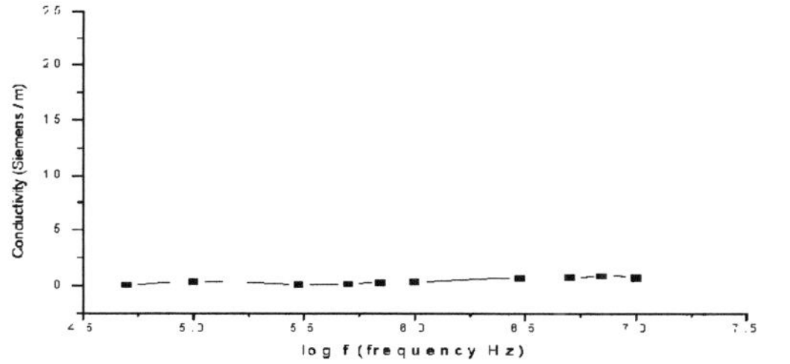

FIGURE 7. Conductivity of PBT ceramics with frequency

CONCLUSION

XRD, SEM, FTIR & TGA analysis shows crystallanity and nano size of the particles and high temperature retainability of the coeramics. The dielectric constant decreases slightly with the frequency , the dielectric constant is almost independent of frequency within a frequency region of 500Hz to 13MHz. The conductivity slightly increases. This indicates that at higher frequencies the mobility of is too slow to contribute to the dielectric constant. This material is opaque to X-rays, casting shadows. PBT ceramics can also be used in Microwave.

REFERENCES

1. A.J Moulson and J.M Herbert, Electro ceramics; Materials properties Application (1990).
2. R. E. Newnham, L. J. Bowen, K. A. Klicker, and L. E. Cross, Matl. Engg , **2**, 107 (1980).
3. Y. Matsuo, and H. Sasaki, J. Am. Ceram. Soc., **48**, 289 (1965).
4. D. A. Buckner and P. D. Wilcox, Am. Ceram. Soc. Bull., **51**, 218 (1972).
5. B.D Cullity, S.R.STOCK. Elements of X-ray diffraction, Prentice Hall, 2001 3rd edition.
6. B Wunderlich, "Thermal Analysis", pp: 40-45, Academic Press, Boston, (1990).
7. M.E Brown,"Introduction to Thermal Analysis: Techniques and Applications," pp 211. First edition Chapman and Hall, New York (1988).

Structural and Optical Properties of In_2S_3 Nano Layers Grown by Close Spaced Evaporation

N. Revathi, P. Prathap and K.T. Ramakrishna Reddy*

Thin Film Laboratory, Department of Physics, Sri Venkateswara University, Tirupati- 517 502, India
e-mail: ktrkreddy@hotmail.com

Abstract. Nanocrystalline In_2S_3 films have been prepared by close spaced evaporation technique. The films were grown on coring 7059 glass substrates at different temperatures in the range, 200 - 350 °C. The structural and surface morphological studies were analyzed by using X-ray diffractometer, scanning electron microscope and atomic force microscope. The optical characteristics of the films were evaluated from the optical transmittance data. All the films showed a strong (103) peak at $2\theta \sim 14.2°$, which correspond to tetragonal structure of β-In_2S_3 and the surface roughness increased with the increase of substrate temperature. The material parameters such as lattice strain and texture coefficient were also determined and the influence of substrate temperature on these parameters was discussed. The optical band gap was found to be direct and the layers grown at 300 °C had a band gap of 2.46eV

Keywords: In_2S_3 thin films, CSE Technique, Structural and morphological properties, Optical properties.

PACS: 73.61.G; 78.66.F; 81.15.E; 82.45.Mp

INTRODUCTION

In_2S_3 is a III-VI compound semiconductor that has potential applications during the past two decade in the field of optoelectronics, green and red phosphors, color television picture tubes and heterojunction for photovoltaic generators. It exists in three forms such as α, β and γ at atmospheric pressure [1] among which β-In_2S_3 is the most stable form. β-In_2S_3 is an n-type semiconductor with a wide energy band gap of 2.6 eV and high transmittance in the visible region. It is a potential substitute to CdS as buffer layer in solar cells and conversion efficiencies >16% have been reported. Thin films of In_2S_3 have been synthesized by different physical and chemical methods. In this study a novel technique, close spaced evaporation has been used for the first time to deposit In_2S_3 films. This technique has proved to be suitable for producing high quality films in a cost effective way. In this technique, material loss during evaporation can be minimized and highly uniform, well adherent films can be prepared. The present paper gives the influence of substrate temperature (T_s) on the structural, morphological and optical properties of nanocrystalline In_2S_3 films grown by close spaced evaporation. The different

CP1004, *Thermophysical Properties of Materials and Devices-NCTP '07*,
edited by P. Predeep, S. Prasanth, and A. S. Prasad

structural and optical parameters were also evaluated and their dependence on the substrate temperature has been discussed.

EXPERIMENTAL DETAILS

In_2S_3 thin films were deposited by close-spaced evaporation technique (CSE) using Hind Hi Vac Box Coater (model BC-300). This technique is slightly different from that of the conventional thermal evaporation in which the distance between the source and substrate was fixed at ~5 cm. The experimental arrangement and other details were reported elsewhere [2]. 5N pure In_2S_3 compound was used to grow the layers and molybdenum boat was used as the evaporation source. The layers were grown at different substrate temperatures, T_s that varied form 200 °C and 350 °C. The deposition was carried out in a vacuum of 10^{-5} mbar on ultrasonically cleaned corning 7059 glass substrates. The deposition rate and thickness of the films were determined using quartz crystal thickness monitor (model QTM-101). The structural properties of the films have been analyzed by Siefert X-ray diffractometer (model 3003TT) with CuK_α radiation source (λ=1.542 A°). A Hitachi (model S: 2400) scanning electron microscope (SEM) and Veco atomic force microscope were used to evaluate the surface morphology and grain size of films. The optical properties were studied using UV-Vis-NIR (Hitachi U-3400) spectrophotometer.

RESULTS AND DISCUSSION

All the deposited layers appeared to be dark red in color and the films had a thickness of ~300nm. The X-ray diffraction analysis of the films deposited at different substrate temperatures revealed that the structure of In_2S_3 films is strongly dependent on the deposition temperature. Fig.1 shows the X-ray diffraction spectrum of In_2S_3 films prepared at a substrate temperature of 300 °C. The X-ray diffraction spectra of the films grown at various substrate temperatures showed several peaks that correspond to the (111), (200), (300), (222), (511), (444) and (551) planes of the cubic β-In_2S_3, and (101), (103), (008), (107), (116), (109), (0012) and (307) orientations, corresponding to the tetragonal β-In_2S_3 structure. Sandoval et al. also reported the existence of both cubic and tetragonal phases in In_2S_3 layers grown by chemical bath deposition [3]. In general, (103) peak is the preferred orientation for most of In_2S_3 films that grew in the tetragonal structure. As the substrate temperature increased to 300 °C, the layers exhibited a strong (103) plane at 2θ~14.2° with an improvement in the crystalline. This might be due to the increase of ad-atom mobility with deposition temperature, which in turn allows the ad-atoms to orient along a particular direction in the crystal. Beyond 300 °C, the intensity of (103) peak decreased, indicating that there is degradation in the crystalline quality of In_2S_3 films, which could be due to sulfur re-evaporation from the films.

The various structural parameters for In_2S_3 films, such as lattice constants, strain, grain size, and texture coefficient were calculated from the XRD data and the variation of these parameters with substrate temperature was also studied. The lattice constants, 'a' and 'c' initially decreased with substrate temperature, reached a minimum value at T_s = 300 °C and increased with further increase of T_s. The evaluated lattice constants for the films grown at 300 °C were a = 7.584 Å and c = 31.977 Å, which are nearer to the bulk values. The change in lattice constants with substrate temperature may be due to the change in the nature of the native imperfections and their density in the films [4]. Sherif et al. reported a similar behaviour in ZnSe films deposited by thermal evaporation method [5].

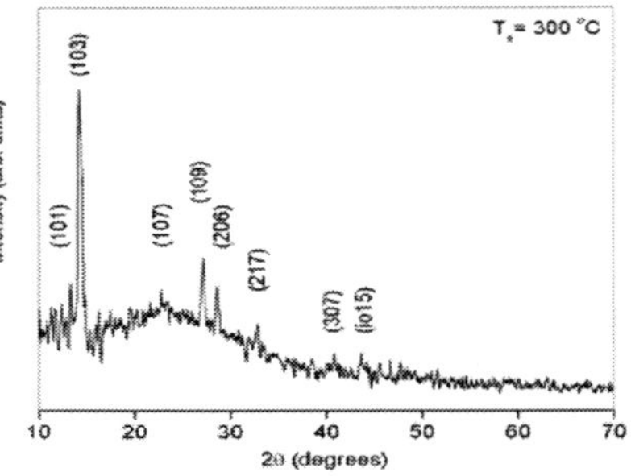

FIGURE 1. XRD spectrum of In_2S_3 layers.

For thin film materials, a variety of textures are possible depending on the deposition conditions used to grow the layers. The texture coefficient (TC) represents the preferential growth of certain plane compared to randomly oriented crystallites. The texture coefficient is given by the relation,

$$TC = \frac{I/I_o}{(1/n)\sum_n I/I_o} \tag{1}$$

where I is the measured intensity of the peak, I_o is its intensity in the JCPDS data and 'n' is number of reflections. If TC = 1, it represents the films with randomly oriented crystallites whereas TC > 1 for the films that are oriented preferentially in a particular direction. The TC and crystallite size increases with increase of T_s upto 300 °C, indicating that the crystallites were highly textured along the (103) plane at T_s = 300 °C.

The lattice strain developed in In_2S_3 films due to the difference in the thermal expansion coefficients of the deposited material and the substrate was calculated using the standard relation [6]. The nature of strain was found to be tensile in the layers prepared at 200 °C and changed to compressive with further increase of deposition temperature. This implies that the films would like to compress parallel to the substrate surface with the increase of T_s. The analysis showed that the grain size increased in the presence of compressive strain. The observed lower crystallite size in the films prepared at < 250 °C, which showed tensile strain, might be due to the retarded crystal growth that arise from the stretched lattice. Ghosh et al. also reported similar results for ZnO films deposited by sol-gel method [7]. This infers that the strain could also have considerable effect on the grain growth in addition to the substrate temperature.

The crystallite size (D) of the films was evaluated using the full width at half maximum of the (103) peak using the Debye-Scherrer formula [8]. It was found that the average grain size of the films initially increased from 10.8 nm to 28.6 nm with increase of substrate temperature, reached a maximum at 300 °C and then decreased to 20.4 nm with further increase of T_s. The increase of grain size might be due to the increase of surface ad-atom mobility, which could also increase the agglomeration of clusters. The decrease of grain size at $T_s > 300$ °C could be due to the re-evaporation of sulfur.

The surface morphology of In_2S_3 films is strongly dependent on the substrate temperature. Fig. 2. (a) shows the SEM micrograph of In_2S_3 films prepared at $T_s = 300$ °C. The SEM picture shows that the sample is continuous and uniform with smooth surface. The image also reveals that β-In_2S_3 particles have a spherical shape with an average grain size of ~ 27.2 nm. Fig. 2. (b) shows the AFM image of the films deposited at $T_s = 300$ °C over 1μ x 1μ scanning range. The average grain size and rms surface roughness of the sample is 28 nm and 1.2 nm respectively. The grain size evaluated from both SEM and AFM are correlated with the crystallite size calculated using the X-ray diffraction analysis and are found to be in good agreement.

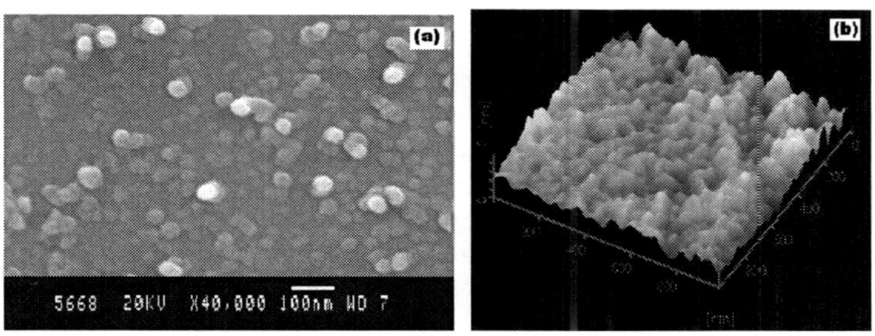

FIGURE 2. (*a*) SEM and (*b*) AFM picture of In_2S_3 films deposited at 300 °C.

The optical properties of In_2S_3 films deposited at different substrate temperatures revealed that the transmittance of the films increased from 60% to 78 % with the increase of T_s from 200 °C to 300 °C and found to decrease beyond 300 °C. Fig.3 shows the transmittance, T versus wavelength, λ spectrum recorded in the wavelength range 500-2500 nm for the films deposited at $T_s = 300$ °C. The decrease of transmittance at higher substrate temperatures, >300 °C might be due to the stoichiometric deviation in the films. The T versus λ plot shows a sharp fall in transmittance near the band edge, indicating that the grown layers had good crystallinity. The absorption coefficient,α which is a function of photon energy, $h\nu$ was evaluated by using the relation,

$$\alpha = -1/d \ \ln(1/T) \tag{2}$$

where d is the thickness of the film and T, the percentage transmittance. The optical energy band gap, E_g of the films was determined by extrapolating the linear portions of the $(\alpha h\upsilon)^2$ vs $h\upsilon$ plot onto the x-axis, as shown in the inset of Fig.3. The evaluated band gap value varied in the range, 2.08 - 2.49 eV with the change of T_s from 200 °C to 350 °C and the layers grown at 300 °C had E_g = 2.46 eV. The change of E_g with T_s is in close agreement with the reported data [9]. The broadening of band gap of the films with T_s might be due to the decrease of mid band gap defects, reported by Dow et al. [10].

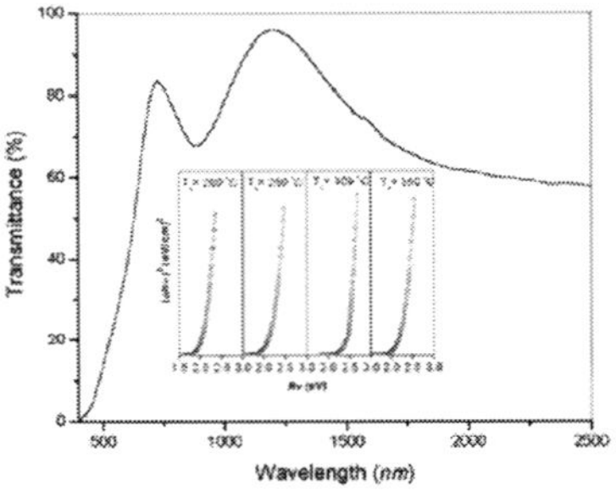

FIGURE 3. Transmittance Vs wavelength spectrum of In_2S_3 films deposited at T_s ~300 °C (Inset shows $(\alpha h\upsilon)^2$ vs $h\upsilon$ plots of the films grown at different T_s)

CONCLUSIONS

Polycrystalline β-In_2S_3 layers have been deposited by close spaced evaporation on corning glass substrates at different substrate temperatures that varied from 200 °C to 350 °C. The films grown at lower T_s consisted of both tetragonal and cubic β structure. The films grown at T_s = 300 °C exhibited a strong (103) peak as the preferred orientation and had tetragonal structure. The structural parameters such as grain size, lattice constants, strain and texture coefficient were found to be highly influenced by the substrate temperature. The morphological studies showed that the surface of the films was uniform, homogenous, void free and the evaluated grain size was ~ 10 - 30 nm. The films deposited at 300°C have shown a better transmittance of ~78% in the visible region compared those deposited at other temperatures. These layers had an optical band gap of ~ 2.46 eV.

REFERENCES

1. R. Diehl and R. Nitsche, *J. Cryst. Growth*, 28 (1975) 306-310.
2. Y.P.V. Subbaiah, P. Prathap, M. Devika, K.T.R. Reddy, *Phys. B*, 365 (2005) 240-246.
3. M.G. Sandoval-Paz, M. Sotelo-Lerma, J.J. Valenzuela-Jauregui, M. Flores-Acosta, R. Ramirez-Bon, *Thin Solid Films*, 472 (2005) 5-10.
4. K. Rrichelt and X. Jiang, *Thin Solid Films*, 191 (1990) 91-126.
5. M.El. Shcrif ct al, *J. Mater. Sci. Mater. Electron*, 7 (1996) 391-395.
6. H.C. Ong, A.X.E. Zhu, G.T. Du, *Appl. Phys. Lett.*, 80 (2002) 941-943.
7. R. Ghosh, D. Basak, S. Fujihara, *J. Appl. Phys.*, 96 (2004) 2689-2692

8. B.E. Warren, X-ray Diffraction, Dover, Newyork, 1990, p.253.
9. M. Calixto-Rodriguez, A.Tiburcio-Silver, A. Ortiz, A. Sanchez-Juarez, *Thin Solid Films*, 480 (2005) 133-137.
10. J.D. Dow, D. Redfield, *Phys. Rev. B*, 5 (1972) 594-610.

Optical Band Gap Study Of Nanocrystalline $NiCr_{0.8}Fe_{1.2}O_4$ Ferrite

Arun S Prasad[a], S N Dolia[a*], M S Dhawan[a] and P Predeep[b]

[a]*Department of Physics, University of Rajasthan, Jaipur 302 004*
[b]*Department of Physics, N. I. T. Calicut*

Abstract. Optical energy band gap of nanocrystalline $NiCr_{0.8}Fe_{1.2}O_4$ ferrite have been studied. The nanocrystalline $NiCr_{0.8}Fe_{1.2}O_4$ ferrite have been synthesized using sol-gel technique. X-ray diffraction pattern confirms the formation of spinel structure in single phase and the average particle size is 4 nm. The energy band gap measurements of nanocrystalline $NiCr_{0.8}Fe_{1.2}O_4$ ferrite in pellet form have been carried out by absorption spectra using double beam spectrophotometer. A pellet of nanoparticle ferrite was made under a load of 10 tons. From the analysis of absorption spectra, nanocrystalline $NiCr_{0.8}Fe_{1.2}O_4$ ferrite have been found to have energy band gap of 3.2 eV at room temperature

Keywords: Nano particles, spinel ferrites, Optical bandgap, Absorption spectra.
PACS: 78.40.-q , 78.66-w, 78.67-n.

INTRODUCTION

Nanocrystalline spinel ferrites currently receive considerable attention during the past several years because of their remarkable electric, magnetic and optical properties, which are considerably different from their bulk counterparts [1,2]. They are largely used as components in recording tape, flexible recording media, microwave devices, communication technology and magnetic fluid as well as biomedical applications. Below a critical size, magnetic particles become single domain in contrast to the usual multidomain structure of the bulk magnetic materials and exhibit interesting magnetic properties such as superparamagnetism, quantum tunneling of magnetization, spin canting, etc. which are used in modern technologies including ferrofluid technology, magnetocaloric refrigeration and magnetic guided drug delivery,[3-6]. In fact, below a critical size, superparamagnetic phenomena occur due to the fact that the thermal energy becomes comparable to the anisotropy energy, which is proportional to the particle volume. Further, high surface area of nanomaterials enhances their absorption properties[7]. The optical energy band gap is more for nanoparticles as compared to their bulk counterpart and increases with decrease in particle size[8].

The nanocrysatlline ferrite is being synthesized by various techniques such as chemical co-precipitation[9], hydrothermal[10], reverse micelle synthesis[11] and combustion[12]. Sol-gel method is being extensively employed as an alternate route to obtain the nanoparticles in larger amounts with narrow size distribution. The present

CP1004, *Thermophysical Properties of Materials and Devices-NCTP '07*,
edited by P. Predeep, S. Prasanth, and A. S. Prasad

work has been taken up to study the optical energy band gap of nanocrystalline $NiCr_{0.8}Fe_{1.2}O_4$ ferrite synthesized by sol-gel method.

EXPERIMENTAL DETAILS

The nanocrystalline $NiCr_{0.8}Fe_{1.2}O_4$ ferrite were prepared by the sol-gel method. In the sol- gel auto-combustion method analytical grade $Ni(NO_3)_2.\ 6H_2O$, $Cr(NO_3)_3.\ 9H_2O$, $Fe(NO_3)_2.\ 9H_2O$ and citric acid ($C_6H_8O_7$) were used as raw materials. The molar ratio of nitrates to citric acid is 1:1 all the nitrates and citric acid solutions were mixed together with continuous stirring. A small amount of ammonia is added to the solution to adjust the PH to about 7. The mixed solution was slowly heated to 90 ^{0}C with steering constantly to transform it into a Xero-gel. Further heating resulted in burning of the gel in a self propagating combustion manner until all the gel gets burnt out completely to form a loose powder. X-ray powder diffraction was performed on the nanocrystalline $NiCr_{0.8}Fe_{1.2}O_4$ ferrite using a Phillips x-ray powder diffractometer PW 1840 employing Fe K_α radiation.

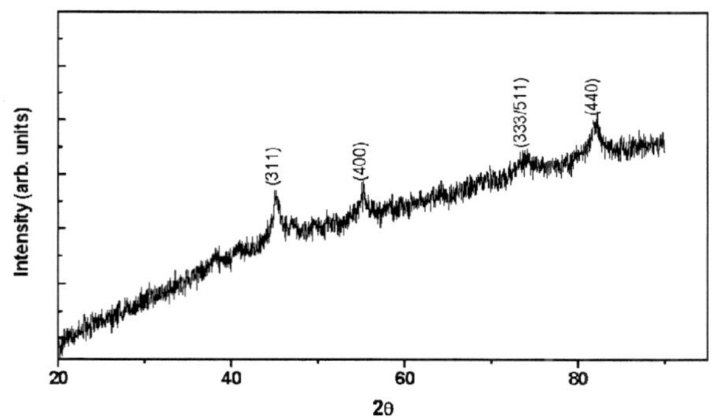

FIGURE 1. X-ray diffraction pattern of the nanocrystalline $NiCr_{0.8}Fe_{1.2}O_4$ ferrite sample of an average particle size 4 nm

The absorption spectrum of nanocrystalline $NiCr_{0.8}Fe_{1.2}O_4$ ferrite was recorded at room temperature with the help of spectrophotometer Hitachi model U-3400, in the wavelength range of 300-800 nm. The lead sulphide detector (PbS) is used for the detection of infrared rays. According to Tauc relation[13], the absorption coefficient for direct band gap material is given by

$$\alpha h\nu = A\ (h\nu - E_g)^n \qquad \ldots(1)$$

where hν is photon energy, E_g the band gap and A is constant which is different for different transitions and n is an index which assumes the values ½,3/2,2 and 3 depending on the nature of the electronic transition responsible for the absorption. n=1/2 is taken for an allowed direct transition. Therefore, by plotting a graph between $(\alpha h\nu)^2$ and hν, a straight line is obtained which gives the value of the direct band gap.

RESULTS AND DISCUSSION

Figure 1 shows indexed X-ray powder diffraction pattern of nanocrystalline $NiCr_{0.8}Fe_{1.2}O_4$ ferrite. The Bragg reflections are indexed in Fe_3O_4 like cubic structure and the estimated cell constant a is 8.38 Å. This confirms that the sample is formed in single phase and the cell constant is very close to that of Fe_3O_4 (a = 8.39 Å). Considerably broadened lines in the XRD pattern are indicative of the presence of nano-size particles. We have used the (311) reflection line in the XRD pattern for obtaining the average particle size with the help of the Debye-Scherrer[14]equation

$$t = \frac{0.9\lambda}{B\cos\theta_B} \quad ; \qquad B = \left(B_M^2 - B_S^2\right)^{1/2} \qquad \ldots(2)$$

where t is the thickness (diameter) of the particle, λ is the X-ray wavelength (1.9373 Å), B_M and B_S are the measured peak broadening and instrumental broadening in radian respectively and θ_B in the Bragg angle of the reflection. The calculated average particle size is 4 nm.

Figure 2 shows the optical absorption spectrum for the pellet of nanocrystalline $NiCr_{0.8}Fe_{1.2}O_4$ ferrite, recorded at room temperature in the wavelength range of 300-800 nm. It is clearly evident from the figure 2 that absorption decreases with the increase in wavelength which indicates the presence of optical band gap in nanocrystalline $NiCr_{0.8}Fe_{1.2}O_4$ ferrite.

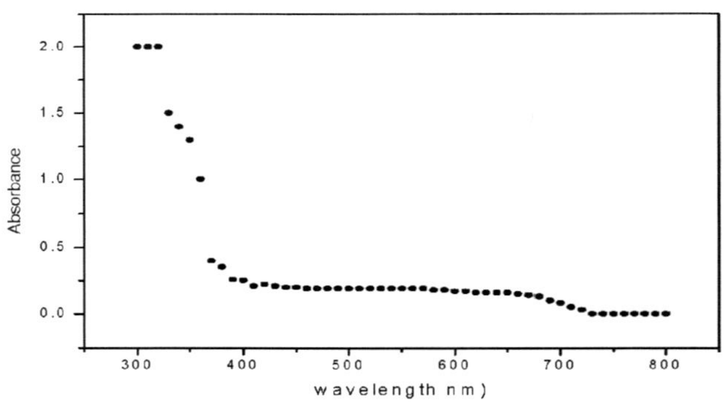

FIGURE 2. Absobance versus wavelength for nanocrystalline $NiCr_{0.8}Fe_{1.2}O_4$ ferrite sample

Tauc relation as given by Eq (1) is used for the determination of direct band gap for nanocrystalline $NiCr_{0.8}Fe_{1.2}O_4$ ferrite. A graph between $(\alpha h\nu)^2$ and hν has been plotted in figure 3. The extrapolation of straight line to $(\alpha h\nu)^2 = 0$ gives the value of direct band gap. From this graph, the value of optical band gap of nanocrystalline $NiCr_{0.8}Fe_{1.2}O_4$ ferrite comes out to be 3.2 eV. The estimated optical band gap of 3.2 eV for nanocrystalline $NiCr_{0.8}Fe_{1.2}O_4$ ferrite having average particle size of 4 nm is high as compared to 2.5 eV for nanocrystalline $NiFe_2O_4$ ferrite having average particle size of 25 nm[15]. This increase in optical band gap from 2.5 eV to 3.2 eV for reduction in average particle sizes from 25 nm to 4 nm attributing to increase in surface to volume ratio for nanoparticles which creates more surface states with higher energies resulting increase in optical energy band gap[16]. However, substitution of Cr^{3+} for Fe^{3+} in $NiFe_2O_4$ ferrite may also be responsible for slight increase in optical band gap.

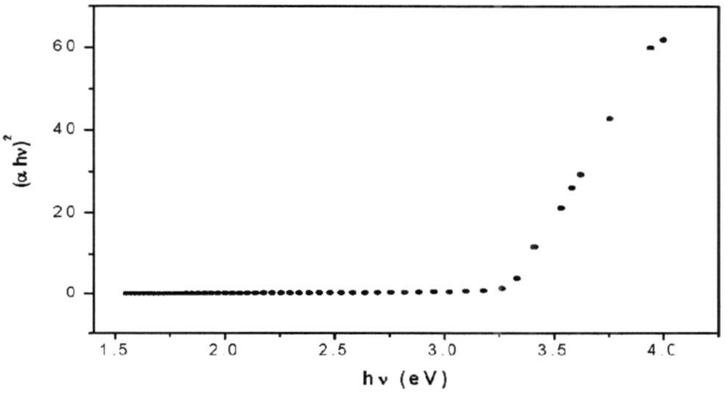

FIGURE 3. $(\alpha hv)^2$ versus photon energy (hv) for nanocrystalline $NiCr_{0.8}Fe_{1.2}O_4$ ferrite sample

CONCLUSION

Nanocrystalline $NiCr_{0.8}Fe_{1.2}O_4$ ferrite has been synthesized using sol-gel method. XRD confirms the formation of spinel cubic structure in single phase and the estimated average particle size is 4 nm. High optical band gap of 3.2 eV for nanocrystalline $NiCr_{0.8}Fe_{1.2}O_4$ ferrite is obtained pertaining to reduction in particle size.

ACKNOWLEDGMENTS

The authors are thankful to Prof. B. K. Srivastava and Prof. A. Krishnamurthy for helpful discussions. We are also thankful to Dr. Y. K. Vijay for providing the spectrophotometer facilities

REFERENCES

1 Ozin G A, *Adv Mater*, **4** (1992) 612.
2 Gleiter H, *Adv Mater* **4** (1992) 474.
3 Ding J, Miao W F, Street R & McCormic, *J Alloys Compd* **281** (1998) 32.
4 Martinez B, Obradors T, Balcell L, Rounanet A & Monty C, *Phys Rev Lett* **80** (1998) 181.
5 Ding J, Reynold T, Miao W F, McCormick & Street R, *Appl Phys Lett* **65** (1994) 7074.
6 Kodama R H, Berkowtiz A E, McNiff E J & Foner S, *Phys Rev Lett* 77 (1966) 395.

7 Verda S R, Parihar M, Negi S C & Kumar N, *Smart materials, structures and MEMs*, (Eds) Proc SPIE, 3321 (1998) 369.

8 Keigo Suzuki & Kazunori Kijima, *Jap J Appl Phys* **44 (4A)** (2005) 2081.

9 S N Dolia & S K Jain, *Hyperfine Interactions* 160 (2005) 219.

10 Chandana Rath, Mishra N C, Anand S, Das R P, Sahu K K, Chandan Upadhayay Chandan & Verma H C, *Appl Phys Lett* **76 (4)** (2000) 475.

11 Rolling B, Happe A, Funke K & Ingram M D, *Phy Rev Lett*, **78** (1997) 2160.

12 Zheng M, Wu X C, Zou B S & Wang Y J, *J Magn Magn Mater*, **183** (1998) 152.

13 Tauc J (Ed.) *Amorphous and Liquid Semiconductors*, Plenum press, New York, (1974).

14 Cullity B D, *Elements of X-ray Diffraction* (Addison-Wesely, Reading, MA) (1978).

15 S N Dolia, Rakesh Sharma, M P Sharma & N S Saxena, *Ind J of Pure & Appl Phys* **44** (2006) 774.

16 . Banerjee S and Chakravorty, *Europhysics Lett* **52 (4)** (2000) 468

* corresponding author, e-mail : sn_dolia@yahoo.com

Chemical synthesis and Characterization of Cu doped ZnS nanoparticles

S. Sambasivam[a], D. Paul Joseph[b], B. K. Reddy[a], C. K. Jayasankar[a*]

[a]*Department of Physics, Sri Venkateswara University, Tirupati – 517 502, India.*
[b]*Materials Science Centre, Department of Nuclear Physics, University of Madras, Guindy Campus, Chennai-600025, India.*
E-mail : ckjaya@yahoo.com (C.K. Jayasankar).

Abstract: $Zn_{1-x}Cu_xS$ nanoparticles with x = 0.0, 0.1, 0.2, 0.3, 0.4 were prepared by chemical co-precipitation from homogeneous solutions of zinc and copper salt compounds, with S^{2-} as precipitating anion, formed by decomposition of thiophenol. XRD studies show the phase singularity of ZnS particles having zinc-blende (cubic) structure. The XRD line broadening indicates that the average crystallite size of the particles is about 7.2 nm. Optical absorption studies indicated a slight blue shift of the fundamental absorption edge with increasing Cu^{2+} doping. The photoluminescence of excited at about 370nm, showed a green emission band at 495 nm corresponding to Cu^{2+} luminescent centre.

Keywords: Nanoparticles, Chemical synthesis method, Optical properties

PACS: 78.67.Bf; 81.16.Be; 74.25.Gz

INTRODUCTION

During recent decades, the preparation and characterization of materials on the nanometer-size scale have provided not only new physics in reducing dimensions, but also the possibility of fabricating novel materials [1-3]. Nanosized ZnS semiconductor (e.g. ~ 3.6eV) is commercially used as phosphor and also in thin-film electroluminescent devices [4-6]. The photophysical and photochemical properties of ZnS semiconductor nanocrystals doped with Mn^{2+}, Ni^{2+}, Ag^{2+} have been studied recently [7-12]. These doped nanoparticles show interesting properties different from those of bulk materials, such as an increased energy band-gap, which results from size quantization confinement [13]. Various authors [14-16] have reported synthesis and photoluminescence properties of ZnS:Cu nanoparticles (~ 2nm) formed by chemical route with different capping agents. These reports have initiated interest in us to use thiophenol as a ligand playing the role of glycine in order to obtain nanosized ZnS:Cu particles and their characterization. The synthesized materials were characterized by XRD, optical absorption, photoluminescence techniques which are systematically analysed and the results are discussed.

CP1004, *Thermophysical Properties of Materials and Devices-NCTP '07*,
edited by P. Predeep, S. Prasanth, and A. S. Prasad

EXPERIMENTAL TECHNIQUES

The samples were prepared by adapting the colloidal chemical co-precipitation method. Zinc acetate and copper acetate starting materials were weighed according to the stoichiometry x = 0.0 0.1, 0.2, 0.3, 0.4 and were dissolved in 100 ml of methanol to make 0.1M solution. The stoichiometric solution was taken in a burette and added in drops with continuous stirring to a mixture of Na_2S (0.1M) + 50 ml of H_2O +2 ml of Thiophenol + 100 ml of methanol until fine precipitates are formed. After complete precipitation, the solution in conical flask was constantly stirred for about 15 h. Then the precipitates were filtered out separately and washed thoroughly with de-ionized water and then calcined at 300°C/2h in vacuum. The XRD measurements were made in Seiferet 3003TT X-ray diffractometer. The optical absorption measurements were made in JASCO-V-560 spectrophotometer. The photoluminescence measurements were recorded in a Jobin Yvon Fluorolog -3 spectrophotometer.

RESULTS AND DISCUSSION

The X-ray diffraction patterns of annealed (300°C/2h) ZnS:Cu samples (x=0.0, 0.1, 0.2, 0.3 and 0.4) obtained using Cu K_α irradiation (λ= 1.5418 Å) in the range 20° - 65° are shown in Fig. 1. It reveals that the particles exhibit a zincblende crystal structure. The diffraction peaks correspond to (111), (200), (220), and (311) planes of the cubic crystalline ZnS, respectively. From XRD analysis, no characteristic peaks of impurity phases have been observed. The average size of particles doped with different dopents, estimated from the Debye-Scheerer formula, is about 7.0-7.5 nm. The average particle-size of ZnS nanocrystals is about 7.2 nm.

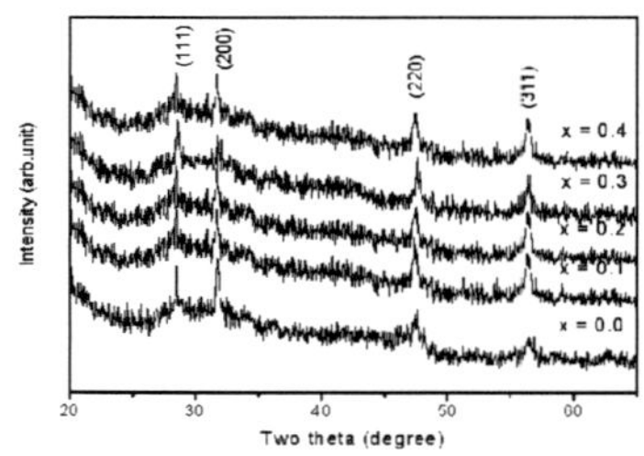

FIGURE 1. XRD pattern of ZnS:Cu (x = 0.0, 0.1, 0.2, 0.3 and 0.4) annealed at 300°C/2h in vacuum.

The optical absorption spectra of ZnS and ZnS:Cu nanoparticles for varying copper dopings are shown in Fig.2 it is well known that bulk ZnS possesses a band gap of 3.5eV corresponding to the absorption peak at 350nm. In the present study, nanoparticles of ZnS:Cu show the absorption peak at 330nm corresponding to an energy gap of ~3.7eV, characterizing the nano- nature, showing a small blue shift with increasing Cu ion content. The trend of band gap variation observed in the present study is similar to that reported by

Wang et al.[17] and Kohsravi et al.[15] in ZnS:Cu nanoparticles of average grain size of 4 nm and 2.1 nm prepared using glycine and sodium hexametaphosphate respectively, as capping agents. Also the band gap values obtained in the present work (average grain size ~7.20 nm) in ZnS:Cu nanoparticles formed using thiophenol as a capping agent are relatively lower compared to the earlier reported band gap values of 4.3 - 4.5 eV with Cu content 0.2 – 2.0 % [10] and 4.1 eV with Cu content 0.06 % [15]. This difference could be attributed to the different capping agents used and also to the difference in the particle sizes. As expected for nanoparticles, in the present study [Fig.2] the absorbance decreases with wavelength beyond the absorption maximum. The decrease in absorbance between 330 and 550nm is assigned to d-d transitions in Cu^{2+}. Above 550nm an increase in absorbance is noticed with increasing wavelength. In undoped ZnS nanoparticles the absorbance increases at a faster rate compared to Cu doped ZnS nanoparticles above 550nm. Thus the observed dependence of absorbance on wavelength reflects the characteristics of zero-phonon absorbance [18].

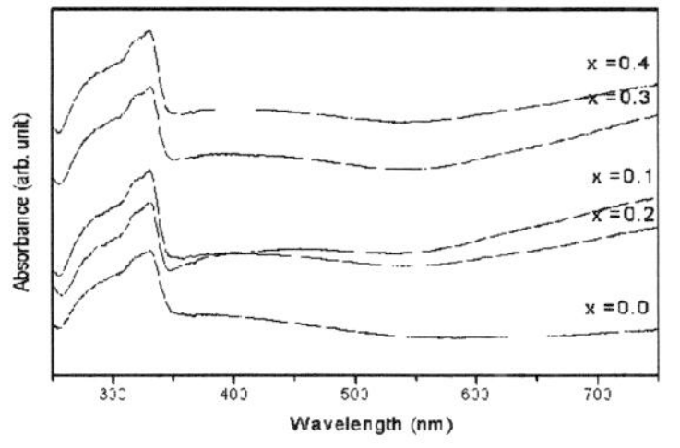

FIGURE 2. Optical absorption spectra of ZnS:Cu (x = 0.0, 0.1, 0.2, 0.3 and 0.4) annealed at 300°C/2h in vacuum.

The excitation spectra of ZnS and ZnS:Cu^{2+} were recorded in Jobin Yvon Fluorolog – 3 spectrophotometer between 320 and 420nm and are shown in Fig.3. The excitation spectrum of undoped ZnS shows a series of excitation maxima. The first excitation maximum is located at 340nm and appears relatively more broad and intense. The second excitation maximum located at 370nm is less intense. However, the peaks appeared at about 393nm and 410nm are poorly resolved for undoped system. This observation suggests different groups of particles with different sizes, each group being characterized by its characteristic and distinguishable band positions. The excitation spectrum of ZnS:Cu^{2+} exhibits single excitation maximum at 370nm indicating the growth of particles with different sizes due to the presence of dopant Cu^{2+}. Its entry into the lattice of ZnS appears to modify the surface free energy of ZnS particles so that the size of particles could be controlled nearly within a narrow range.

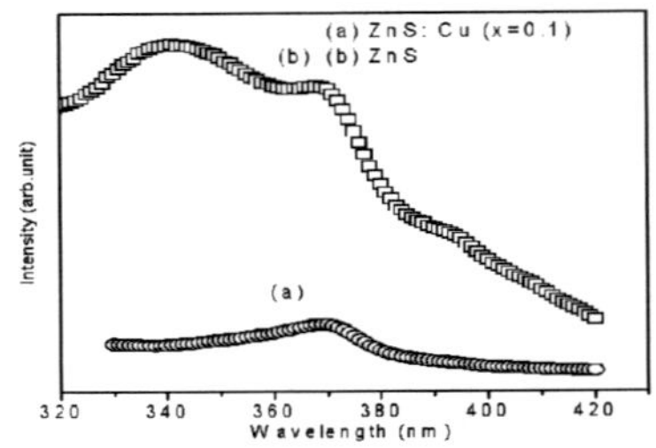

FIGURE 3. PL excitation spectra of ZnS and ZnS:Cu (x=0.1) nanoparticles annealed at 300^{0}C/2h in vacuum.

The emission spectra (exciting at 370nm) of ZnS:Cu (x=0.0, 0.1, 0.2, 0.3 and 0.4) samples are shown in Fig. 4. Undoped ZnS gives an intense broad emission band covering the major part of the visible region peaking at 491nm. Such green emission in ZnS:Cu nanoparticles has also been reported by others [15,17]. The emission spectra of ZnS:Cu (0.1) and ZnS:Cu(0.2) also appear similar to that of undoped ZnS, but the gradual decrease in intensity with the increase in copper doping is an obvious evidence for entry of copper into the lattice of ZnS which was also evident from the XRD spectra which showed no extra peaks. No appreciable shift in emission maximum is noticed with increase in Cu content. The decrease in intensity with increasing Cu content is attributed to repeated excitation within the copper sites of the samples. Such a decrease in intensity with increasing copper loading was also reported by Wang et al. [17].

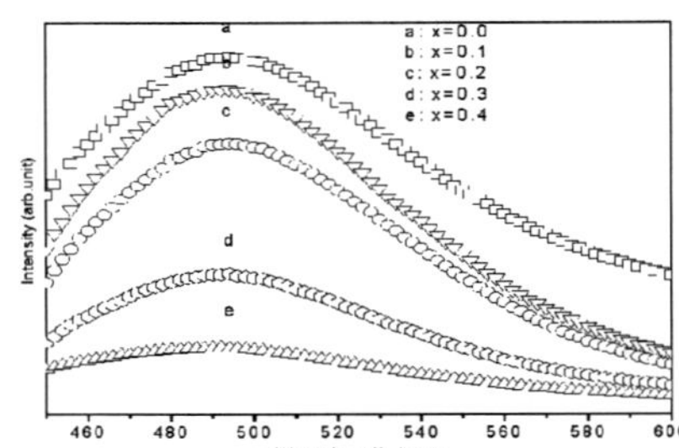

FIGURE 4. PL emission spectra of ZnS and ZnS:Cu (x= 0.0, 0.1, 0.2, 0.3 and 0.4) nanoparticles annealed at 300^{0}C/2h in vacuum.

CONCLUSIONS

ZnS:Cu nanoparticles with Cu concentrations 0 – 0.4 were successfully synthesized by the colloidal chemical co-precipitation method with thiophenol as the ligand. XRD measurements confirmed the formation of substitutional single zinc blende phase in the entire range of composition. The function of the ligand, thiophenol can be summarized as providing solubility leveling and surface protecting effects. Optical absorption studies indicated a slight blue shift in the absorption edge with increasing Cu^{2+} doping concentration. The photoluminescence of ZnS:Cu excited at about 370nm, showed green emissions. No appreciable shift in the emission wavelength is noticed with increasing Cu content although the emission intensity decreases considerably with broadened emission bands.

ACKNOWLEDGMENTS

The authors are grateful to the University Grant Commission, New Delhi, Government of India, for providing financial support.

REFERENCES

1. P. Calvert: Nature 383, 300 (1996).
2. R. Pool: Science 248, 1186 (1990).
3. A. P. Alivisators: Science 271, 933 (1996).
4. G.F. Garlick, A.F. Gibson: J. Opt. Soc. Am. 39, 935 (1949).
5. M. Ohring: The Materials Science of Thin Films (Academic Press, San Diego 1992).
6. I.P. McClean, C.B. Thoms: Semicond. Sci. Technol. 7, 1394 (1992).
7. J. Yu, H. Liu, Y. Wang, EE. Fernandez, W. Jia: J. Lumin. 76-77, 252 (1998).
8. R.N. Bhargava, D. Gallagher, T. Welker: J. Lumin. 60-61, 275 (1994).
9. K. Sooklal, B.S. Cullum, M.S. Angel, C.J. Murphy: J. Phys. Chem. 100, 4551 (1996).
10. G.N. Lvanova, V.A. Kasiyan, N.D. Nedeoglo, D.D. Nedeoglo: J. Lumin. 82, 277 (1999).
11. P. Yang, M.Lu, D. Xu, D. Yuan: Appl. Phys. A 74, 257 (2002).
12. D. Denzler, M. Olschjewski, K. Sattler: Appl. Phys. Lett. 84, 2841 (1998).
13. L. Brus: J. Phys. Chem. 90, 2555 (1986).
14. K. Manzoor, S.R. Vadera, N. Kumar, T.R.N. Kutty, Mater. Chem. Phys. 82, 718 (2003).
15. Ali Azam Khosravi, Manisha Kundu, Lalita Jatwa, and S.K. Deshpande, Appl. Phys. Lett. 67, 2702 (1995).
16. Lingdong Sun, Changhui Liu, Chunsheng Liao, Chunhua Yan, Solid State Commun. 111, 483 (1999).
17. Mingwen Wang, Lingdong Sun, Xuefeng Fu, Chunsheng Liao, Chunhua Yan, Solid State Commun. 115, 493 (2000).
18. Shionoya, W. M. Yen, Phosphor Handbook, CRC press, 1998, pp. 248 – 249.

Preparation and Characterization of SnO Nanoplatelets by Microwave Innovative Technique

T. Krishnakumar[a], Nicola Pinna[b], K. Perumal[a] And R. Jayaprakash[a]*

[a]*Nanotechnology Lab, Department of physics, Sri Ramakrishna Mission Vidyalaya College of Arts and Science, Coimbatore, Tamilnadu - 641 020 India*
[b]*Department of Chemistry and CICECO, University of de Aveiro, Campus Universitario de Santiago, 3810-193 Aveiro, Portugal*

Abstract. Tin oxide (SnO) nanoplatelets have been synthesized by a Microwave innovative technique with an operating frequency of 2.45 GHz in just few minutes. The crystalline size and structure was evaluated from XRD pattern. The SEM and TEM analysis showed the single crystal platelets together with many small particles attached to its surface. The FT-IR and the electrical conductivity of the samples have also been investigated.

Keywords: Microwave innovative, Tin oxide, Tetragonal Romarchite Structure, Nanoplatelets, Lithium-ion secondary battery
PACS No: 61.82RX

INTRODUCTION

Tin oxide belongs to a class of materials that combines high electrical conductivity with optical transparency and thus constitutes an important component for optoelectronic applications. SnO has been used as a potential anode material for the lithium-ion secondary battery and that can store Li ions with a Coulombic capacity reaching that of hydrogen-storage alloys, ensuring protection against dendritic Li formation. The Li_2O is formed, when Li ions react with the SnO according to the following reaction[1].

$$6.4e^- + 6.4\ Li^+ + SnO_{(s)} \rightarrow Li_2O_{(s)} + Li_{4.4}\ Sn_{(s)}$$

The Li_2O helps to stabilize the anode structure of Lithium-ion secondary battery. For this application SnO powder is successfully synthesized by minimize the oxidation of Sn^{2-} into Sn^{4+}[2].

Many methods have been developed in order to synthesize tin oxide nanoparticles. Recently microwave assisted solution route has emerged as a new method for the production of nanosized material with highly crystalline properties, shortened reaction time, rapid volumetric heating increased reaction rates and high purity. In the present work we have studied the effect of microwave irradiation on hydrated tin oxide precipitate in aqueous medium and formation of metal sub-oxides material SnO with high crystalline rate but not the oxidation rate.

CP1004, *Thermophysical Properties of Materials and Devices-NCTP '07*,
edited by P. Predeep, S. Prasanth, and A. S. Prasad

EXPERIMENTAL

A 0.1M solution of tin(II) chloride in deionized water is prepared. Then pH of the solution is maintained between 7 and 9 using liquid ammonia diluted with water. The resulting precipitate was washed with water no chlorine ions are detected. The resulting precipitate is irradiated with household microwave oven for 10minutes. The radiation wavelength is 2.45GHz and power up to 1 kW.

RESULTS AND DISCUSSION

Fig. 1 show that the XRD pattern of tin oxide powder before and after microwave treatment for pH 7 and 9. The XRD pattern before, after microwave treatment shows the existence of hydroxyl group ($Sn_6O_4(OH)_4$ and SnO nanocrystals respectively. Microwave radiation causes change hydroxyl group into oxide group. It is evident that the diffractions match perfectly with the SnO Tetragonal Romarchite structure. The following Miller indexes [101], [110], [002], [200], [112], [211], [103], [220] were assigned [3].

The deduced lattice parameter values (a = b = 3.7873 Å and c = 4.8352 Å) of tin oxide (SnO) nanoplatelets have been in good agreement with the standard values (a = b = 3.8029 Å and c = 4.8382 Å)[4]. The average crystalline size of SnO crystal was calculated by using Debye Scherer's formula and calculated as 26nm, 29nm for pH-7, 9 respectively.

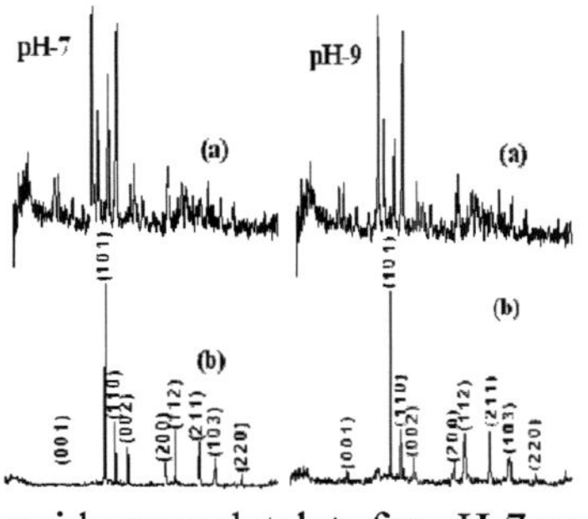

FIGURE 1. XRD patterns of tin oxide nanoplatelets for pH-7 and 9 (a)Before (b)After Microwave treatment

Fig 2(a) shows the morphologies of the microwave synthesized tin oxide nanoparticle as platelet like morphology with an average width ranging from 3 to 8 μm and thickness of around 100 nm. However, the particle size estimated from XRD is much smaller than the one deduced from SEM micrograph measurements. This could be explained by the fact that together with the nanoplatelets many smaller particles are present and are clearly visible in micrographs.

(a) **(b)**

FIGURE 2. (a)SEM images of tin oxide nanoplatelets (b)TEM images of SnO nanoplatelets

In order to answer to this question transmission electron microscopy (TEM) image of the SnO is acquired together with selected area electron diffraction (SAED). Fig. 2(b). shows a SnO single crystal platelet together with many small particles attached to its surface. The SAED shows bright spots, which can be indexed to the Tetragonal Romarchite structure in the [001] zone axis and rise from the single crystal nanoplatelet. The diffraction shows also low intensity rings due to the diffraction caused by small (SnO) nanoparticles, which can be clearly seen on the micrograph. Thus, the sample is a mixture of platelets and small nanoparticles. Hence, their presence can explain the small particle size extracted from the XRD measurements.

Fig. 3(A). shows that FTIR spectrum of tin oxide before microwave treatment (dried precipitate), after microwave treatment. The adsorption band at 540 cm^{-1} can be ascribed to the terminal oxygen vibration (ν_{Sn-OH})[5] for the dried precipitate. The absorption band located around 480 cm^{-1} for the sample after microwave treatment, has been ascribed to the stretching frequency of oxide group (ν_{SnO})[6].

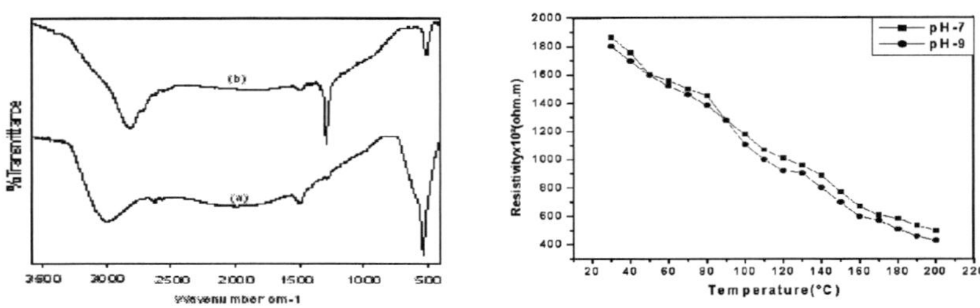

(A) (B)

FIGURE 3. (A) FTIR spectrum of tin oxide nanoplatelets(a) Beforeand (b)afterMicrowave treatment (B) Resistivity of tin oxide nanoplatelets for pH-7 and pH-9

Fig. 3(B). shows that the resistivity of the tin oxide nanoplatelets as a function of temperature for pH-7 and pH-9. Also, at low temperature the conductivity can occur due to the contribution of migration of charged species like O_2^{2-}, O^-, OH^-, H_3O^+[7]. With the increase in temperature, the desorption of physically adsorbed water should lead to a decrease in the resistance due to the release of carriers. Hence, in the low temperature range of 200 °C, the resistance of SnO decreases with increasing the temperature. The resistivity of tin oxide is decreased by increasing the pH of the solution, which is mainly due their grain size[8].

CONCLUSION

In summary, the crystalline SnO nanoplatelets have been successfully prepared by microwave technique. The SEM results show that the microwave irradiation during the powder formation process affected the grain shape due to the volumetric heating and it showed as nanoplatelets. The crystalline sizes are same in both XRD and TEM results, which confirm the presence of single crystalline nanoplatelets. The nanostructured platelets have been further characterized by FT-IR and electrical conductivity.

REFERENCES

1. Yoshio Idota, Tadahiko Kubota, Akihiro Matsufuji, Yukio Maekawa, Tsutomu Miyasaka , Science, **276**, 1395-1397 (1997).
2. Dien-Shi Wu, Chih-Yu Han, Shi-Yu Wang, Nae-Lih Wu, I.A. Rusakova - Mater. Latter, **53**, 155-159 (2002).
3. Kwang-Min Lee, Doh-Jae Lee, Hoon Ahn, Mater. Lette, **58**, 3122-3125 (2004).
4. J. Pannetier, G. Denes - Acta Crystallogr., Sect. B: Struct. Crystallogr. Cryst. Chem. **36**, 2763-2765 (1980).
5. Jianrong Zhang, Lian Gao, Inorga. Chem. Commu. **7**, 91-93 (2004).
6. Quan Li, Xundao Yuan, Guangfu Zang, Shiquan Xi - Mater. Chem. and Phys, **47**, 239-245 (1997).
7. U. Weimer, W. Gopal, Sens. and Actuat. B **26**, 13-18 (1995).
8. A.Cobot, J.Arbiol, J.R.Morante, U.Weimar, N.Barsan and W.Gopal - Sens. and Actuat. B **70**, 87-100 (2000).

DEVICES AND APPLICATIONS

Structure and Electrical Characteristics of Hot Wall Deposited $CuInSe_2$ Thin Films and Study on Back Contact Layers in $CuInSe_2$ Based Solar Cells

S.Agilan[a,b*], D.Mangalaraj[b], Sa.K.Narayandass[b], G.Mohan Rao[c]

[a] *Department of physics, Coimbatore Institute of Technology, Coimbatore, India*
[b] *Department of Physics, Bharathiar University, Coimbatore, India*
[c] *Department of Instrumentation, Indian Institute of Science, Bangalore, India.*

Abstract. $CuInSe_2$ thin films were prepared on to well cleaned glass substrates by a hot wall deposition technique. The X-ray diffraction studies revealed that all the deposited films are polycrystalline in nature, single phase, and exhibiting chalcopyrite structure The crystallites were found to have a preferred orientation along the (112) direction. The resistivity is found to decrease with increase in temperature. Two types of conduction mechanisms are present in the hot wall deposited $CuInSe_2$ films. In the temperature region below 215 K the conduction is due to variable range hopping mechanism and in the temperature region above 215 K the conduction is due to thermally activated process. It is observed that the solar cell with molybdenum as back contact has low series resistance (R_s), high shunt resistance (R_{sh}) and large fill factor (FF) when compared with $CuInSe_2$ based solar cells with other back contact material layers.

Keywords: Hot wall deposition, $CuInSe_2$ thin films, Structure, Electrical conductivity, Solar Cell and Back Contact.
*Corresponding Author; e-mail: agils70@yahoo.co.in (S.Agilan) Ph: (0)999 490 999 3
PACS: 68.55ag

INTRODUCTION

$CuInSe_2$ (copper indium di selenide) is a promising material for thin film solar cells because of its extraordinary radiation stability[1]. $CuInSe_2$ films possess certain exceptional material characteristics like band gap, absorption coefficient and minority carrier diffusion length which are particularly suitable for photovoltaic applications. The relation between the energy gap and the lattice constant of isovalent chalcopyrite materials and the material requirements for $CuInSe_2$ solar cells are analysed by Konovalov [2]. In the present study, stoichiometric $CuInSe_2$ thin films have been deposited on to well cleaned glass plates by hot wall deposition technique. Deposited $CuInSe_2$ thin films are structurally characterized and dependence of conduction mechanism with temperature have been studied. An attempt is made to study the effect of back contact layer in the fabricated $CuInSe_2$ based solar cells.

CP1004, *Thermophysical Properties of Materials and Devices-NCTP '07*,
edited by P. Predeep, S. Prasanth, and A. S. Prasad

EXPERIMENTAL DETAILS

The prepared bulk $CuInSe_2$ compound was evaporated and deposited onto well cleaned glass substrates by hot wall deposition technique. The schematic diagram of the hot wall experimental set up is shown in Fig. 1. $CuInSe_2$ thin films were deposited onto well cleaned glass substrates, using quartz tube of optimized length of 0.11m, with the source and wall temperatures around 1270 K. The structure of the films was studied using X-ray diffractometer Model Philips PW 3710 with CuK_α radiation. Chemical constituents present in the deposited films were estimated using an Energy Dispersive X-ray Analysis (EDAX) Model 7060 Oxford Instruments

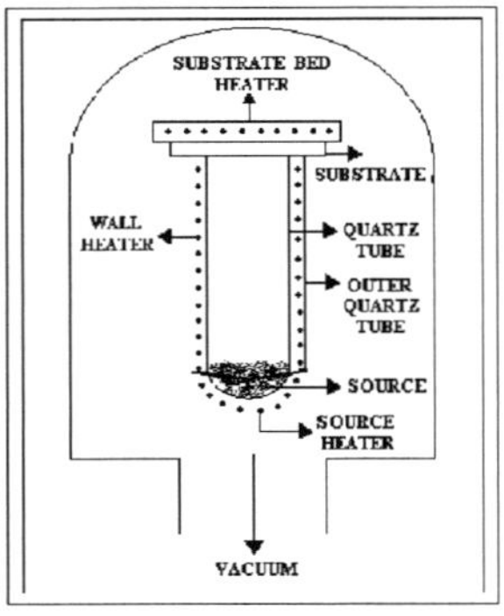

FIGURE 1. The schematic diagram of the hot wall experimental set up.

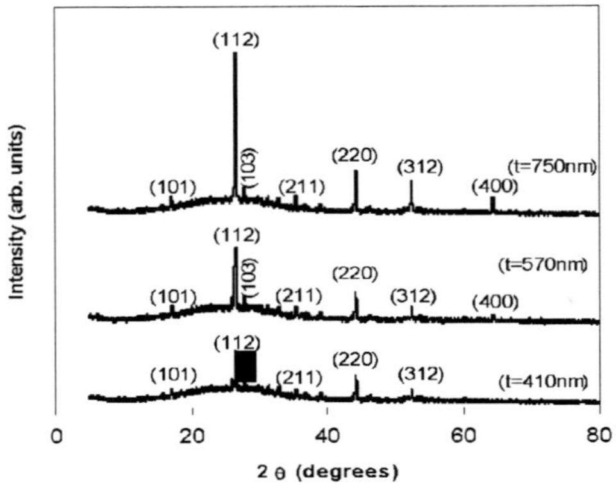

FIGURE 2. The X-ray diffraction pattern of $CuInSe_2$ thin films of different thicknesses.

(Cambridge). The electrical resistance was measured in the temperature range 80 to 340 K (under vacuum of 10^{-3} Torr) using a Keithley 614 electrometer as an ohm-meter. The intensity of illumination on the solar cell was measured using a suryamapi (CEL, India). The current and voltage were measured using optical power meter [70310, Oriel Instruments, USA] and an electrometer (Keithley 2001) respectively.

RESULTS AND DISCUSSION

Structural Properties of $CuInSe_2$ Thin Films

The X-ray diffraction pattern of $CuInSe_2$ thin films of different thicknesses is shown in Fig. 2. All the deposited $CuInSe_2$ films are found to be polycrystalline in nature exhibiting chalcopyrite structure. The micro structural parameters of the deposited $CuInSe_2$ films have been evaluated and are shown in Table 1. It is observed that the grain size increases but the strain and dislocation density decrease with the increase of film thickness[3,4]. The presence of (101), (103) and (211) diffraction peaks confirm the chalcopyrite structure in the deposited films [5,6].

It is found that the elemental composition in the bulk is almost maintained in the deposited film, as copper:24.653 atomic%, indium:24.288 atomic % and selenium: 51.059 atomic%. $CuInSe_2$ thin films with Cu/In ratio greater than one have been reported by Menna et al. [7] as suitable for high efficiency devices.

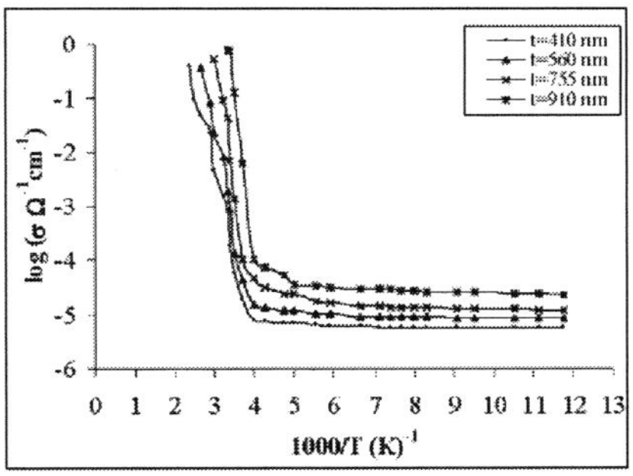

FIGURE 3. Plot of log σ versus 1000/T of $CuInSe_2$ thin films of different thicknesses.

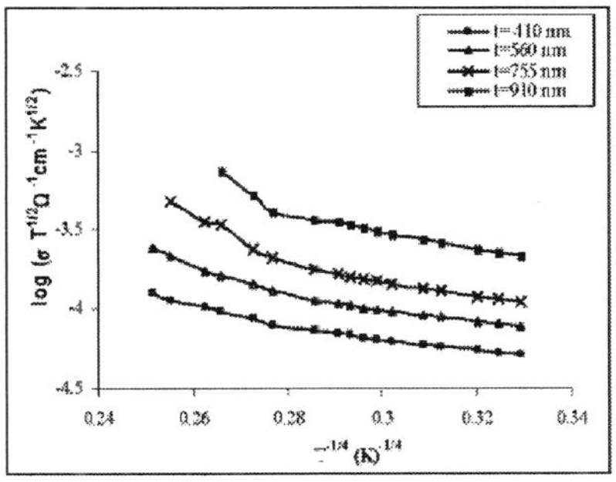

FIGURE 4. Plot of log(σ $T^{1/2}$) versus $T^{-1/4}$ of $CuInSe_2$ thin films of different thicknesses.

Film thickness (nm)	Lattice parameter a Å	Lattice parameter c Å	c/a ratio	Tetragonal Distortion 2-c/a	Grain size D (nm)	Dislocation Density $x10^{14}$ lin/m^2	Strain ε $x10^{-4}$
410	5.779	11.623	2.0112	-0.0112	89.37	1.252	4.507
570	5.780	11.621	2.0105	-0.0105	162.75	0.378	2.475
750	5.785	11.613	2.0074	-0.0074	243.68	0.168	1.653

TABLE 1. The micro structural parameters of the deposited $CuInSe_2$ films of different

Temperature dependence of DC conduction

The resistivity of the film is found to decrease with increase in temperature which confirms the semiconductor nature of the deposited $CuInSe_2$ films. Figure 3 shows the variation of electrical conductivity of $CuInSe_2$ films with the inverse of absolute temperature. The plot is a linearly varying one and therefore the measured experimental data fits to the relation,

$$\sigma = \sigma_o \exp(-Ea/k_B T). \tag{1}$$

The presence of two regions with different slopes in the plot suggests that there are two types of conduction mechanism present in the $CuInSe_2$ films deposited by a hot wall deposition technique. The point of inflection is found to be around 215 K. In the temperature region below 215 K, the conduction is due to variable range hopping mechanism and in the temperature region above 215 K, it is due to thermally activated process. From the plot it is clear that for temperatures greater than 215 K, the electrical conductivity varies linearly with temperature and so the electrical data fits into the relation

$$\sigma = \sigma_o \exp(-\Delta E/k_B T), \tag{2}$$

therefore the conductivity is attributed to thermal excitation of charge carriers from grain boundaries to the neutral region of grains.

The activation energy associated with the hot wall deposited $CuInSe_2$ films at two different temperatures are shown in Table 2. The two different activation energies of low and high value obtained for temperatures 135 K (below 215 K) and 313 K (above 215 K) can be attributed to the hopping conduction mechanism and thermally activated process respectively. Similar characteristics have been reported for $CuInSe_2$ films by earlier workers [8,9].

From Table 2 it is observed that the activation energy associated with hot wall deposited $CuInSe_2$ films decreases considerably with the increase in film thickness and thereby it proves the semiconducting behavior of the deposited films. The decrease in activation energy as the thickness increases suggests that the grain boundary scattering contribution reduces significantly as thickness increases.

Thickness (nm)	Temperature (K)	Activation energy (cV)
410	135	3.26×10^{-3}
	313	0.706
560	135	1.98×10^{-3}
	313	0.545
755	135	1.59×10^{-3}
	313	0.381
910	135	1.33×10^{-3}
	313	0.155

TABLE 2. The activation energies of hot wall deposited $CuInSe_2$ films at two different temperatures.

The plot of log $(\sigma T^{1/2})$ versus $T^{-1/4}$ shown in Fig. 4 represents the conductivity below 215 K. As the plot is a linearly varying one, the experimental data fits into the relation given by Mott [10]

$$\sigma = \sigma_0 T^{-1/2} \exp[-(T_0/T)^{1/4}]. \tag{3}$$

Since T_0/T values are found to be greater than 1, the conductivity mechanism in the hot wall deposited $CuInSe_2$ films is variable range hopping in the temperature range below 215 K [11-13].

Effect of back contact in $CuInSe_2$ based solar cells

The contact layer must function as barrier that hinders diffusion of impurities from the substrate into the absorber [14]. A high optical reflectance is necessary to minimize optical losses. p-$CuInSe_2$/n-CdS solar cells of area 50 mm^2 are fabricated on the metallic back contact layers Mo, W, Ta and Cr (≈1μm) separately. Figure 5 shows the I-V characteristics of the fabricated solar cell with different back contacts. On comparing the cell parameters given in Table 3, it is observed that $CuInSe_2$ based solar cells with molybdenum as back contact possess a low series Resistance (R_s), high shunt resistance (R_{sh}), maximum Power P_{max} and high fill factor (FF) values than the other cells with different metallic back contact layers.

More over the deviation of the characteristic curve from the ideal curve near the open circuit region is highly reduced in the case in which molybdenum is used as back contact. Hence the molybdenum is the ubiquitous choice for back contact in $CuInSe_2$ based solar cell device fabrication.

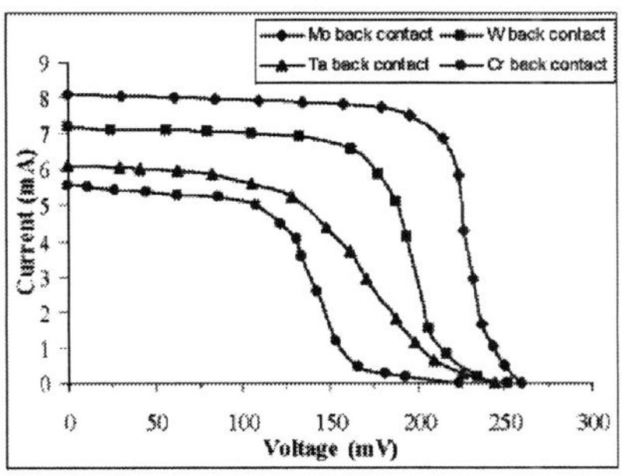

FIGURE 5. I-V characteristics of the fabricated solar cell with different back contact layers.

Metallic back contact layer	V_{oc} mV	I_{sc} mA	R_{sh} Ω	R_s Ω	P_{max} mW	Fill Factor
Mo	259.84	8.09	955	37	1.474	0.701
W	251.50	7.20	860	48	1.068	0.589
Ta	244.35	6.12	780	49	0.674	0.451
Cr	223.41	5.56	395	106	0.548	0.441

TABLE 3. Solar cell parameters of the fabricated solar cells with different back contact layers.

CONCLUSION

$CuInSe_2$ thin films were prepared on to well cleaned glass substrates by hot wall deposition technique. The X-ray diffraction studies revealed that all the deposited films are polycrystalline in nature, single phase, and exhibiting chalcopyrite structure. Two types of conduction mechanisms are present in the hot wall deposited $CuInSe_2$ films. The activation energy is found decreasing as the thickness of the film increases. It is found that molybdenum is the ubiquitous choice for back contact in $CuInSe_2$ device fabrication.

ACKNOWLEDGEMENT

One of the authors (S.A) expresses his gratitude to "Tamil Nadu State Council for Science and Technology-Chennai" for the award of "Young Scientist Fellowship" to him.

REFERENCES

1. A.V.Postnikov, M.V.Yakushev, Thin solid films 451-452 (2004) 141
2. I.Konovalov, Thin solid films 451-452 (2004) 413.
3. K.Senthil, D.Nataraj, K.Prabakar, D.Mangalaraj, Sa.K.Narayandass, N.Udhayakumar, N.Krishnakumar, Materials Chemistry and Physics 58 (1999) 221.
4. M.Dhanam, R.Balasundaraprabhu, S.Jayakumar, S.Gopalakrishnan, M.D.Kannan, Physica status solidi (a) 191 (2002) 149.
5. R.Pal, K.K.Chattopadhya, S.Chandhuri, A.K.Pal,Solar Energy Materials and Solar Cells 33 (1994) 241.
6. M.Nishitani, T.Negami, M.Terauchi, T.Hirao, Japanese Journal of Applied Physics 31 (1992) 192.
7. P. Menna, A.Perretta, M.Pellegrino, L.Quenica, M.L.Addonizio, Solar Energy Materials and Solar Cells 35 (1994) 165.
8. A.H.Moharam, M.M.Hafiz and A.Salem, Aplied Surface Science 172 (2001) 61.
9. Y.D.Tembhurkar and J.P.Hirde, Thin Solid Films 215 (1992) 65
10. N.F.Mott and E.A.Davis, Electronics Processes in Non Crystalline Materials, Clarendron, Oxford, 1971.
11. R.Dewsberry, Journal of Physics D: Applied Physics, 8 (1975) 1797.
12. D.Lemoine and J.Mendolia, Physics Letters, A82 (1981) 418.
13. A.J.Lewis, Physical Review B, 13 (1976) 2565.
14. K. Orgassa, H.W. Schock, J.H. Werner, Thin Solid Films 431–432 (2003) 387.

Dye Sensitized Solar Cell Based On Pyronin G Dye and TiO_2

P. Balraju[a], Yojana Janu[b], M.S. Roy[b*] and G. D. Sharma[a]

[a] *Department of Physics, Jai Narain Vyas Univeristy, Jodhpur, Rajasthan – 342 005, India*
[b] *Defence Laboratory, Jodhpur, Rajasthan – 342 011, India*

Abstract. Dye sensitized solar cell (DSSC), was fabricated with nano-crystalline TiO_2 absorbed Pyronin G (PYR G) dye using polar solvent treated poly(3,4 ethylene dioxythiophene):poly (sturene sulfate) (PEDOT:PSS) coated conductive glass (Florine doped tin oxide, FTO) as counter electrode. Their absorption spectra, photoaction and photovoltaic properties were studied. Impedance analysis was also employed to analyse the charge transfer kinetics at the electrolyte/counter electrode interface. Using the solvent treated PEDOT:PSS coated FTO as counter electrode, a 1.2% power conversion efficiency of DSSC was obtained under 73mW cm^{-2} irradiation.

Keywords: Dye sensitized, Impedance analysis, charge transfer kinetics
PACS: 72.40.+w Photoconduction and photovoltaic effects

INTRODUCTION

Dye-sensitized solar cells (DSC) have been a topic of intensive research as a next-generation solar cell since in 1991 O'Reagan and Grätzel could obtain high conversion efficiencies with such cells for the first time [1]. In the last decade, after their first report on the dye-sensitized solar cell, a number of investigations on this type solar cell have been carried out[2-6]. Many advantages such as low production cost and simple fabrication process have made DSC promising as next-generation photovoltaic devices. The DSC is a photoelectrochemical device that employs a light-absorbing dye to harvest solar energy in the initial form of electronic and vibrational energy, which is converted into electrical power. The principle of DSCs is considered similar to that of photosynthesis in a leaf, as, unlike in conventional solar cells. The process is thought to be as follows: electrons are injected from the photoexcited dye into the conduction band of the semiconductor oxide, from where they pass through the nanoparticles to the transparent conducting oxide current collector and into the external circuit. The sensitizer is regenerated by electron transfer from a donor, typically iodide ions, which are dissolved in the electrolyte that is present in the porous semiconductor. The triiodide ions formed during the reaction diffuse to the counter electrode, where they are reduced back to iodide by the conduction band electrons that have passed though the external circuit that performs the electrical work.

Dye-sensitized solar cells are an option for a low-cost production of solar cells. Investment costs for fabrication are low, and materials of the cells may become cheap

CP1004, *Thermophysical Properties of Materials and Devices-NCTP '07*,
edited by P. Predeep, S. Prasanth, and A. S. Prasad

in a large-scale production. At present, costs are significant for cell elements like the dye or the, so far required Pt catalyst at the back contact. Our work focuses both on the fundamental, and on the up scaling issues required for industrialization of Dye-sensitized solar cells. In the present approach we have used modified PEDOT-PSS electrode instead of platinum counter electrode and PYR G dye which is available commercially at more economical cost.

EXPERIMENTAL PROCEDURE

Preparation of TiO_2 film

TiO_2 (Degusa P25) porus film was deposited by doctor blade technique on FTO glass substrate having the sheet resistance 10 Ohm/sq, 80% optical transmittance in the visible region. After the TiO_2 layers were dried at 80°C, then the film was sintered at 450°C for 1 h in air to improve the electronic contact not only between the particles and support but also among the particles [7]. For adsorbing photo sensitized dye on the TiO_2 surface, TiO_2 electrode was immersed in a solution of Pyronine G for 15 h at room temperature, and rinsed with ethanol.

Preparation of counter electrode

Counter electrode was prepared by using modified poly(3,4-ethylene dioxythiophene): Poly(styrene sulfonate) film as reported elsewhere [8]. The PEDOT-PSS aqueous solution and the organic solvent DMSO were mixed by the volume ratio of 3:1. The mixed solution of PEDOT-PSS and organic solvent was filtered and stirred continuously for 6h at room temperature. A small amount of carbon black (0.1 wt% with respect to the solvent treated PEDOT-PSS aqueous solution) is also added to improve the conductivity and roughness of the films. The solvent treated PEDOT-PSS films were formed on FTO by dip coating the above-mentioned polymer adhesive solutions. The film was dried at 80 °C for 30 min in air.

Solar cell assembly

Dye-adsorbed TiO_2 electrode and counter-electrode were assembled as a sandwich type cell, and then the assembled cell was filled with the redox electrolyte

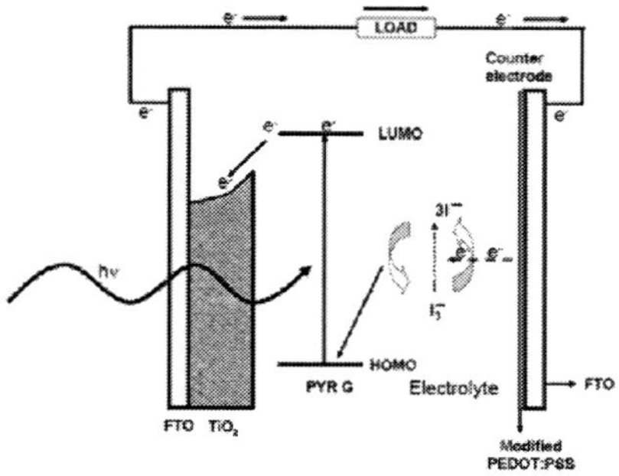

FIGURE.1. Schematic diagram of a dye sensitized solar cell

containing I^-/I^{-3} redox couple. The electrolyte introduced into the cell by capillary effect through thin gap and the gap was sealed with thermal plastic. Schematic diagram of dye-sensitized solar cell is shown in Fig. 1.

RESULTS AND DISCUSSION

Fig.2 represent the photocurrent-voltage (J-V) curve of DSC fabricated using modified PEDOT: PSS counter electrode and PYR G as a dye sensitizer, under irradiation with white light from a 100 W metal halide lamp in ambient atmosphere. From the figure the values of short circuit current (J_{sc}), open circuit voltage (V_{oc}) of the DSC can be obtained directly, which is 4.08 mA and 0.6 V respectively. The maximum power (P_{max}) of the DSC is confirmed based on P_{max} = (J X V) at the point M on Fig.2, which is 120 μW.

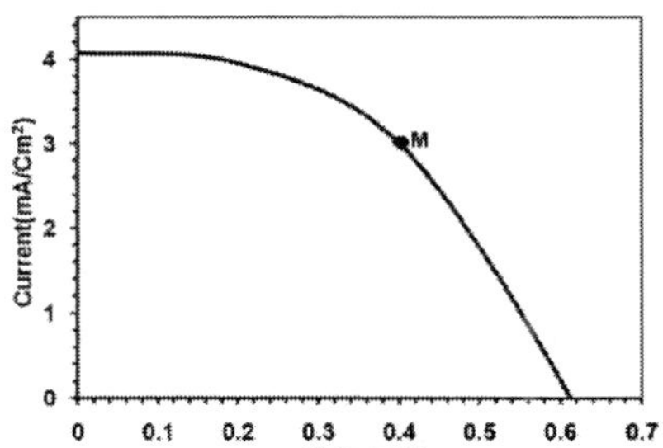

FIGURE 2 Photocurrent–voltage characteristic of the solar cell based on TiO_2 sensitized by PYR G dye using modified PEDOT: PSS counter electrode

The fill factor (FF) (related to series resistance for a practical solar cell) is calculated using the formula

$$FF = \frac{J_{max} \times V_{max}}{J_{sc} \times V_{oc}}$$

in Fig.2, which is 0.33. The overall efficiency η of the DSC was calculated from the expression

$$\eta = \frac{J_{sc}.V_{oc}.FF}{P_{in}}$$

where P_{in} is the intensity of the incident light. The power conversion efficiency of 1.2% was obtained.

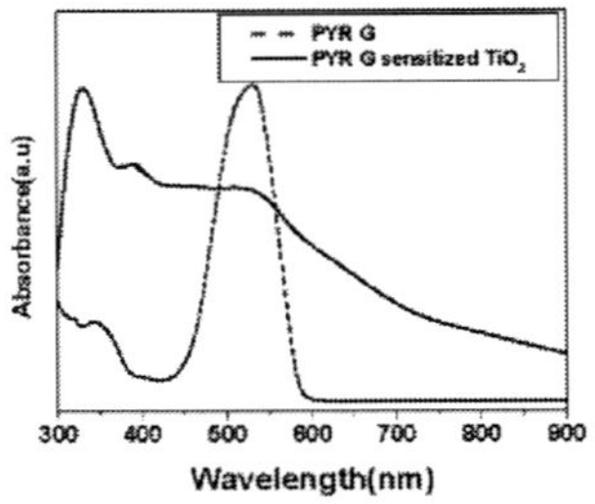

FIGURE 3. Absorption spectra of Pyronine G and Pyronine G sensitized TiO_2

The optical spectrum of the PYR G and PYR G dye sensitized TiO_2 is recorded and shown in Fig.3. The dye sensitized TiO_2 electrode exhibit strong absorption in the visible range confirming the binding of the sensitizer to the semiconductor surface. It is seen in the figure that PYR G absorbs in window extending from 450 nm to 590 nm and sensitization of TiO_2 with PYR G dye imparts enhancement in extent of absorption with a shifts towards longer wavelength leading to reduction in optical band gap of the material.

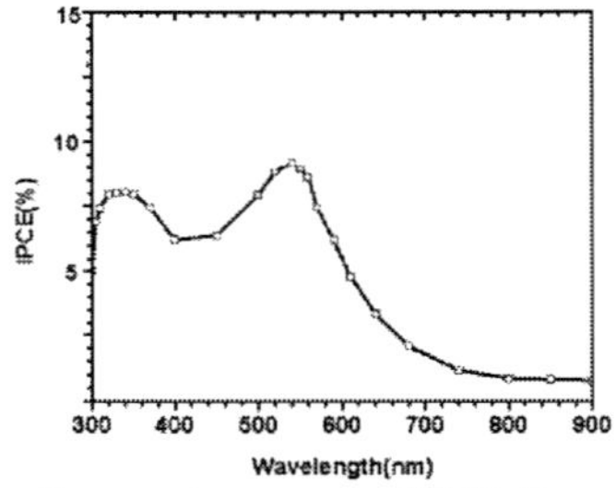

FIGURE 4. Photoaction spectra of PYR G sensitized DSC

Fig. 4 represents the photoaction spectra of the DSC in visible light range. The close match between the photocurrent action spectrum and the absorption spectrum indicates that photosensitization mechanism is operative in generation of photocurrent. IPCE represents the percentage of incident photons that are converted to electrons at a certain wavelength and is defined by the formula

$$IPCE(\lambda) = 1240\left(\frac{J_{sc}}{\lambda\phi}\right)$$

where λ is the wavelength (in nanometers) and Φ is the incident radiative flux (in units of Wm^{-2}) [9]. The spectrum is relatively broad, exhibiting a peak at 550 nm with a maximum of 9% due to electron injection in the TiO_2 conduction band, which is characteristic of the dye absorption spectrum. The peak between 320 and 350 nm with 8% of conversion efficiency in the IPCE is assigned to the direct excitation of the TiO_2 band gap.

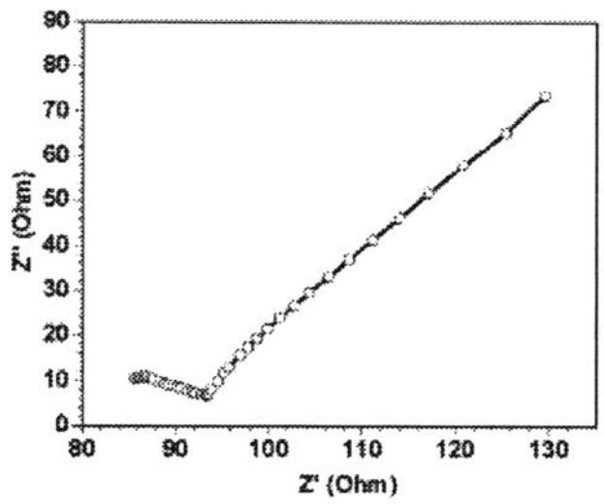

FIGURE 5. Electrochemical impedance spectrum of DSC

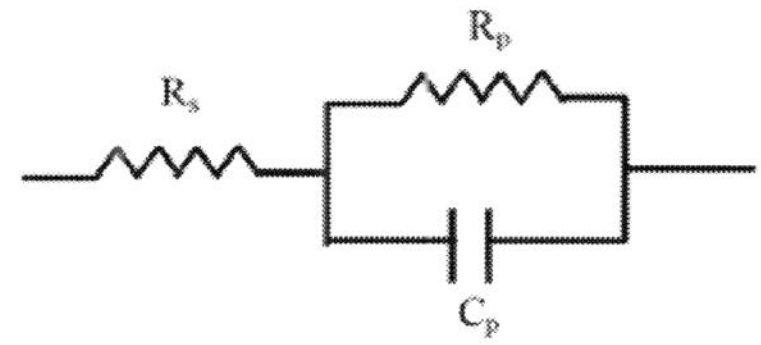

FIGURE 6. Equivalent circuit model of the experimental cell

Impedance function is a complex function: the real part of impedance, Z' characterizes the long-range transport (irreversible process like energy dissipation) while the imaginary part of impedance, Z" characterizes polarization or magnetization (reversible process like energy storage) in materials [10]. Impedance measurement for dye-sensitized solar cells is an effective method for investigating the contribution of each internal resistance of cells. Impedance analysis of the cell was carried out by plotting Cole-Cole plot in between real and imaginary impedance with illumination under the frequency range from 100 Hz to 1 MHz using a Galvanostat/ Potentiostat (Autolab, Eco Chemie Netherlands). All measurements were carried out in room temperature. The corresponding equivalent circuit is given in Fig.6. One part is parallel connection of resistance (R_p) and capacitance (C_p) for interfacial reaction. The other part is series connection of electrolyte resistance, electrode resistance, and interfacial resistance between electrode and TCO. Among other parameters obtained from the fitting, the series resistance was estimated as $R_S = 90\ \Omega$. The Rs is resistance related to charge-transfer processes occurring at the modified PEDOT-PSS counter electrode and electrolyte. It should increase with increase in the surface area of the counter electrodes. The poor fill factor of the cell is also due to the high series resistance. This can be minimized by increasing conductivity and increasing roughness factor[11] of the modified PEDOT-PSS counter electrode.

CONCLUSION

DSSC fabricated by adopting Doctor Blade technique using modified PEDOT-PSS electrode instead of high cost platinum counter electrode and commercially available PYR G dye. The new class of dye, whose energy matches the TiO_2 for DSSC applications, provides an alternative over conventional ruthenium complex-based DSSCs. The cell shows reproducible photo response in terms of J_{sc} 4.08 mA; V_{oc} 600 mV; FF 0.33 and the conversion efficiency η =1.2 percent. The incident photon to current conversion efficiency IPCE was found to be maximum of 9% at 550 nm. Impedance spectrum of the DSSC was also analyzed by plotting equivalent circuit.

REFERENCES

1. B. O'Reagan, M. Grätzel, Nature **353,** 737-470 (1991).
2. A. Hagfeldt, M. Gratzel, Acc. Chem. Res. **33,** 269-277 (2000).
3. M.K. Nazeerudden, A. Kay, I. Rodicio, R. Humphry-Baker, E. Muller, P. Liska, N. Vlachopoulos, M. Gratzel, J. Am. Chem. Soc. **115,** 6382-6390 (1993).
4. U. Bach, D. Lupo, P. Cornte, J.E. Moser, F. Weissortei, J. Selbeck, H. Spreitzer, M. Gratzel, Nature **395,** 583-585(1998).
5. M. Gratzel, J. Photochem. Photobiol. A **164,** 3-14 (2004).
6. M.K. Nazeerudden, P. Pechy, T. Renouard, S.M. Zakeeruddin, B.R. Humphry, P. Comte, P. Liska, L. Cevey, E. Costa, V. Shklover, L. Spiccia, G.B. Deacon, C.A. Bignozzi, M. Gratzel, J. Am. Chem. Soc. **123,** 1613-1624 (2001).
7. K. Kalyanasundaram, M. Gratzel, Coord. Chem. Rev. **177,** 347-414 (1998).
8. J.G. Chen , H.Y. Wei , K.C. Ho, Sol. Ener. Mat. & Sol. Cells. **91**, 1472-1477 (2007).
9. C.J. Barbé, F. Arendse, P. Comte, M. Jirousek, F. Lenzmann, V.Shklover, M. Grätzel, J. Am. Chem. Soc. **80,** 3157-3161 (1997).
10. Z.Shi, M.Liu, Electrical and electrochemical analysis of nanophase materials, Z.L.Wang (Ed.), Characterization of Nanophase Materials, Wilwy-VCH, Weinheim, 2000, p.172.
11. N. Koide, A. Islam, Y. Chiba, L. Han, J. Photochem. Photobiology A: Chemistry **182,** 296-305 (2006).

Activated Reactive Evaporated Nano Crystalline WO_3 Thin Films for Electrochromic Windows

K. Hari Krishna[a] and O.M. Hussain[a*]

[a] *Thin Film Laboratory, Department of Physics, Sri Venkateswara University, TIRUPATI – 517 502, India.*

Abstract: Tungsten trioxide thin films were deposited by using home built activated reactive evaporation (ARE) technique on ITO coated substrates at an optimized conditions. The microstructural, optical and electrochromic properties of the as deposited WO_3 films were investigated. The X-ray diffraction pattern of the film displayed characteristic triplet of (200), (020) and (002) Bragg reflections, in which (200) orientation is predominant and attributed to the monoclinic structure. The films were found to be smooth, uniform and composed with roughly spherically shaped nano grains of size in the range 60 - 70 nm. The as deposited WO3 films exhibited 85% of optical transmittance with evaluated optical band gap of 3.46 eV. The electrochromic studies were carried out for these WO_3 films and observed to be 69% of optical modulation with colouration efficiency of 120 mC/cm^2 at the wavelength of 550 nm.

Keywords: Activated reactive evaporation technique, Nano crystalline tungsten trioxide thin films, Microstructural, optical and Electrochromic properties.

PACS : 72.80.Ga, 78.70.Ck, 74.25.Gz, 76.30.Mi, 78.20.Jq.

INTRODUCTION

Electrochromic materials with semiconducting nature are capable of changing their optical properties under the action of an applied electric field or an electric current In this direction, transition metal oxides have been exhibiting promising and persistent electrochromic properties [1]. Among all the transition metal oxides, tungsten trioxide (WO_3) has received the considerable attention by the material science researchers from the birth of the electrochromism due to it's significant electrochromic properties such as large optical modulation, colouration efficiency and good durability [2]. These properties inturn depend upon the type of deposition technique and respective film processing parameters. Several physical and chemical vapor thin films deposition techniques have been employed to deposit WO_3 thin films such as electron beam evaporation, pulsed laser deposition, sputtering, sol-gel and potentiostatically electro deposition techniques [3-7].

The activated reactive evaporation technique (ARE) is one of the plasma assisted physical vapor deposition techniques to prepare better quality nano crystallineWO_3 thin films of nearly stochiometric, well adherent and large film area with good uniformity. Hence in the present study, the WO_3 thin films were prepared by using ARE technique on

CP1004, *Thermophysical Properties of Materials and Devices-NCTP '07*, edited by P. Predeep, S. Prasanth, and A. S. Prasad

ITO coated glass substrates in controlled environment in the deposition chamber. The structural, surface morphological, optical and electrochromic properties were investigated for their effective utilization in advanced smart window technology.

EXPERIMENTAL

Tungsten trioxide thin films were deposited by using a home built activated reactive evaporation system [8] on ITO coated glass substrates maintained at constant substrate temperature of 573 K during the film depositions. A Seifert computerized X–ray diffractometer (model 3003 TT) was employed for the structural analysis of the as deposited films. A CuK_{α} target was used as the X–ray source with wavelength λ = 1.5418 Å. Atomic force microscopy (AFM) (Digital Instrument: Dimension 3100 series) was used to study the surface morphology of the films in a single contact mode of operation. The optical measurements were carried out by Hitachi U-3400 UV-VIS-NIR double beam spectrophotometer in the wave length range of 300 – 1500 nm. Lithium insertion studies were investigated by a dry method in which lithium niobate ($LiNbO_3$) was heat-treated under high vacuum to give off lithium atoms for insertion in the exposed WO_3 films. The degree of such lithiation was measured by noting the charge in the quartz crystal thickness (the effective mass) and calibrated against the electrochemical insertion.

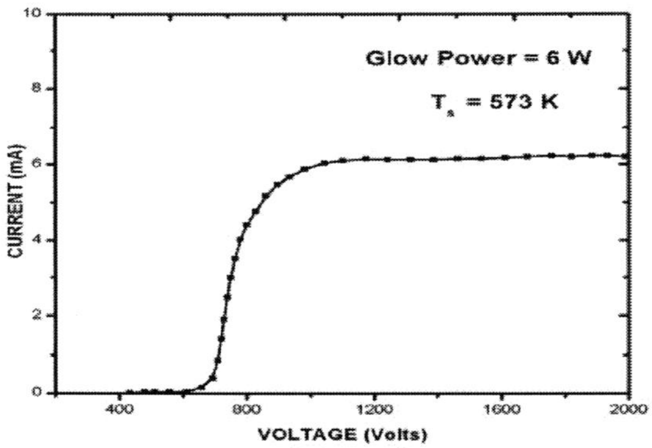

FIGURE. 1 Current-Voltage characteristics of the glow discharge at an oxygen partial pressure of 5 X 10^{-3} Torr.

The current–voltage (C-V) characteristics of the discharge were studied in order to understand the characteristics of the plasma. The glow discharge is established when the voltage across the electrodes reaches the breakdown value. Initially, at low voltages the discharge covers only the area near the rim of the cathode. When the power is raised, the current increases and the discharge extends to cover the whole surface of the cathode. Figure.1 shows C-V characteristics of the glow discharge at an oxygen partial pressure of

5 X 10^{-3} Torr. As voltage increases the current also increases and at a certain voltage, the current reaches to a saturation value. This is because at certain voltages, the production of electrons and ions are the same and some of these electrons and positive ions will recombine to produce neutral species. The parameter that defines the density of the charged particles in the plasma is the degree of ionization of the gas. It specifies the fraction of the particles in the gaseous phase which are ionized. The degree of ionization for sustained plasma in low pressure discharges is typically in the range 10^{-6} to 10^{-3}. However, if the electric discharge is assisted and confined by an additional magnetic field, the degree of ionization can reach values of 10^{-2} or higher. The degree of ionization increases by the increase of glow power and at higher glow power a sustained glow can be observed. Also the degree of ionization increases with the increase of oxygen partial pressure. Therefore from the above examination of C-V characteristics of the glow discharge, the high density of stabilized plasma provided with high degree of ionization at optimized glow power of 6 W and oxygen partial of 5 X 10^{-3} Torr. Hence in the present investigation WO_3 thin films were prepared at optimized conditions as mentioned above.

RESULTS AND DISCUSSIONS

The structural characterization for the as deposited WO_3 thin film at $T_s = 573$ K was carried out by using X-ray diffraction technique as shown in the fig.2. The XRD pattern of the film exhibited broad characteristic Bragg reflections (200), (020) and (002) at 23.1^0, 23.6^0 and 24.1^0 diffraction angles in which (200) orientation is noticed to be predominant. Along these characteristic triplet, additionally (004) and (400) diffraction peaks were identified at 46.5^0 and 48.6^0 respectively. The presence of broad diffraction

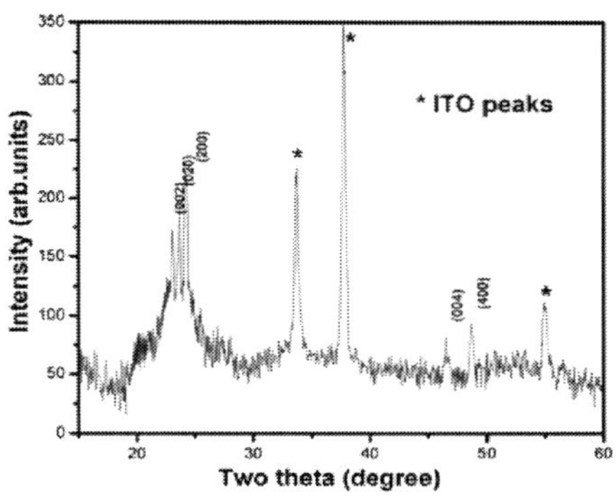

FIGURE.2 The X-ray diffraction pattern of the tungsten trioxide films prepared at $T_s = 573$ K.

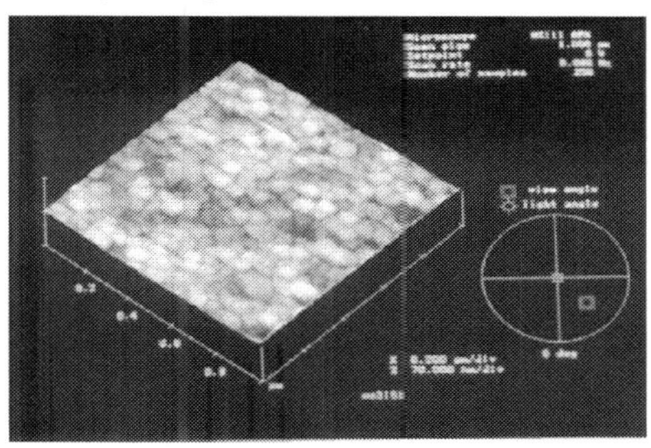

FIGURE. 3 AFM micrograph of the tungsten trioxide thin films deposited at $T_s = 573$ K.

peaks at respective diffraction angles reveal the existence of nano crystallites in the amorphous matrix of the as deposited WO_3 films. The presence of characteristic triplet (200), (020) and (002) attribute to the monoclinic structure of the WO_3 film and corresponding evaluated lattice constants are a = 7.309 Å, b = 7.601 Å, c = 7.711 Å and β = 91.131°, which are in good agreement with JCPDF data file (no.83-0951). The Scherrer's formula was used to calculate the average crystallite size of the crystallites present in the film and found to be in the order of 70 nm. The Surface morphological features of the as deposited WO_3 film at T_s = 573 K, were studied using atomic force microscopy as shown in the fig.3. From AFM micrograph it is clear that, the film was observed to be composed of spherically shaped nano grains in the range 64 nm and the individual grains are clearly visible. The evaluated root mean square roughness of the films is 11 nm. The regular distribution of grains all over the surface is attributed to the uniform temperature gradient maintained on the substrate. The nano crystalline WO_3 film showed characteristic and extremely interior porous nature. Visibly, the film is highly transparent showing no tint and colour variation with respect to viewing angle. However, such films do present a slightly cloudy appearance due to a small component of their diffuse reflectance arising from the nano grains.

The optical transmittance spectrum were recorded in the wave length range 300–1500 nm for nano crystalline WO_3 thin films deposited at T_s = 573 K as shown in fig. 4. From the figure it is observed that, the WO_3 film exhibited 85% of optical transmittance in the visible region. The sharp decrease in the optical transmittance with most of the radiation absorbed for incident photons in the wavelength range 300 – 500 nm is associated with the fundamental absorption.

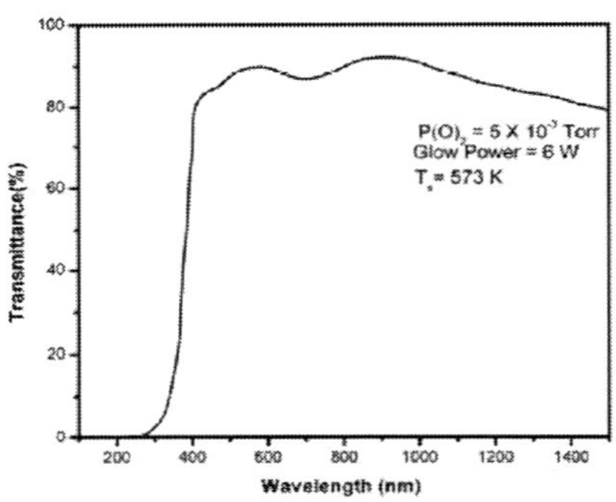

FIGURE. 4 Optical transmittance spectrum of the nanocrystalline WO_3 thin film prepared at T_s = 573 K

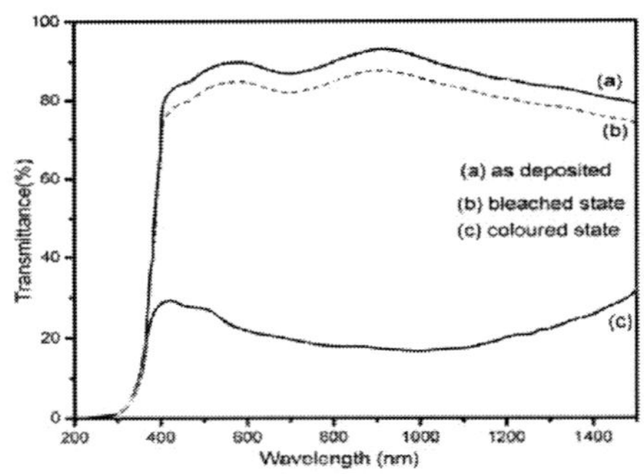

FIGURE. 5 Optical transmission spectra for nano rystalline WO_3 film both in coloured and bleached states.

The optical absorption coefficient (α) of the ARE tungsten trioxide thin films is calculated from the obtained optical transmittance of the films. The experimentally measured optical absorption coefficient for tungsten trioxide films is found to give a better fit to the relation $(\alpha h\nu)^{1/2} = B(h\nu - E_g)$, where hv is the energy of the incident photon, B the absorption edge width parameter and E_g the optical band gap. The optical band gap was estimated by extrapolating the linear region of the plot to zero absorption and noticed that as deposited nano crystalline WO_3 film prepared at T_s = 573 K possessed larger optical band gap value of 3.46 eV, which is in good agreement with the previous reported value. The electrochromic (EC) studies were carried out for the nano crystalline tungsten trioxide thin film prepared at Ts = 573 K. Figure. 5 represents the variation of the optical transmittance of WO_3 films in bleached state, it is unambiguous that ARE deposited tungsten trioxide thin films exhibited higher optical modulation of 69%. In the visible region the optical transmittance is found to be significantly different for the tungsten trioxide films in the colored state, this effect could make them useful in EC display devices and smart window applications. The higher optical modulation was observed in the visible region of the spectrum and lower optical modulation in the infrared region. This may be due to the maximum optical absorption of WO_3 films in their coloration and bleaching process in the visible region. For the nano crystalline WO_3 films the optical density (OD) was evaluated at the wavelength range of 300-900 nm The CE of WO_3 films is evaluated by using relation E = OD/Q_{in}, where OD is the optical density and Q_{in} (mC cm^{-2}) is the charge injected during the colouration cycle. The nano crystalline WO_3 film deposited at T_s = 573 K has CE value of 120 cm^2/C at the wavelength of (λ) 550 nm.

CONCLUSIONS

Nanocrystalline Tungsten trioxide thin films deposited by activated reactive evaporation technique were observed to be well adherent with uniform thickness over considerable area on the ITO coated substrates. The XRD pattern of the films deposited at optimized conditions of T_s = 573 K, $P(O)_2$ = 5 X 10^{-3} Torr and glow power of 6W respectively, has exhibited broad characteristic triplet, in which (200) orientation is predominant. These results supported the existence of monoclinic structure. AFM investigation revealed the cloudy and highly interior porous nature of surface the films, composed of nano grains in the order 64 nm.The ARE deposited WO_3 thin films showed 85% of optical transmittance and demonstrated better optical modulation of 69% in the visible region. Therefore, we conclude that, activated reactive evaporated tungsten trioxide thin films of significant electrochromic properties, can be adopted as active layers in advanced electrochromic windows.

REFERENCES

1. C.G. Granqvist, Hand book of Electrochromic Materials (Amsterdam: Elsevier), 1995.
2. Yoji Yamada, Kenji Tabata, Solar Energy Materials & Solar Cells **91**, 29-37 (2007).
3. R. Siva Kumar, R. Gopal Krishnan, Optical Materials **29**, 679-687 (2007).
4. O.M. Hussain, A.S.S.Smitha, Applied Physics A **81,** 1291-1297 (2005).
5. E. Washizu, A. Yamamoto, Solid State Ionics **165,** 175-180 (2003).
6. P.K. Biswas, N.C. Pramanik, Materials Letters **57**, 4429-4432 (2003).
7. A.K. Srivastava, M. Deepa, Solid State Ionics **176**, 1161-1168 (2005).
8. O.M. Hussain, K.S. Rao, Materials Chemistry and Physics **9862**, 1-9 (2003).

Mobile based Appliances switching using Bluetooth

Dr Sureshchandra J.Gupta[a], Kalp Desai [a], Deepak Gaikawad [a], Vijay N. Pawar [a] and Devendranath R. Gangal[a]

[a] *University Department of Physics, University of Mumbai, Kalina, Mumbai -400 098. INDIA..*

Abstract. How many times do you have to get up from your desk to switch on your Air conditioner or fan when you are completely into your table work? How many times do you feel lazy to get off your comfort to switch on/off your home appliances in different rooms? How much energy do you lose in a day for operating your appliances? The solution is either a large amount of manual work - or the idea that is presented over here: APP-CON (APP-CON stands for appliances control). Here the ordinary cell phone with bluetooth capability acts as remote designed in such a manner that it acts as a helping hand to human by reducing its manual work and therefore saving human energy. The cell phone control of APP-CON units lets you access many of your home appliances situated in different rooms by using just a single remote from distance. Electronics hobbyists would love to make such a remote control themselves. But they find it difficult due to complex circuitry rather than the high cost because of using a number of frequency counting techniques and decade counters. The APP-CON system given here overcomes the aforesaid problems by using a single microcontroller and moreover a simple program or software for bluetooth enabled cell phone and employing simple coding and decoding of remote signals. Here the mobile based remote control is used to operate a number of home appliances basically consists of Bluetooth technology. The unit consists of a transmitter and a receiver consisting of a microcontroller. The importance of bluetooth technology is that the signal to be transmitted from transmitter to the receiver is done without requiring line of sight.

Keywords: Bluetooth, Switching, Appliances.
PACS: 84.40.x

INTRODUCTION

Inside App-Con

APP-CON stands for Appliances control using Bluetooth. The APP-CON unit allows user to access many of the home appliances like Fans, Tube light, Lamps, etc using Remote. APP-CON provides user-friendly access and has software based operation. This APP-CON consists of two phases, Mobile phone Remote and Receiver. The foresaid remote with Bluetooth capability operates with the help of designed software which senses the keys specified for switching ON/OFF different appliances. The mobile device transmits data whenever the key is pressed. This data together with start stop and default bit stream is transmitted through Bluetooth device

CP1004, *Thermophysical Properties of Materials and Devices-NCTP '07*,
edited by P. Predeep, S. Prasanth, and A. S. Prasad

from the mobile phone. The second phase of APP-CON is receiver. The bit stream transmitted by remote is recepted/sensed by Bluetooth receiver module which is based on the principle of RF reception/detection. The received bit stream is given to Microcontroller for processing. The Microcontroller first confirms whether data transmitted is from the remote of the same device, if yes it further checks the request is from which channel (Appliance), if no it continues with the scanning for the request. In case of failure of the remote the Appliances can be controlled by manual switches.

Features

- Software based operation.
- Mutli-channel Transmission.
- Accurate frequency transmission.
- Bluetooth based principle of operation.
- Unique remote control based operation.
- Remote/Switch based operation of devices.
- LED indication of state of devices.

Bluetooth communication

Bluetooth is a method for data communication that uses short-range radio links to replace cables between computers and their connected units. Industry-wide Bluetooth promises very substantial benefits for wireless network operators, end workers, and content developers of exciting new applications. The functional overview and applications of Bluetooth, and deals with the development of a model for recording, printing, monitoring, and controlling of eight process variables at the same time, using a distributed control system. The industrial automation via Bluetooth using IISS. Industrial automation is one of the major applications of Bluetooth technology. Industrial automation, in terms of controlling or monitoring a factory, office, or industrial process, means to install machines that can do the work instead of human workers. Industrial plants consists of many devices interconnected in different ways ranging from simple data collection units (I/O) to more intelligent devices such as sensors, one-loop controllers, or programmable controllers, and a supervisory system used as a human–machine interface (HMI) for data logging and supervisory control. An IISS is a controlling device that monitors the devices in accompany. It basically communicates via the interface card in the PC; the hardware is connected parallel across the device, and it is interfaced with the PC via a transceiver. The device can be accessed both manually via the switches and remotely via the PC. A simulation of connecting a PC with the machines in a company was executed. Also, a software program using C language; will show how the remote monitoring takes place between the control room and the PC. Bluetooth is an open standard for wireless connectivity with supporters mostly from the PC and cell phone industries. Not surprisingly, its primary market is for data and voice transfer between communication devices and PCs. In this way, it is similar in purpose to the IrDA protocol. Bluetooth, however, is a radio frequency (RF) technology utilizing the unlicensed 2.4 GHz industrial, scientific, and medical (ISM) band. Target applications include PC and peripheral networking, hidden computing, and data synchronization such as for address books and calendars. Other applications could include home networking and home appliances of the future such as smart appliances, heating systems, and entertainment devices. Bluetooth attempts to provide significant advantages over other data transfer technologies, such as IrDA and HomeRF, vying for similar markets. Despite comments from the Bluetooth SIG indicating that the technology is complementary to IrDA; it is clearly a competitor for PC-to peripheral connection. IrDA is already popular in PC peripherals, but is severely limited by the short connection distance of 1 m and the line-of-sight requirement for communication. This limitation eliminates the feasibility of using IrDA for hidden computing, where the communicating devices are nearby but not visible to one another. Due to its RF nature, Bluetooth is not subject to such limitations. In addition to wireless device connections up to 10 m (up to 100 m if the transmitter's power is increased), devices need not be within line of sight and may even connect through walls or other nonmetal objects. This allows for applications such as a cell phone in a pocket or a briefcase acting as a modem for a laptop or PDA. Bluetooth is designed to be low cost, eventually under $10/unit. On the flip side, however, are the limited connection distance and, even more damaging, the transmission speeds. Bluetooth supports only 780 kb/s, which may be used for 721 kb/s unidirectional data transfer (57.6 kb/s return direction) or up to 432.6 kb/s symmetric data transfer. These rates are comparable to the 1–2 Mb/s supported by HomeRF and, although live digital video is still beyond the capability of any RF technology, perfectly adequate for file transfer and printing applications. Finally, Bluetooth's main strength is its ability to simultaneously handle both data and voice transmissions. It is capable of supporting one asynchronous data channel and up to three synchronous voice channels, or one channel supporting both voice and data. This capability combined with ad hoc device connection and automatic service discovery make it a superior solution for mobile devices and Internet applications.

This combination allows such innovative solutions as a mobile hands-free headset for voice calls, print to fax capability, and automatically synchronizing PDA, laptop, cell phone address book, and other applications.

BLOCK DIAGRAM

Transmitter

Mobile phone: The Mobile phone with Bluetooth enabled capabilities is used with software program for controlling the appliances loaded on the phone memory. The numerical keys on the phone acts as the keys for switching the appliances ON/OFF on the receiver side. The software scans which key is pressed and accordingly transmits the signal to the Bluetooth module through a single port line.

Bluetooth Module: It is used for RF transmission of stream of data using the ISM band frequency of 2.4GHz depending upon signals given by mobile phone

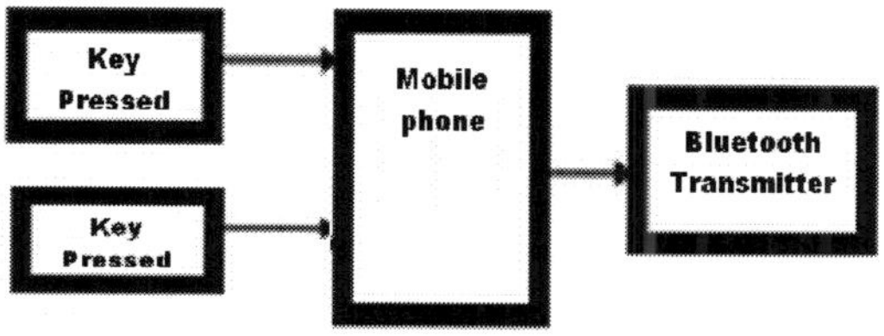

FIGURE 1. Block diagram of Transmitter Section

Receiver

Power Supply: It is a regulator circuit that supplies constant input voltage of 5V for the circuit operation. It is the heart of the receiver stating that its beats are required for operation of receiver.

Bluetooth receiver: It receives the bluetooth signal that is transmitted by the Bluetooth device of the remote operating at frequency of 2.4GHz.

Micro controller: It is the brain of receiver. It detects the signal sensed by the bluetooth sensor. It turns devices on or off devices depending upon the input signals received at the port line P3.0.

Driver Stage: It is used to increase the current driving capability of the circuit for operation of relays. It amplifies the current up to 12 A for the smooth operation of relays. Switches: There are totally 8-switches for eight different devices connected to the eight different port lines of microcontroller for manual operation. Relay & LED: It is an electromagnetic switch that turns devices on or off device depending upon output of microcontroller and the status of manually operated switch. LED is used to indicate the state of devices. Diodes are connected at the input terminals of the relays to protect it from the back EMF of the relay's coil.

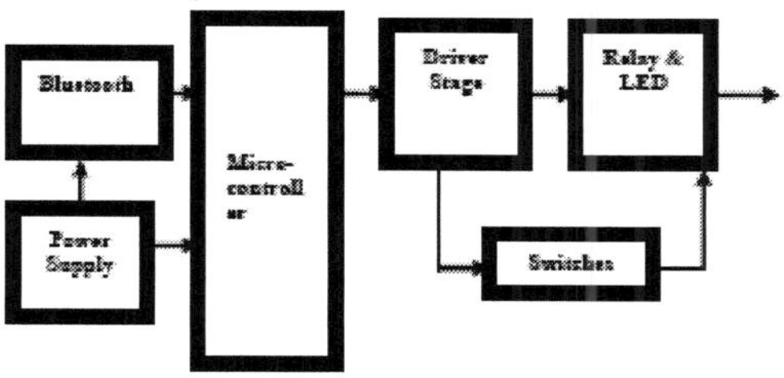

FIGURE 2. Block diagram of Receiver Section

WORKING OF CIRCUIT

APP-CON is Appliances Control; with help of remote controls various appliances. It comprises of two sections:

1. REMOTE
2. RECEIVER

Remote

The software loaded on the phone memory starts after switching the bluetooth ON in the phone. Numeric keys on the phone pad can be assigned the value or the appliances name for switching it ON/OFF. Once the key are assigned the values the program when started continuously scans the keys and when detects pressed key, transmits stream of data. The stream of output data is given to the Bluetooth transmitter. This Bluetooth module transmits the data at frequency of 2.4GHz ISM (Industrial Scientific and Medical) band.

Receiver

The receiver circuit operates on 5V given by the regulator circuit. The regulator circuit uses IC 7805, which generates constant 5V DC voltage. The active stream of data transmitted by the transmitter remote is sensed by the bluetooth sensor module. The Bluetooth sensor works on the principle of RF frequency detection. The output of the bluetooth sensor is given to the input port line of microcontroller (P3.0). The microcontroller checks for the MSB code, and confirms that the received data is from APP-CON remote only. If the data matches, then the LSB code is checked and compared with the set values for different channels and accordingly the data is send to output Port1 (P1.0 – P1.7). The eight different amplifier circuits are connected to the four port lines of microcontroller (P1.0 – P1.7). The amplifier circuit comprises of a relay and a transistor to drive the relay. The relays connected operate at 12V, 5A. The transistor (BD 139) is connected along with the resistor to amplify the current to drive the relay. The diodes are connected to input of the relays, to protect the relays from the back emf. The LEDs are connected to indicate the status of the channels. The switches are connected for operation of the channels during the failure of remote.

Software

The whole operation of the system which is performed by micro-controller is obtained by writing the program in assembly language software.

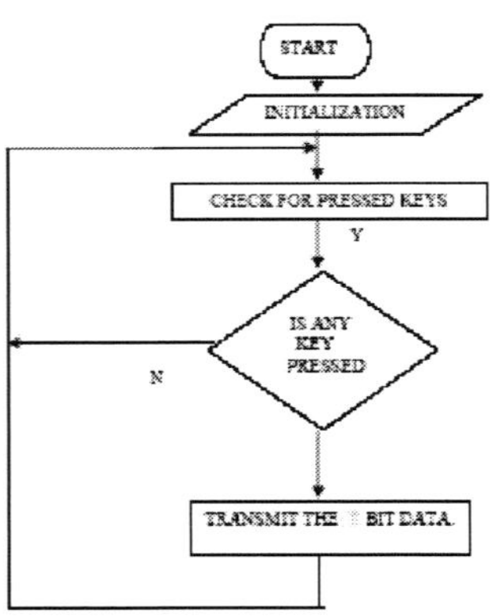

FIGURE3: Flowchart for Remote

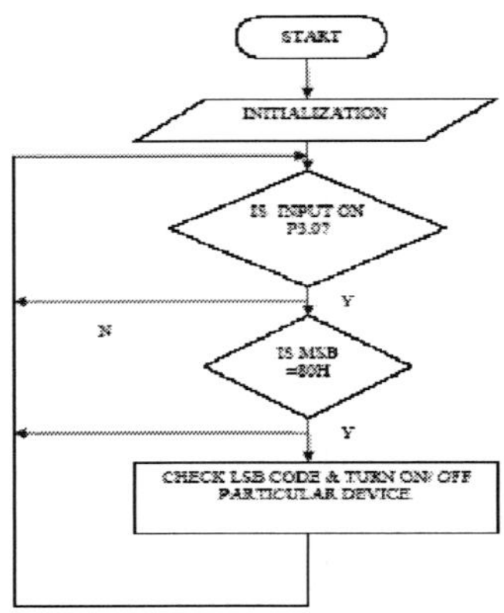

FIGURE4: Flowchart for Receiver.

CONCLUSION & FUTURE SCOPE

Earlier remote control were made using a complex frequency counting technique and various types of counters resulting into complex circuitry due to non-availability of Microcontrollers and Programmers for the unit. The remote can function up to a distance of 10 meters. Due to Bluetooth Technology used in APP-CON remote control can be operated from a anywhere within the distance of 10meters overcoming the line of sight problem with operating frequency of 2.4GHz for transmission. APP-CON remote is more advantageous due to simple program or software employing simple coding and decoding techniques. In APP-CON the transmitter section that is the remote control consists of a Mobile phone and a simple program or software which checks for the key pressed.

After identifying the key pressed it will generate data in form of bits through its one of the ports. Also the receiver section consists of a Microcontroller and a program which checks for the input bits. Now with the help of initially stored code it will check and compare whether the data is of same system. After all it will identify the appliance to be controlled and finally switches that appliance. This is all simply done by Microcontroller avoiding the complex circuitry.

Future Advancements

As in APP-CON eight appliances are interfaced to the receiver so as for the purpose of switching but further, it can be expanded to as many appliances. Here APP-CON is operated to switch on/off the appliances connected to the device but further we can configure it in such a manner that it can control the device by varying its other parameters like speed, intensity, etc. Further by using PC-based connection more functionality and industrial applications can be controlled and operated.

REFERENCES

1. Behrouz, Forouzan A, "Data Communications and Networking, TMH", Edition 2000.
2. Halsall, Fred, "Data Communications, Computer Networks and Open Systems", 4th Ed, MA: Addison-Wesley, 1995.
3. Stallings, William, "Data and Computer Communication", 5th Ed, NY: Prentice Hall, 1995
4. K. V. S. S. S. S. Sairam, N. Gunasekaran, S. Rama Reddy, "Bluetooth in Wireless Communication", IEEE Communication Magazine, June 2002.
5. Brag, J. and Sturman, C., "Bluetooth: Connect Without Cables", NJ: Prentice Hall, 1992.
6. Davis, P. T. and Macguffin, C. R., "Wireless Local Area Networks", NY: McGraw Hill, 1995.
7. Stallings, Williams, "Wireless Communication and Networks", Pearson Educations Ltd. 2004.

TCAD for SiGe MOSFET: Modeling of Diffusion Kinetics and Device Simulation

Shahul Hameed[a] and Sanjay S. Mane[b]

[a]*Dept. of Electronics and Communication Engineering, T.K.M.College of Engineering, Kollam-691005, India*
[b]*Dept. of Electronics and Communication Engineering , Sardar Patel College of Engineering, Mumbai*
Corresponding Author's E-mail: shahul.iitkgp@gmail.com

Abstract: Considering a semi-infinite surface, fitting the solution of diffusion equation for SiGe so that prediction of the annealed profile in terms of its concentration and depth might be a pointer for researchers. An analytical model of dopant diffusion in SiGe is presented. Excellent matches have been observed between the reported data and simulation results. The model parameters included in device simulation software and device characteristics of n channel SiGe MOSFET have been observed

Keywords: SiGe, TCAD, nMOSFET, ATHENA/ATLAS, constant dose diffusion, constant source diffusion
PACS: 02.60.Cb, 02.70.-c

INTRODUCTION

The device trend towards smaller and faster versions leads the researchers for newer materials other than silicon, where $Si_{1-x}Ge_x$ is proved to be a natural choice. The easiness of incorporating into existing technologies and the lattice matching are the key features of this alloy. Diffusion, a much-discussed process in silicon technology has been extensively used in SiGe process as well. So far, analytical model for the diffusion profile in SiGe is not seen and our attempt is to introduce such a model for arsenic, thus to predict the enhancement of depth in this advance material. We have simulated the model using MATLAB and fitted the results with the reported data as well as the process simulation results from ATHENA. Excellent matches have been observed under identical implant and anneal conditions.

To cope with explosive development costs and competition in the semiconductor industry today, Technology Computer Aided Design (TCAD) methodologies are extensively being used in the development. In this paper, we report on the ATHENA/ATLAS simulation of typical device characteristics for n-channel SiGe MOSFETs incorporating the presented model of arsenic diffusion kinetics.

CP1004, *Thermophysical Properties of Materials and Devices-NCTP '07*, edited by P. Predeep, S. Prasanth, and A. S. Prasad

SIMULATION METHODOLOGY

Model of the Arsenic Diffusion Kinetics in SiGe

Fick's diffusion equation, gives the basis of continuum theory of diffusion. The numerical solution of the equation cannot be transformed into the profile and hence an analytical solution for extended initial and boundary conditions would be the right approach towards the attempt for a dopant profile in SiGe [1]. We have the diffusion equation as

$$\frac{\partial N(x,t)}{\partial t} = D\frac{\partial^2 N(x,t)}{\partial x^2} \tag{1}$$

Since it is a second-order partial differential equation, a specific problem requires two boundary conditions and an initial condition to specify a unique solution. The constant-source diffusion is carried out with a source for the diffusing material being continuously present at the surface of the silicon wafer. The effect is to hold the concentration at the surface of the wafer at a fixed value. The concentration drops off from that value inside the wafer. The boundary and initial conditions are $N(x=0,\ t) = No$, $N(x \to \infty, t) = 0$, $N(x \neq 0, t = 0) = 0$. With these conditions, the solution to the diffusion equation takes the form

$$N(x,t) = N_0 erfc(\frac{x}{2\sqrt{Dt}}) \tag{2}$$

The diffusion of impurities will not be ideally downwards and lateral diffusion is a universally accepted phenomenon to be considered in any practical systems. Taking into account such diffusion sideways, eqn. (2) can be rewritten as

$$N(x,t) = \frac{No}{2\sqrt{1+\frac{2Dt}{\Delta R_p^{\ 2}}}} erfc[\frac{\frac{x}{4Dt} - \frac{x_p}{2\Delta R_p^{\ 2}}}{\sqrt{1/2\Delta R_p^{\ 2} - 1/4Dt}}] \tag{3}$$

where x_p is a function of the projected range and ΔR_p is the lateral straggle.

In constant-dose diffusion, a fixed amount of dopant is introduced at the surface of the wafer and then allowed to diffuse into the body of the wafer. Since the total dose is fixed, the surface concentration must decrease as the dopant moves into the wafer. The dose of dopant in a wafer can be defined as the integral of the doping profile. In general, the integral will extend from the surface through the entire thickness of the wafer as in eqn. (4),

$$Q = \int_0^\infty N(x)dx \tag{4}$$

In using the definition (4), we are assuming that doping profile is the same everywhere on the surface of the wafer. The initial condition and one of the boundary conditions for the constant-dose diffusion are same as for the constant-source diffusion- $N(x \neq 0, t=0) = 0$, $N(x \to \infty, t) = 0$.

The other boundary condition is determined through the constant-dose definition

$$Q = \int_0^\infty N(x,t)dx = \text{constant} \tag{5}$$

Using these conditions in the diffusion equation leads to the constant-source diffusion profile:

$$N(x,t) = \frac{Q}{\sqrt{\pi D t}} \exp\left(\frac{-x^2}{4Dt}\right) \tag{6}$$

Eqn. (6) has the general form of a Gaussian function. The surface concentration of the profile is given by $N(x=0,t) = \dfrac{Q}{\sqrt{\pi D t}}$ (7)

In this case also, the lateral diffusion is considered and a term associated with this phenomenon is introduced as in eqn.(3),

$$N(x,t) = \frac{Q/\sqrt{\pi D t}}{2\sqrt{1+2Dt/\Delta R_p^{2}}}[\exp[-\frac{(x-x_p)^2}{2\Delta R_p^{2} - 4Dt}] \tag{8}$$

Though in all practical diffusion systems, unlimited source is used, the annealed profile will be showing a dip in the surface concentration as observed in constant dose diffusion system and the truncated gaussian will have a channeling effect towards the proximal end. By introducing a parameter k which represents the different boundary conditions at the sample surface (x=0) and combining the effects in 2 and 8, the annealed profile shall be redefined by a product function of complementary error function and truncated gaussian,

$$N(x,t) = \frac{Q/\sqrt{\pi D t}}{2\sqrt{1+2Dt/\Delta R_p^{2}}}\{erfc[-\frac{\frac{x_p}{2\Delta R_p^{2}} - \frac{x}{4Dt}}{\sqrt{1/2\Delta R_p^{2} - 1/4Dt}}]\exp[\frac{(x-x_p)^2}{2\Delta R_p^{2} - 4Dt}] \tag{9}$$

$$+k.erfc[-\frac{\frac{x_p}{2\Delta R_p^{2}} + \frac{x}{4Dt}}{\sqrt{1/2\Delta R_p^{2} - 1/4Dt}}]\exp[\frac{(x+x_p)^2}{2\Delta R_p^{2} - 4Dt}]\}$$

The parameter k = -1 corresponding to a surface acting as a perfect sink and k = +1 corresponds to an ideally reflecting surface. When t = 0 and k = 0, the eqn (9) reduces to gaussian.

The behavior of *n type* dopants such as arsenic and phosphorus in $S_{1-x}Ge_x$ shows an enhancement of depth due to high effective diffusivity. The enhancement in depth is included in the eqn.(9) so as to make it a model for the annealed profile of n type dopants in $S_{1-x}Ge_x$

$$N(x,t) = \frac{Q/\sqrt{\pi.D.t}}{2\sqrt{1+2D.t/\Delta R_p^{2}}}\{erfc[-\frac{\frac{x_p}{2\Delta R_p^{2}} - \frac{x}{4D.t.h}}{\sqrt{1/2\Delta R_p^{2} - 1/4D.t.h}}]\exp[\frac{(x-x_p)^2}{2\Delta R_p^{2} - 4D.t.h}] \tag{10}$$

$$+k.erfc[-\frac{\frac{x_p}{2\Delta R_p^{2}} + \frac{x}{4D.t.h}}{\sqrt{1/2\Delta R_p^{2} - 1/4D.t.h}}]\exp[\frac{(x+x_p)^2}{2\Delta R_p^{2} - 4D.t.h}]\}$$

where h is the factor of enhancement in depth.

The peak of the profile x_p is extracted from the implanted profile (before annealing), which in turn is obtained from the equation

$$N(x) = Q/\sqrt{\pi.D.t}[\exp(-(x-R_p)^2/2\Delta R_p^2] \qquad \text{for } x>0 \qquad (11)$$

where Rp is the projected range.

Device Modeling using Athena/Atlas

The values of diffusivity, projected range and straggle as used in implementing the diffusion kinetics model described in above section are incorporated in SILVACO via C-interpreter function, for simulating the nMOSFET. The current characteristics have been studied for its performance.

SIMULATION RESULTS AND DISCUSSION

Arsenic Diffusion Profile for Small Anneal Times

The model of the profile presented in eqn. (10) is implemented for different cases of arsenic diffusion in SiGe with varying germanium contents We made an extensive comparison with a number of published results[2]-[4] and the error rate was within negligible tolerance of 2-5%.

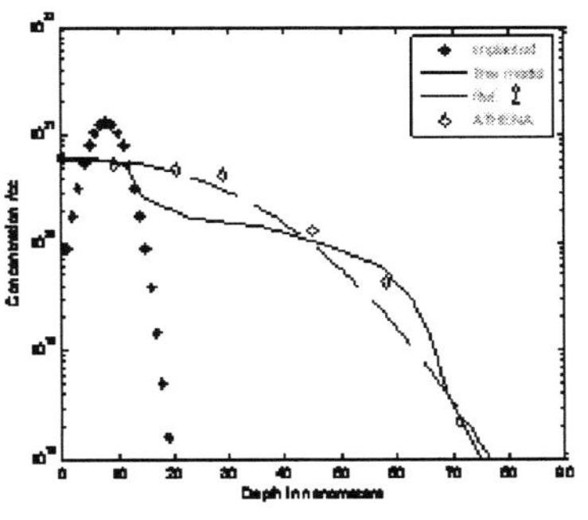

FIGURE1. Simulated profile for 15 keV $1X10^{15}/cm^2$, 1000°C for 10 seconds

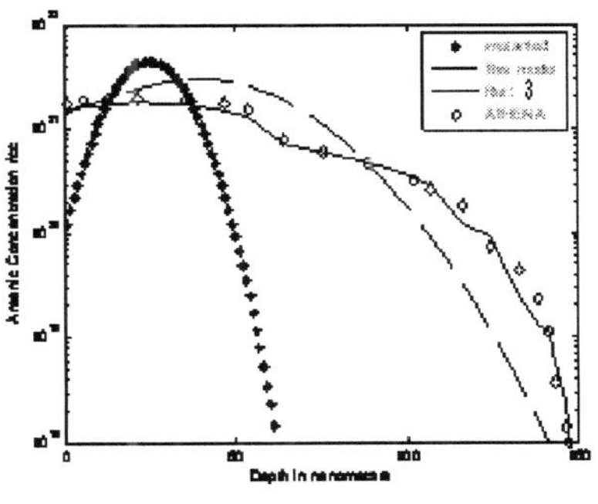

FIGURE 2. Simulated Profile for 40keV $1X10^{16}/cm^2$, 1025°C for 30 seconds

Fig.1 and 2 show verifications of the simulation results of the model presented in eqn. (1) for 20% and 5% of Ge mole fractions at different implant and anneal conditions. The results are compared with identical experimental values and ATHENA results also substantiate its validity.

Simulation of nMOSFET using ATHENA/ ATLAS

The model presented in eqn. (10) is incorporated in ATHENA and the resulting device structure is shown in fig (3-a), which is successively used in ATLAS for device characterization. An enhancement in simulated drain current at gate bias of 1-2.5 V@0.5, as shown in fig.3-b. This supports the relevance of the presented diffusion kinetics model in the device level implementations.

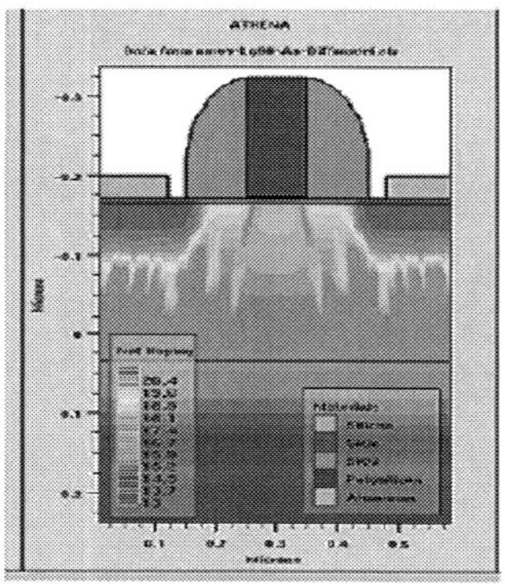

FIGURE3 a. Structure of the nMOS device

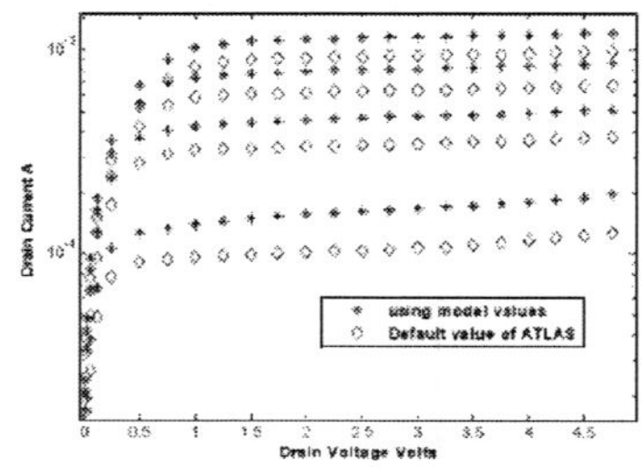

FIGURE3-b. I-V Characteristics of the device

CONCLUSIONS

Prediction of the depth of arsenic profile in SiGe, after implantation, shall be useful for validating the experimental results and hence shall further enhance the research in shallow implantations. This is made possible by incorporation of enhancement in depth, in the solution of diffusion equation. Considering a semi-infinite surface, fitting the solution of diffusion equation for SiGe so that prediction of the annealed profile in terms of its concentration and depth might be a pointer for researchers. The device thus simulated shows an improvement in current for low gate bias, explains its relevance in low power operation.

REFERENCES

1. R.Ghez, G.S.Oehrlein, T.O.Sedgwick, F.F.Morehead and Y.H.Lee, *Applied Physics Letters* 45(8) October 1984
2. S.Eguchi, J.J.Lee, S.J.Rhee, D.L.Kwong, M.L.Lee, E.A.Fitzgerald, I.Aberg and J.L.Hoyt, *Applied Surface Science* 224 (2004) 59-62
3. M.J.Mitchell, P.Ashburm, J.M.Bonar and P.L.F.Hemment, *J. Applied Physics*, Volume 93, Number 8 2003
4. Suresh Uppal, J.M.Bonar, Jing Zhang, A.F.W.Willoughby, *Material Science and Engineering B* 114--115 (2004) 349-351

MATERIALS AND CHARACTERIZATION

Structural and optical characterization of (PEO+PVAc) polymer blend electrolyte films

N.Reddeppa, T.J.R. Reddy, V.B.S.Achari, V.V.R.N. Rao, A.K. Sharma*
Department of Physics, S.V. University, Tirupati-517502, INDIA

Abstract. Solid polymer blend electrolyte films based on polyethylene oxide (PEO) and polyvinyl acetate (PVAc) were prepared in various concentrations by solution-cast technique. The structure of these blend electrolyte films was determined using X-ray diffraction (XRD) studies. The composition of the blend was confirmed by Fourier transform infrared (FTIR) spectroscopy. Optical absorption studies were made in the wavelength range 200-600 nm. Optical properties like absorption edge and optical band gaps were estimated from their optical absorption spectra. It was found that the energy gaps and band edge values shifted to lower energies on blending with PVAc.

Key words: Polymer ectrolyte films, XRD, optical properties.
PACS: 78.66.Qr, 78.70.Ck, 74.25.Gz

INTRODUCTION

Since the discovery of high electrical conductivity from blending poly (ethylene oxide) with potassium salts by Fenton et al.[1] polymer electrolytes have attracted a lot of interest, especially because of their potential use in thin film batteries[1]. Polymer electrolytes consist of polar polymers and ionizable salts. PEO is the most popular polymer used, due to its high solvating power with metal ions, good processability, and outstanding mechanical properties [2-6].The polymer blends often exhibit properties that are superior compared to the properties of each individual component polymer[7-12]. The main advantages of the blend systems are simplicity of preparation and ease of control of physical properties by compositional change [13,14]. However, the superior properties depend upon the miscibility of homopolymers on the molecular scale. In the present study we have chosen poly (ethylene oxide) -poly (vinyl acetate) blend system because of the excellent heat resistance and abrasion resistance of its constituents. These constituents are mainly used in the manufacture of shatterproof glass, electrical insulation coatings and adhesive compositions.

Although the ionic conductivity of PEO polymer electrolytes at 100°C is about to 10^{-4}-10^{-3} S/cm, at room temperature it is usually less than 10^{-6} S/cm. Therefore, much effort has been devoted to improve the ambient ionic conductivity of the polymer electrolytes by improving of the cation migration in the amorphous domain or in the helical structure of the crystal. Since ions primarily transport in the amorphous phase,

CP1004, *Thermophysical Properties of Materials and Devices-NCTP '07*,
edited by P. Predeep, S. Prasanth, and A. S. Prasad

many investigations, which aimed to improve ionic conductivity, have focussed on increasing the amorphous phase of the material.

The investigation of the optical properties of these systems is still limited and much experimental work needs to be done in this direction. Ultraviolet absorption is generally believed to be associated with electron (or anion)transitions from the top of the valence band to bottom of the conduction band [15-17]. This report is concerned with the optical properties of poly(ethylene oxide)PEO/PVAc blends in the UV-visible region. In this study, PVAc was blended with PEO and the results of such a study on the structural and optical properties of the polyblend films are reported.

EXPERIMENTAL DETAILS

In the present study films of polyvinyl acetate (PVAc) blended with polyethylene oxide (PEO) in various compositions of (90:10), (80:20), (70:30), (60:40), (50:50) by wt% were prepared using the solution–cast technique. PEO and PVAc were dissolved in methanol (CH_3OH) and the mixture was stirred at room temperature for 6-8 hours. The stirred solution was then cast onto polypropylene dishes and evaporated slowly at room temperature. The films were then dried in a vacuum of 10^{-3} torr to eliminate all the traces of the solvent. The X-ray diffraction studies were performed by means of SEIFERT X-ray diffractometer in the range 10 - 65°. IR spectra profiles were obtained using a Perkin - Elmer FTIR spectrophotometer in the range 500-4000 cm^{-1} in the transmittance mode. The optical absorption spectra of pure and polyblend films were recorded at room temperature in wavelength range 200-600 nm using UV-VIS-NIR (Model UV-3100) spectrophotometer. The absorption coefficient (α) was calculated from these spectra and the optical constants such as band edge, optical band gaps (both direct and indirect) were determined.

RESULTS AND DISCUSSION

XRD analysis is very useful in knowing the structure of a material. These patterns enable one to find out whether a material is crystalline or amorphous. The diffraction patterns for crystalline materials show well defined peaks while for amorphous materials the pattern does not show any peaks.

In the present study, the structure (crystalline or amorphous) was determined by taking XRD patterns on pure PEO, PVAc and their blends to correlate their structure with other properties. It was observed from the diffractograms (Fig. 1) that while PEO had fairly large crystalline content, PVAc was essentially amorphous. As PVAc content in PEO was increased, the crystalline content in the sample decreased as suggested by the decreasing intensity and sharpness of the peaks (Fig.1 b, c, d, e, f).

The complexation of PEO with PVAc was confirmed by IR studies. The IR spectra of pure PEO, PVAc and (PEO+PVAc) polymer blends are shown in Fig.2. It is clear that the IR spectra of polymer blend exhibits several changes in comparison with the IR spectra of pure PEO and PVAc.

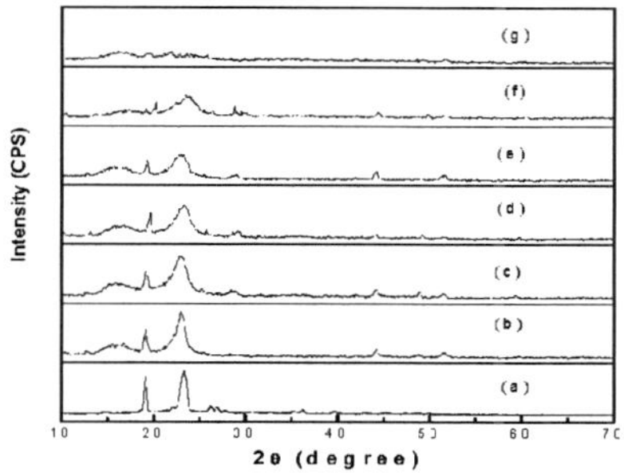

FIGURE 1. XRD patterns of (a) Pure PEO (b) PEO+PVAc (90:10) (c) PEO+PVAc (80:20) (d) PEO+PVAc (70:30) (c) PEO+PVAc 60:40 (f) PEO+PVAc (50:50) (g) Pure PVAc

IR spectra showed that the chemical structure was maintained as all constituent groups were present in the spectra for PEO, PVAc and their blends. The spectrum of polyblends exhibited characteristic absorption peaks of both monomeric units. There was a gradual change in intensity of absorption bands as the percentage of PAVc was increased in the blend. These observations indicate clearly the miscibility of PVAc with PEO polymer.

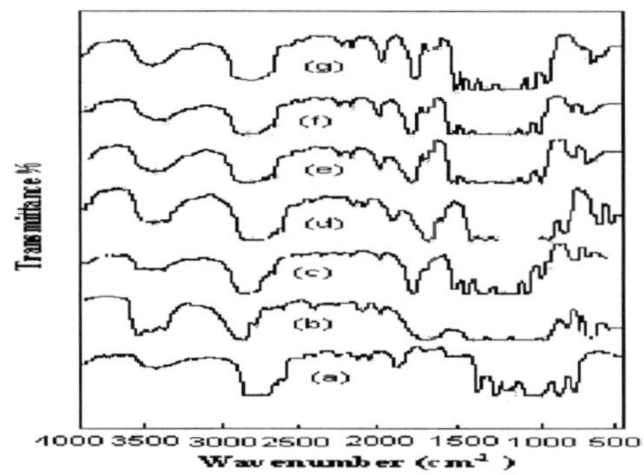

FIGURE 2. FTIR plots of (a) Pure PEO (b) PEO+PVAc (90:10) (c) PEO+PVAc (80:20) (d) PEO+PVAc (70:30) (e) PEO+PVAc 60:40 (f) PEO+PVAc (50:50) (g) Pure PVAc

Optical absorption studies of pure and polyblend films were carried out to determine the optical constants such as optical band gap (E_g) and the position of the fundamental band edge. The absorption coefficient was determined from the spectra using the formula,

$$\alpha = A/d$$

where A is the absorbance and d the thickness of the film.

When a direct band gap exists, the absorption coefficient has the following dependence on the energy of the incident photon [18,19]

$$\alpha h\upsilon = C\,(h\upsilon - E_g)^{1/2}$$

where E_g is the band gap, 'C' a constant dependent on specimen structure, υ, the frequency of incident light and 'h' is Planck's constant.

Thus, a plot of $(\alpha h\upsilon)^2$ versus photon energy as axis should be linear. The intercept on the energy on extrapolating the linear portion of the curves to zero absorption value may be interpreted as the value of the band gap. For pure PEO polymer film the direct band gap lies at 5.09 eV, while, for blend films the values are found to decrease with increasing PVAc concentration (Table 1).

TABLE 1. . Activation energies, absorption edge and band gaps of poly blend system.

Polymer electrolyte	Optical band gaps (eV) Direct	Indirect	Absorption edge (eV)
Pure PEO	5.09	5.01	4.61
PEO:PVAc(90:10)	4.91	4.96	4.41
PEO:PVAc(80:20)	4.80	4.83	4.26
PEO:PVAc(70:30)	4.75	4.75	4.19
PEO:PVAc(60:40)	4.65	4.69	4.12
PEO:PVAc(50:50)	4.58	4.62	4.06

The indirect band gaps were obtained from the plots of $(\alpha h\upsilon)^{1/2}$ versus photon energy as shown in fig.3. For pure PEO polymer film the indirect band gap lies at 5.01 eV while for blend films the values are found to decrease which are given in Table.1

The position of absorption edge values was calculated by extrapolating the linear portions of the α versus $h\upsilon$ plots as shown in fig.4 to zero absorption value. For pure film, the absorption edge lies at 4.61eV. Where for blend films the values are 4.41 ,4.26,4.19,4.12, and 4.06 eV respectively.

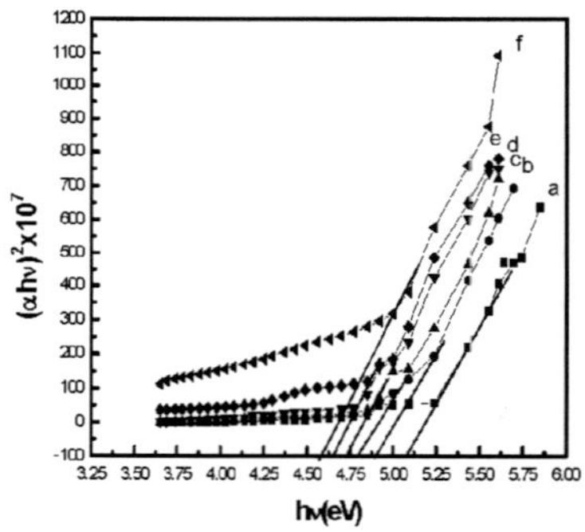

FIGURE 3. $(\alpha h\nu)^2$ vs $h\upsilon$ plots of pure PEO and various compositions of PVAc blend films

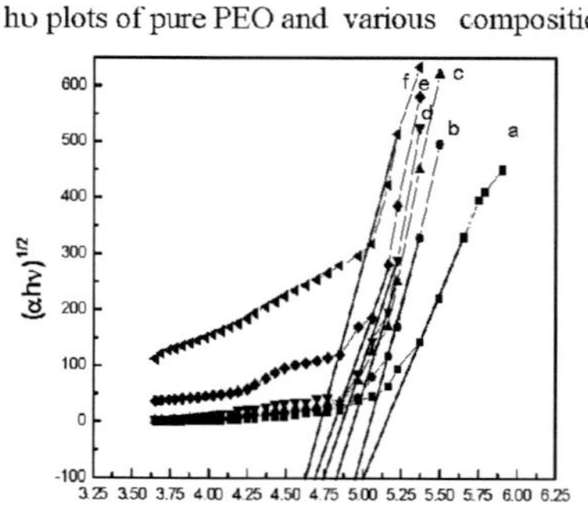

FIGURE 4. .$(\alpha h\nu)^{1/2}$ vs $h\upsilon$ plots of pure PEO and various compositions of PVAc blend films

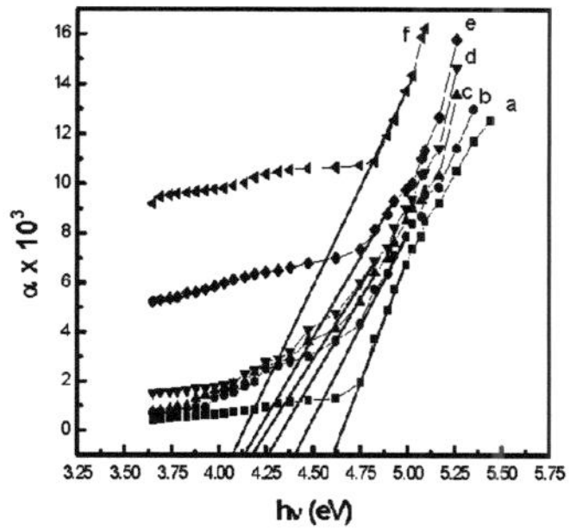

FIGURE 5. α vs hυ plots of of pure PEO and various compositions of PVAc blend films

CONCLUSIONS

X–ray analysis of various polyblend samples showed that the amorphous content of the polyblend increased with the increase in the percentage of PVAc in the polyblend. IR analysis of the blends have shown that there was gradual variation in the intensity of absorption peaks with blend ratio suggesting that the two constituent polymers have mixed fairly well. Optical absorption edge and optical energies gap (both direct and indirect) showed a decreasing trend with increasing PVAc concentration.

REFERENCES

1. B.E.Fenton, J.M. Parker and P.V. Wright, Polym. J., **14**, 589(1973).
2. M.L. Kaplan, I.E. Reitman and R.J. Cava, Polym. J. **30**,504(1989).
3. M. Gauthier, M. Armond and D. Muller, Aprotic polymer electrolyts and their applications, in: A. Terje, . Skotheim(eds), Electroresponsive Molecular and polymeric system, Marcel Dekker, New York, 1988.
4. R.G. Linford, Electrochemical Science and Technology of Polymers, vol. I, Elsevier, New York, 1987.
5. J.R.Maccallum,C.A.Vincent, Polymer Electrolyte Re-views I, Elsevier, New York, 1987.
6. J.R. Maccallum, C.A. Vincent, Polymer Electrolyte Reviews II, Elsevier, New York,1989.
7. M.C. Wintergrill, J.J. Fontanella, M.K. Smith, S.S. Greenbaum, K.J. Adam and C.C. Andeen, Polym. J. **28**,633(1987)
8. J. Procharski, W. Wiecxorek, J. Przyluski and K. Such, Appl. Phys.A **49**,55(1989)
9. S.K. Al-Ani, A.A. Hiigozy and J. Mater. Sci. **26,** 3670(1991).
10. S.K. Al-Ani, I.H. Al-Hassany and Z.T. Al-Dahan. J. Mater.Sci. 30 (1985).
11. Y. Al-Ramadin, M. Ahmad, A. Zihlif and E. Martuscelli, Polym.Comm. **40**,3877(1999).
12. S.J. Al-Ani, Y. Al-Ramadin, A. Zihlif and G. Ragosta, Polym. Testing: Mater. Prop. **18**, 611(1999).
13. F. Urbach, Phys. Rev. **92**, 1324(1953)
14. N.F. Mott and E.A. Davis, Electronic Processes in Non-Crystalline Materials, second ed Clarendon, Oxford, 1979.
15. Y. Al-Ramadin, A.Zihlif and Al-Ani, Opt. Matter. **5,** 69(1969).
16. S.K. Al-Ani, H.O. Al-Hassany and Z.T. Al-Dahahan, J. Mater. Sci. 30 (1995).
17. E.A. Davis and N.F. Mott, Phil. Mag. **22**, 903(1970).
18. D.S. Davis and J.S. Shalliday. J.Phys. D.Appl.Phy.**9**, 1020(1976).
19. G.M. Thutupally and S.G. Tomlin. Phys. Rev. **118**, 1639(1960).

* Corresponding author: profaks_svu@yahoo.co.in

FTIR, Micro-Raman and Ellipsometry Studies on Silicon Oxynitride Layers Synthesized by Low Energy Ion Implantation

Alka R. Chauhan*, A.D.Yadav and S.K. Dubey

Department of Physics, University of Mumbai, Vidyanagari Campus, Santacruz (E), Mumbai - 400 098, India.

Abstract. Single crystal silicon were implanted at room temperature with $^{16}O_2^+$ and $^{14}N_2^+$ 30 keV ions in 1:1 ratio with fluences of 2.5x 10^{17}, 5x 10^{17}, 7.5x 10^{17} and 1x10^{18} ions-cm^{-2} to form silicon oxynitride layers. Rapid thermal annealing (RTA) of these samples was carried out at temperature of 623, 873 and 1173 K in nitrogen ambient for 5 min. Fourier transform infrared (FTIR), Micro-Raman and ellipsometry measurements were performed on RTA samples to investigate the structure and phase. The FTIR studies show that the structures of ion-beam synthesized oxynitride layers are strongly dependent on total ion-fluence and annealing temperature. The Micro-Raman studies on annealing revealed formation of partially amorphous oxygen and nitrogen rich silicon oxynitride structures for lower and higher ion fluences respectively with crystalline silicon beneath it. From the ellipsometry studies, the refractive index of the ion beam synthesized layers were found to be in 1.54 - 1.96 range.

Keywords: Silicon oxynitride, ion implantation, Fourier Transform Infrared, Micro-Raman.
PACS: 61.82.Ms; 61.72.uf; 33.20.Ea; 78.30.Er.

INTRODUCTION

Silicon oxynitride ($Si_xO_yN_z$) has become a very important and promising material in semiconductor application due to its excellent properties [1-3]. The silicon oxynitride ($Si_xO_yN_z$) layer provides a large degree of freedom in integrated optics as it incorporates the positive feature of both silicon oxide (SiO_2) and silicon nitride (Si_3N_4) [4]. Ion implantation has an edge over other conventional techniques as it is a highly controllable process and can be done at relatively low temperature [5, 6]. The formation of ultra-thin and buried silicon oxynitride layers at very low energy (20 eV/5 keV) and high energy respectively has been reported earlier [7-8]. This work reports the experimental results of oxynitride formation by molecular oxygen ($^{16}O_2^+$) and nitrogen ($^{14}N_2^+$) ion implantation at low energy (30 keV) to the ratio of 1:1. The quality of the $Si_xO_yN_z$/Si structures and bonds formed were analyzed from Fourier transform infrared spectroscopy (FTIR) and Micro-Raman spectroscopy. For these silicon oxynitride films, the thickness and refractive index were determined using ellipsometry.

CP1004, *Thermophysical Properties of Materials and Devices-NCTP '07*,
edited by P. Predeep, S. Prasanth, and A. S. Prasad

EXPERMENTAL

The silicon oxynitride layers were formed on single crystal <111> n-type silicon wafer by implanting of 30 keV $^{16}O_2^+$ and $^{14}N_2^+$ ion beams sequentially at room temperature to the ratio of 1:1 with ion fluences of $2.5x10^{17}$, $5x10^{17}$, $7.5x10^{17}$, and $1x10^{18}$ ions-cm^{-2}. Ion-beam current density of 3-8 μA cm^{-2} was used for implantation. FTIR measurements were carried out by using FTIR JASCO model 610 spectrometer with substrate background corrections. The Rapid Thermal Annealing (RTA) of implanted samples was carried out in dry nitrogen ambient at 673, 873 and 1173 K for 5 min. Micro-Raman spectroscopic measurements were performed on 1173K annealed samples using LabRAM HR microscope at room temperature using 514.5 nm line of 4W argon ion laser available at IIT, Bombay. The thickness and refractive index of 1173 K annealed samples were measured using GAERTNER ellipsometer.

RESULTS AND DISCUSSION

Fourier Transform Infrared (FTIR) studies

The FTIR spectra for silicon samples implanted with molecular oxygen ($^{16}O_2^-$) and nitrogen ($^{14}N_2^+$) to the ratio of 1:1 with different fluences indicate the formation of silicon oxynitride ($Si_xO_yN_z$) layers. The spectra of silicon samples implanted with oxygen and nitrogen to a fluence of $2.5x10^{17}$ ions cm^{-2} and annealed at different temperatures for 5 min are given in Fig. 1(a). The spectra for unannealed sample show peak at ~978 cm^{-1} indicating formation of nitrogen rich oxynitride structure. RTA treatment at 673 and 873 K does not give significant change in the IR spectra except for the shift in the peak towards the higher wavenumber with slight increase in the intensity. The spectrum show significant change on annealing at 1173 K. It shows prominent peak of Si-O and a hump corresponding to Si-N bands at ~1051 cm^{-1} and ~996 cm^{-1} respectively. The concentration of the absorbing bond was calculated for all fluence levels (9). The total number of bonds increases from $7.69x10^{22}$ for as implanted sample to $8.92x10^{22}$ cm^{-3} for 1173 K annealed sample. On annealing, rearrangement of atoms in the implanted layer occurs which leads to formation of distinct phases of SiO_2 and Si_3N_4 with some complex structures. The peak shifting is observed from lower wavenumber to higher wavenumber due to the release of band strain and rearrangement of atoms. The FTIR spectra of silicon samples implanted with oxygen and nitrogen to a fluence of $5x10^{17}$ and $7.5x10^{17}$ ions cm^{-2} and annealed at different temperatures for 5 min are shown in Fig. 1(b) and Fig. 1(c) respectively. For unannealed sample, the spectra show double peak corresponding to Si-O and Si-N bonds.

The peak corresponding to Si-O bond is seen to be prominent along with a small peak or hump corresponding to Si-N. The layer formed is found to be oxygen rich silicon oxynitride structure with less contribution from nitrogen. The RTA treatments of these layers show band enhancement with increasing peak intensity. The IR spectra of annealed samples show formation of SiO_2, Si_3N_4 and some complex structures. The peak separation indicates structural saturation of silicon oxynitride which results in phase separation. The intensity of these peaks is seen to be increasing with RTA temperature. The total number of bonds increases on annealing from 8.10×10^{22} to 10.44×10^{22} cm^{-3} for a fluence of 5×10^{17} and from 8.87×10^{22} to 12.08×10^{22} cm^{-3} for a fluence of 7.5×10^{17} ions cm^{-2} respectively.

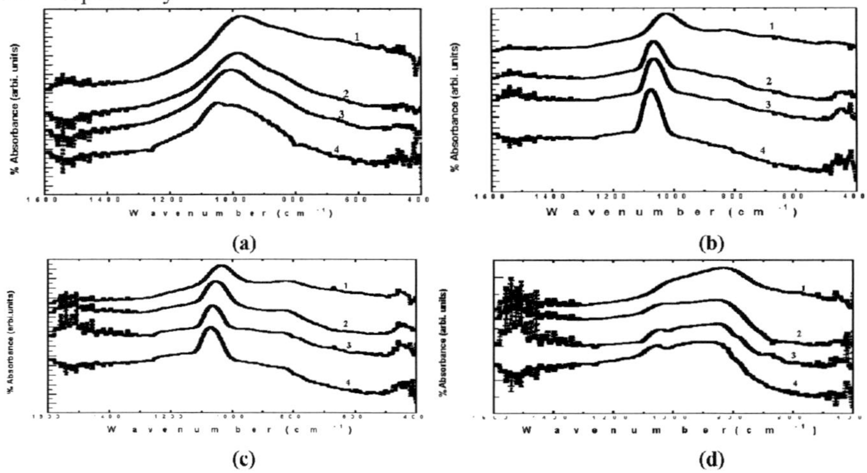

FIGURE 1. FTIR spectra of silicon samples implanted at 30 keV with molecular oxygen and nitrogen in the ratio 1:1 at fluence level of (a) 2.5×10^{17}; (b) 5×10^{17}; (c) 7.5×10^{17}; (d) 1×10^{18} ions cm^{-2} and further annealed for 5 min at temperatures: (1) as-implanted, (2) 673 K, (3) 873 K and (4) 1173 K.

For a fluence level of 1×10^{18} ions-cm^{-2}, the FTIR spectra of silicon samples implanted with oxygen and nitrogen in the ratio of 1:1 and annealed at different temperatures for 5 min are shown in Fig. 1(d). The FTIR spectrum of the as-implanted sample shows a complex structure with peak heading towards nitride at 838 cm^{-1} corresponding to Si-N stretching vibration in silicon oxynitride layer. The RTA treatment at a temperature of 673 K for 5 min. shows a significant change in the spectrum with main peak shifting to 858 cm^{-1} corresponding to Si-N stretching bond. It also indicates the formation of some other complex structures of silicon oxynitride. The splitting of the peak at 851 and 1040 cm^{-1} corresponding to Si-N and Si-O stretching bonds respectively is observed after annealing the sample at 873 K. A further annealing at higher temperatures makes main peak prominent and gives rise to a broad peak of nitrogen as well as indicates layers to be nitrogen rich. The total number of bonds increases from 8.96×10^{22} for as implanted sample to 14.14×10^{22} cm^{-3} for 1173 K annealed sample. The total number of bonds seems to saturate at this level which shows homogenous structure formation due to sputtering effect of concentration profile of implanted atoms.

Micro-Raman studies

The Micro-Raman studies were carried out to estimate stress by observing the band shift of bulk silicon and to investigate the compositions of formed structures. Fig.2 (a). compares the Raman spectra of bare silicon sample and the samples implanted with oxygen ($^{16}O_2^+$) and nitrogen ($^{14}N_2^+$) in the ratio of 1:1 with different fluences ranging from $2.5x10^{17}$ to $1x10^{18}$ions-cm^{-2} after RTA at 1173 K for 5 min in nitrogen

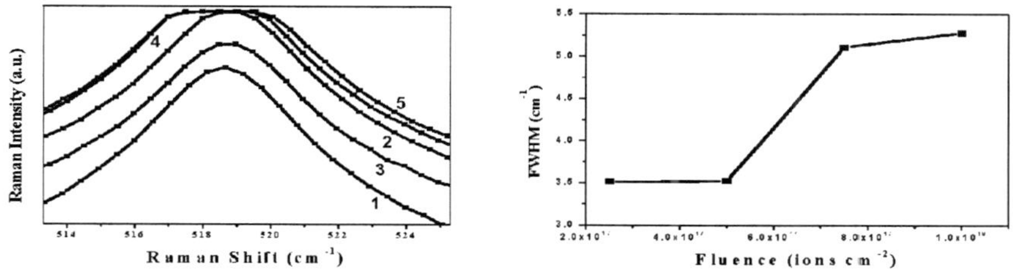

(a) (b)

FIGURE 2. (a) Raman Intensity versus Raman Shift (cm^{-1}) spectra of silicon samples implanted at 30keV with molecular oxygen and nitrogen in the ratio 1:1 at different fluence levels; (1) $2.5x10^{17}$, (2) $5x10^{17}$, (3) $7.5x10^{17}$ and (4) $1x10^{18}$ ions cm^{-2} and annealed at 1173 K for 5 min in nitrogen ambient.and **(b)** FWHM of the crystalline silicon peak as a function of ion fluence.

ambient. The Raman spectrum for the bare silicon sample has a significant crystalline band centered at 520 cm^{-1}[10].The prominent band shift is seen in case of samples implanted at various fluences and RTA at 1173 K for 5 min because of the stress introduced due to implantation. The spectra for samples implanted with fluence $2.5x10^{17}$ and $5x10^{17}$ ions-cm^{-2} show that the band width of the c-Si peak increases with increasing the ion fluence. A further increase in the fluence to $7.5x10^{17}$ ions-cm^{-2} shows shifting of c-Si peak to 517 cm^{-1}. For the sample implanted with fluence of $1x10^{18}$ ions cm^{-2} and annealed at 1173 K for 5 min, the Micro-Raman spectra shows silicon peak with increased FWHM and broad amorphous nitrogen rich silicon oxynitride band with increased intensity in the region 820-1100 cm^{-1} [11] .The spectral intensity and the peak position of the samples show the progressive formation of crystalline silicon structure along with prominent partially amorphous complex structure of oxide and nitride (silicon oxynitride). The shifting of the prominent crystalline silicon band at 520 cm^{-1} after annealing at 1173 K, with increased intensity for different ion fluences indicates recovery of the damaged silicon layer. Fig. 2(b). shows the plot of FWHM of silicon band versus ion fluence for implanted samples The FWHM of the silicon band is found to be increasing from 3.25 cm^{-1} to 5.28 cm^{-1} with respect to increasing ion fluence indicating broadening of peak with increasing ion fluence.

Ellipsometric studies

Ellipsometric studies were carried out to measure thickness and refractive index of silicon oxynitride thin films annealed at 1173 K. For the fluence level of $2.5x10^{17}$ ions cm^{-2} formation of a thin layer of thickness 70 nm with a refractive index of 1.54 is observed, which indicates formation of oxygen rich silicon oxynitride structure as the refractive index of oxygen is 1.47. The FTIR measurements have also given the similar results showing formation of oxygen rich silicon oxynitride layer with prominent peak at ~ 1051 cm^{-1}. On further increasing the ion fluence to $5x10^{17}$ ions cm^{-2} and $7.5x10^{17}$ ions cm^{-2} the layer thickness increases to 77 and 97 nm with refractive index of 1.55 and 1.59 respectively and are found to be oxygen rich. The similar results have been observed in FTIR measurements with prominent Si-O peak and a small hump for Si-N. The layers are found to be nitrogen rich silicon oxynitride at a fluence of $1x10^{18}$ ions cm^{-2} since the refractive index of the layer is found to be 1.96 which is closer to the refractive index of silicon nitride ~ 2.0. The FTIR results for this particular fluence have shown nitrogen rich silicon oxynitride structure with a prominent peak of Si-N at ~ 851 cm^{-1}. The Ellipsometry results are in conformity with our FTIR results. It has been reported that the more incorporation of oxygen in silicon oxynitride films leads towards the formation of layer with lower refractive index and the increase in nitrogen content in silicon oxynitride film is accompanied by a proportional increase in refractive index [12].

CONCLUSIONS

The FTIR studies show that at an annealing temperature of 1173K oxygen rich or nitrogen rich silicon oxynitride film are formed depending on the implanted ion fluence levels. The Micro-Raman spectroscopic studies show that the FWHM of the silicon band is found to be increasing from 3.25 cm^{-1} to 5.28 cm^{-1} with respect to increasing ion fluence indicating broadening of peak with increasing ion fluence. Ellipsometric studies show that the refractive index and thickness of silicon oxynitride thin film are highly dependent on implanted fluence levels. For lower fluences refractive index is closer to silicon oxide whereas for higher fluence it is found to be closer to silicon nitride indicating oxygen and nitrogen rich silicon oxynitride layer.

ACKNOWLEDGEMENT

We are thankful to University Grants Commission, New Delhi for financial assistance through major research project " F.10-47/2001 (SR-I) dated 29.03.2001.

REFERENCES

1. G.A. Al-Jumaily, T.A. Mooney, W.A. Spurgeon and H.M. Dauplaise, *J. Vac. Sci. Technol.* A7(1989) 2280.
2. J. Finster, E.-D. Klingenberg and J. Heeg. *Vacuum* 41 (1990), pp. 1586.
3. P. V. Bulkin, P. L. Swart, and B. M. Lacquet, *J. Non-Cryst. Solids* 187 (1995), pp.484–488.
4. D.Cariado, I. Pereyra, M.I. Alayo, *Materials Characterization* 50 (2003) pp.167-171.
5. W.J. M.J. Josquin, *Nucl. Instr. and Meth.* 209/210 (1983) 581.
6. T.D.M. Salgado, F.C. Stedile, C.Krug, I.J.R. Baumvol, C.Radtke, *Nucl.Inst and Meth. B.* 148 (1-4) (1999) 252.
7. A.I.Belogorokhov, V.T. Bublik, K.D. Scherba chev,Yu.N. Parkhomenko, V.V. Makarov, A.B. Danilin, *Nucl. Instr. and Meth. B* 147 (1999) pp.320.
8. J.A. Diniz, A.P. Sotero, G.S. Lujan, P.J. Tatsch, J.W. Swart. *Nucl. Instr. Meth. B* 166-167 (2000) pp. 64-69.
9. J.C.Rostaing, Y.Cros,S.c. Gujrathi, S.Poulain, *J. Non-Cryst. Solids* 97-98, (1987) 1051.
10. M. Monlinari, H. Rinnert, M. Vergnatt and P. Weisbecker, *Materials science and Engineering* B101 (2003) pp.186.
11. M. Molinari, H.Rinnert, M.Vergnat, P. Weisbecker, *Material Science and Engineering* B101 (2003) pp.186-189.
12. M.I. Alayo, I.Pereyra, W.L.Scopes, M.C.A. Fatini, *Thin Solid Films* 402 (2002) pp. 154.

Hydration Characteristics of Metakaolin Admixtured Cement using DTA, XRD and SEM Techniques

D. Govindarajan*, R. Gopalakrishnan
Department of Physics, Annamalai University, Annamalai Nagar, 608 002, India
e_mail: degerajan@rediffmail.com

Abstract. The paper aims to investigate hydration and pozzolanic reaction in Portland cement paste with different replacement percentages (0%, 10%, 20% and 30%) of metakaolin. The compressive strength of the metakaolin admixtured cement was measured at 1 day, 1 week and 4 weeks. The compressive strength developments of the metakaolin admixtured cement are compared with Portland cement. It is found that metakaolin contributes significantly to strength development as an accelerating admixture for Portland cement. The pozzolanic reactions and the reaction products were determined by DTA, XRD and SEM.
Keywords: Metakaolin, Compressive strength, DTA, XRD, SEM.
PACS: 61.10.NZ, 68.37.Hk

INTRODUCTION

Pozzolanic materials along with Portland cement are increasing, as this mixture offer cost reduction, energy savings, superior products and also one way to reduce greenhouse gas and hence global warming. One such pozzolan is metakaolin (MK), obtained by dehydrating kaolin at temperature from 650 to 850°C [1]. Many studies have focused on the use of MK as cement replacements [2-5]. An attempt has been made to understand their role in physical, chemical and mechanical properties of metakaolin blended cement pastes [6]. As far as we are aware, the studies on metakaolin admixtured Indian cement paste with different hydrated periods are meager. The purpose of this research is to examine the hydration products and to explore pozzolanic reaction mechanism of Indian ordinary Portland cement (OPC) replaced with MK. In this study DTA, XRD and SEM techniques helps quantify and visualize the hydration products that are formed.

MATERIALS AND METHODS

Metakaolin and Ordinary Portland cement is a commercial one. Present investigation, Ordinary Portland cement and metakaolin admixtured cement paste was prepared by mixing double distilled water in water to cement ratio (W/C) of 0.4. OPC is partially replaced with 0%, 10%, 20% and 30% of MK by weight. Samples hydrated for periods 1 day, 1 week and 4 weeks were subjected to acetone [7].

Compressive strength has been measured on OPC and metakaolin admixtured cement paste in a water to cement ratio of 0.4 [8] and reported in Table. 1.

CP1004, *Thermophysical Properties of Materials and Devices-NCTP '07*,
edited by P. Predeep, S. Prasanth, and A. S. Prasad

The Thermal curves of the OPC and metakaolin cement pastes were taken in a Perkin Elmer thermal analyzer. Hydrated products of OPC and metakaolin admixtured cement pastes were identified by Scintag X-Ray diffractometer with CuKα radiation and 2θ range of 5-70°.For microstructure study, the hydrated fractured surfaces of metakaolin admixtured samples were clicked using JEOL SEM model JSM-5610-LV.

RESULTS AND DISCUSSION

Compressive Strength

Table. 1 shows the compressive strength of cement pastes containing different % of MK as a function of curing time. The results indicate that the compressive strength increases with increasing hydration periods for all replacement % of MK. Initially MK particles fill the space between cement particles, accelerate the cement hydration and pozzolanic reaction [10]. As a result, more C-S-H gel and other hydration products are formed and deposited with in pore system with hydration time increases. This leads to an increase in compressive strength.

TABLE. 1. Compressive strength of metakaolin admixtured cement.

Samples	Compressive strength (MPa)		
	1 day	1 week	4 weeks
MK0%	10.4	32.6	47.7
MK10%	18.4	41.2	55.3
MK20%	21.4	52.4	69.3
MK30%	27.4	57.4	74.2

Differential Thermal Analysis

Fig. 1 shows the DTA thermograms of OPC pastes at different hydration periods. At 1 day hydrated cement paste, four endothermic peaks are observed. The first peak at 115°C which is mainly due to the presence of calcium silicate hydrate. The second endothermic peak observed at 130°C represents decomposition of ettringite. The third peak at 450°C represents the decomposition of calcium hydroxide. The fourth endothermic peak at 670°C is due to decomposition of calcium carbonate. From 1 week and 4 weeks DTA curves, these four endotherms are increased in size with increasing time periods and slightly shifting to a higher temperature are observed. An identical observation was reported earlier in Simona Medvescek et al. [11].The DTA curves of the MK admixtured cement pastes (Figs. 2-4), the intensity of the $Ca(OH)_2$ peaks gradually decreases when the replacement % MK increases. Due to this pozzolanic reaction, additional C-S-H developed and hence intensity of C-S-H increases. These findings are totally agreed with previous results [12].

X-Ray Diffraction Analysis

The XRD patterns of the OPC, hydrated with 1 day, 1 week and 4 weeks are presented in Fig. 5. For all hydrated cement pastes, ettringite (2θ =9.1°, 23.1°, 41.2°), C-S-H (2θ = 29.5°), $Ca(OH)_2$ (2θ = 18.1°, 33.5°) and $CaCO_3$ (2θ= 38.6°) crystallised phases are appeared [22-25]. The intensity of peak corresponding to C_3S (2θ =32.2°, 51.5°, 62.4°) and C_2S (2θ = 32.5°, 56.1°) decreases and peak at 29.5° due to CSH phase increases with hydration proceeds [13]. The peaks at 38.6° are due to $CaCO_3$ which is produced at the time of hydration with atmospheric carbondioxide. The crystalline $Ca(OH)_2$ product was formed appreciable amount at 1 day. At 1 week and 4 weeks, the intensity of $Ca(OH)_2$ are higher. These results are confirmed through DTA (Fig. 1). From the XRD, Figs. 6-8, due to pozzolanic reaction of MK, a systematic decrease in intensity of $Ca(OH)_2$, C_3S, C_2S with increase in C-S-H peak intensity are observed. These trends are higher when replacement percentage increases.

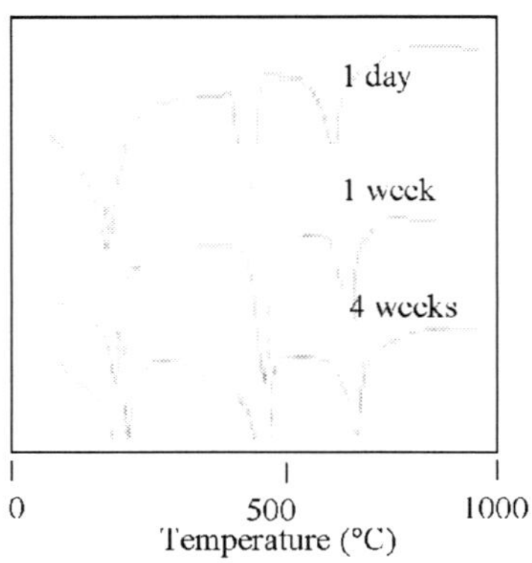

FIGURE . 1 DTA curves of hydrated OPC paste.
MK

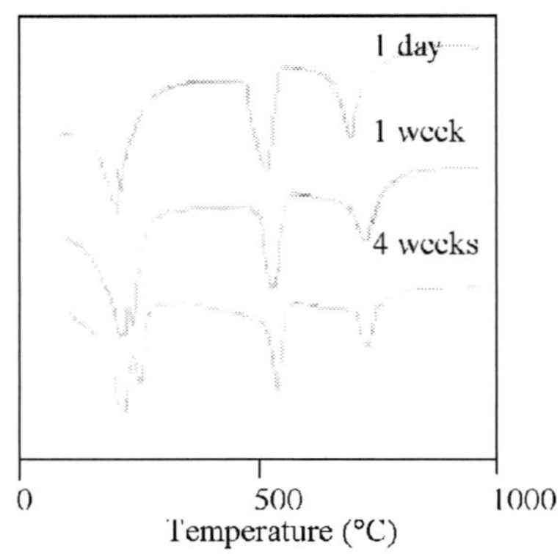

FIGURE. 2 DTA curves of hydratedOPC+10%
paste

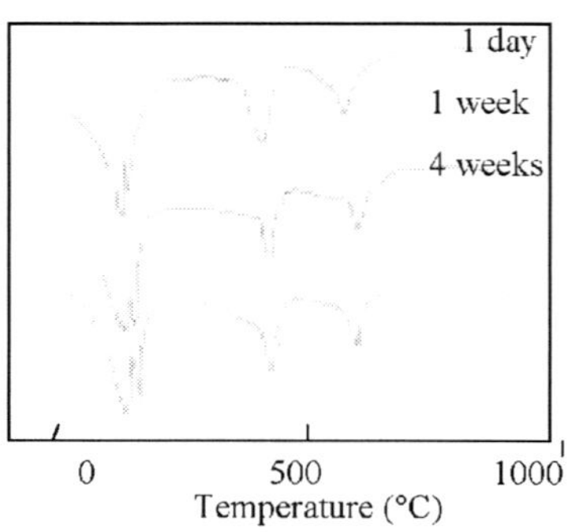

FIGURE .3. DTA curves of hydrated

OPC+20% MK paste

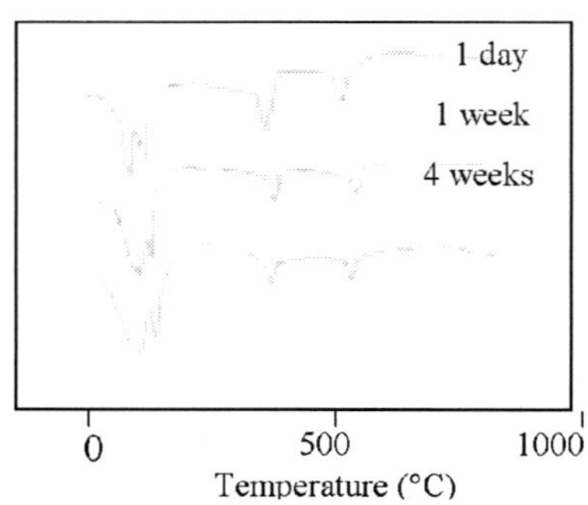

FIGURE 4 DTA curves of hydrated OPC+30%
MK paste

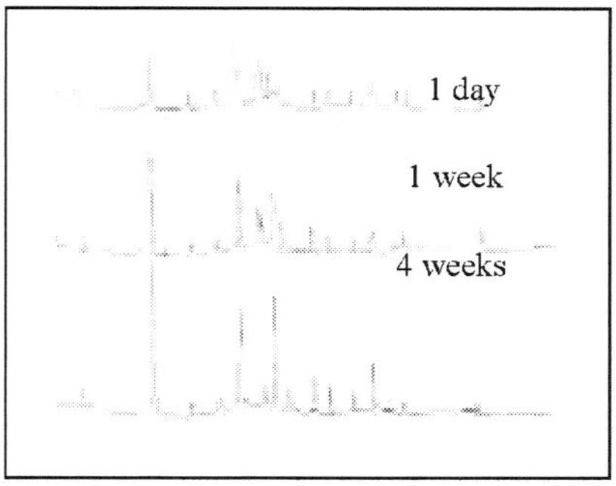

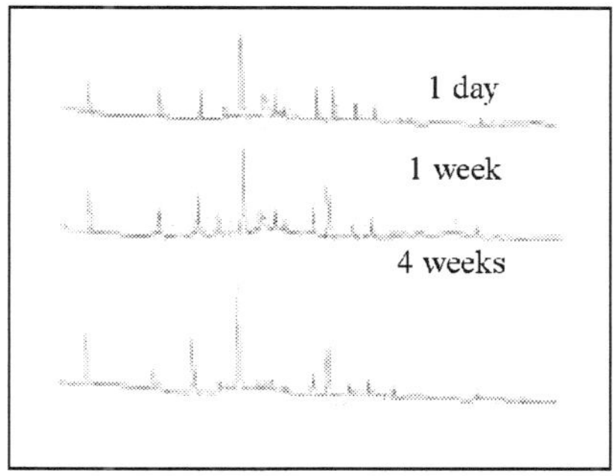

FIGURE . 5 XRD pattern of hydrated OPC paste

FIGURE 6. XRD pattern of hydrated OPC+10% MK paste

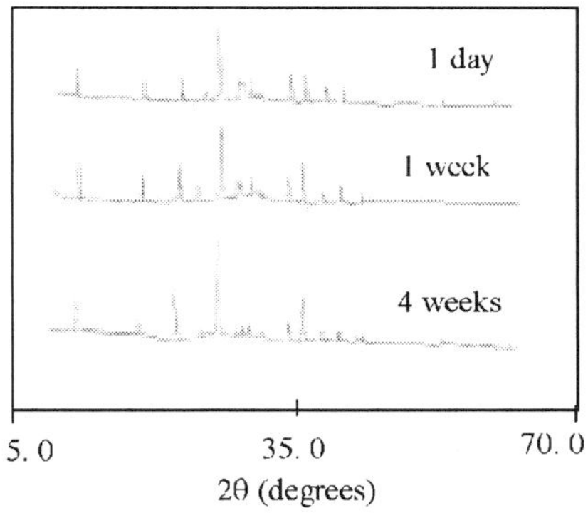

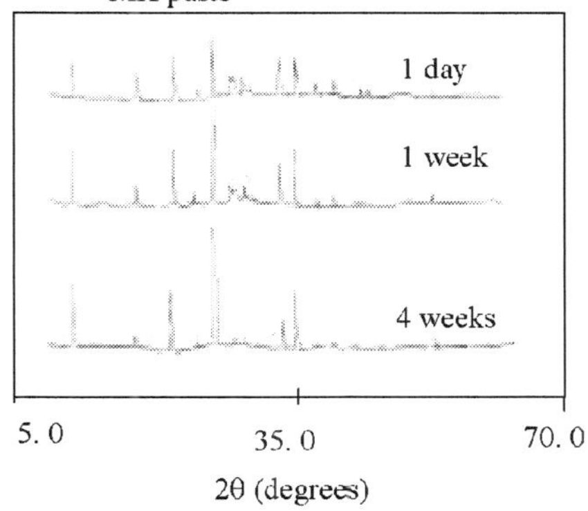

FIGURE 7. XRD pattern of hydrated OPC+20% MK paste

FIGURE . 8. XRD pattern of hydrated OPC+30% MK paste

SCANNING ELECTRON MICROSCOPY

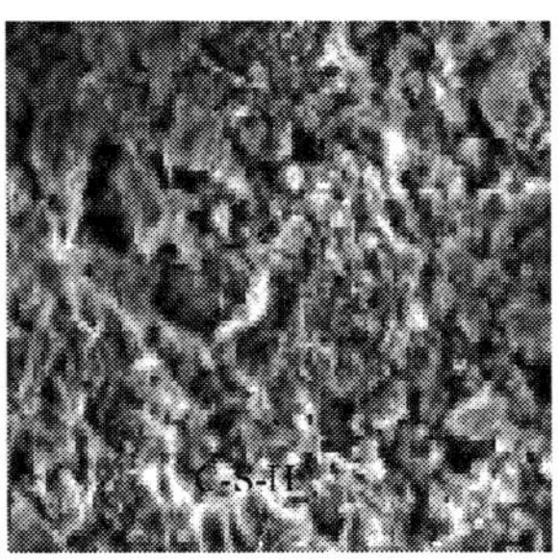

FIGURE. 9. Hydrated microstruture OPC 1 day

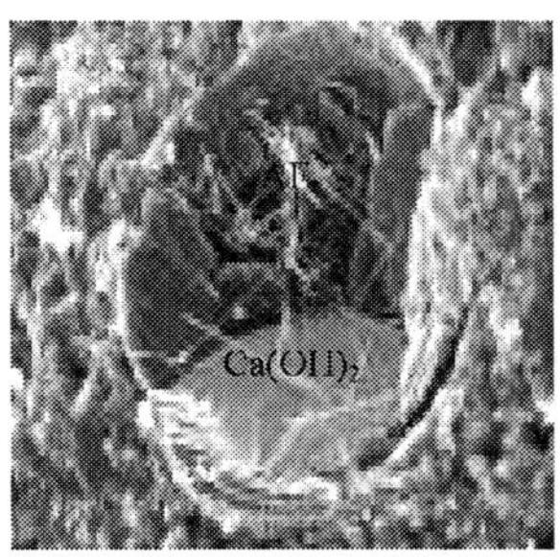

FIGURE. 10 Hydrated micro-studture OPC 1 week

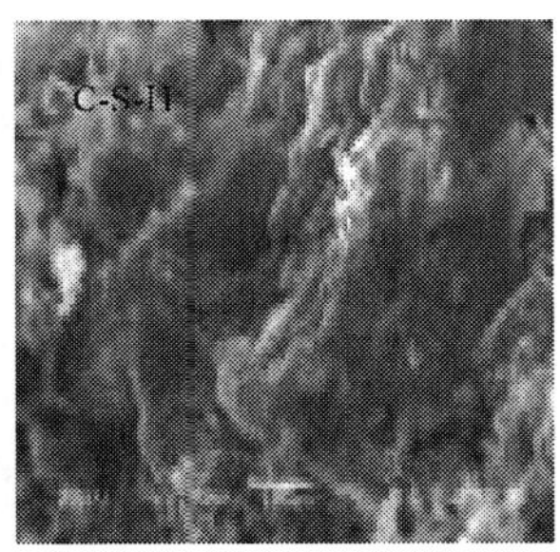

FIGURE 11 Hdrated micro-structure OPC 4 weeks

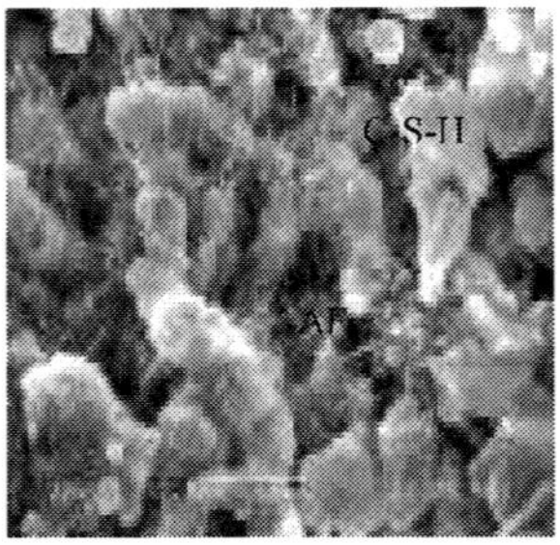

FIGURE. 12 Hydrated microstruture OPC+30%MK 1 day

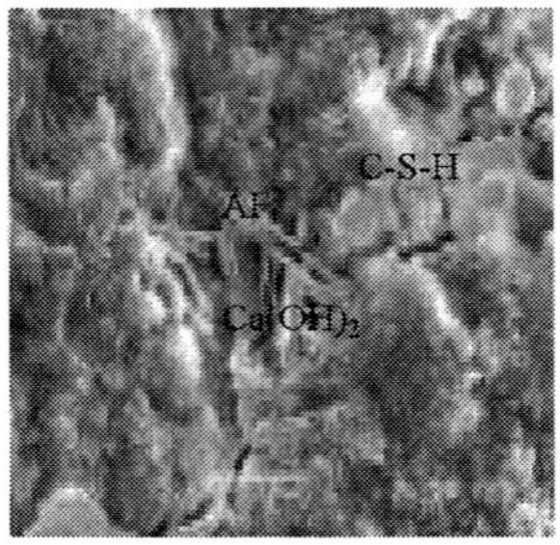

FIGURE. 13 Hydrated micro-studturc+30%MK OPC 1 week

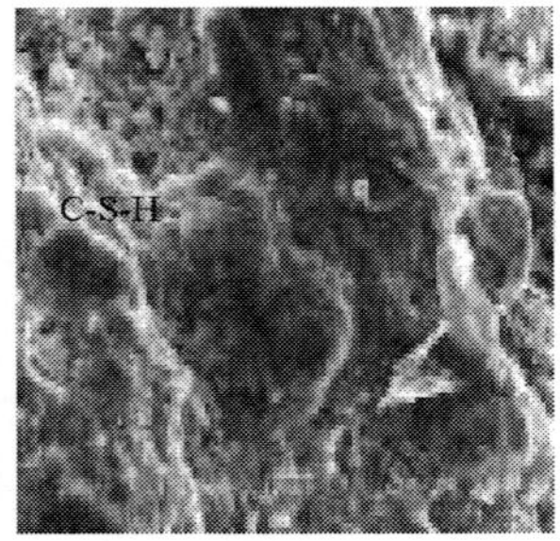

FIGURE 14 Hdrated micro-structurc+30%MK OPC 4 Weeks.

Figs. 9-11 and Figs. 12-14 are, respectively, SEM microstructure of OPC and OPC in combination of 30% MK at the ages of 1 day, 1 week and 4 weeks. It was shown in Fig 9-11, main hydration products of OPC are C-S-H gel, AFt rods and $Ca(OH)_2$ are observed [14]. As hydration continued, a gradual decrease in the unhydrated cement can be seen. At 1 day 30% MK replaced cement paste (Fig.12) almost the whole cement surface was covered with AFt rods and CSH gel. Thus the 30% MK accelerate the pozzolanic reaction results the more and quick formation of AFt and C-S-H gel and are accumulated in sufficient amounts to interlock or zipping (Figs. 13 -14) and to exhibit higher strength.

CONCLUSION

From the results presented in this paper, the following conclusions emerge:

1. The replacement of metakaolin in cement shortens the initial and final setting time and this indicate the MK accelerates the cement hydration.
2. For all curing periods and all replacement percentages of metakaolin admixtured samples shows higher compressive strength than that of OPC.
3. DTA, XRD and SEM results shows that the $Ca(OH)_2$ content decreases consequently the amount of C-S-H increases. As a result the paste becomes more homogeneous and denser in structure. This shows that pozzolanic reactions were enhanced the hydration reaction by increasing the replacement percentages of metakaolin.
4. The effect of metakaolin in ordinary Portland cement is not only environmental benefits alone but also beneficial for the improvement of microstructure and compressive strength of the cement.

REFERENCES

1. N. J. Saikia, P. Sengupta, P. K. Gogoi and P.C. Borthakur, *Cem. Concr. Res* **32,** 1717-1724, (2002).
2. Chi-Sun Poon, Salman Azhar and Mike Anson, *Cem. Concr. Comp* **25,** 83-89, (2003).
3. E. Badogiannis, V.G. Papadakis ,E. Chaniotakis and S. Tsivilis, *Cem. Concr. Res* **34,** 1035-1041, (1999).

4. S. Wild, J. M. Khatib and A. Jones, *Cem. Concr. Res* **26(10),** 1537-1544, (1996).
5. Zhang Yunsheng, Sun We, Jin Zuquan and Yu Hongfa, *Mater. Lett* **61,** 1552-1557, (2007).
6. N. J. Coleman and W. R. Mcwhinnic, *J. Mater. Sci* **35,** 2701-2710, (2000).
7. N. Voglis, G. Kakali, S. Chaniotaki and S. Tsivilis, Cem. Concr. Comp **27,** 191-196, (2005).
8. M. S. Shetty, Concrete Technology, S. Chand and Co.Ltd. New Delhi (2004).
9. L. Zongjin and Zhu Ding, Cem. Concr. Res **33,** 579-584, (2003).
10. B. B. Sabir, S. Wild and J. Bai, *Cem. Concr. Res* **23,** 441-454, (2001).
11. Simona Medvescek, Roman Gabrovsek, Venceslav Keucic and Anton Meden , *Acta. Chim. Slov* **53,** 172-179, (2006).
12. Moises Frias Rojas and M. I. Sanchez de Rojas, *Cem. Concr. Res* **33,** 643-649, (2003).
13. Meenu Sharma and Vishwamittar, K. S. Harchand and D. Raj, Cem. Concr. Res **21,** 484-488, (1991).
14. B. J. Dalgleish, P. L. Pratt and E. Toulson, *J. Mat. Sci* **17,** 2199-2207, (1982).

Electrical, Optical and Structural Characterization of Pure and Potassium Doped PVC Based Polymer Electrolyte Films

T. Janaki Rami Reddy, N. Reddeppa, V.B.S.Achari, V.V.R. N. Rao, A.K. Sharma*

Department of Physics, S.V. University, Tirupati-517502, INDIA.

Abstract. The electrical conductivity and optical absorption of pure and $KClO_3$ doped polyvinyl chloride (PVC) films were studied under different conditions. The features of complexation of the electrolytes were studied by X-ray diffraction (XRD). The electrical conductivity increased with increasing dopant concentration, which is attributed to formation of charge transfer complexes. Optical properties like direct and indirect band gap and optical absorption edge were estimated for pure and doped films from their optical absorption spectra in the wavelength region 200-550 nm. It was found that the energy gap and band edge values shifted to lower energies on doping with $KClO_3$ salt

Key words: PVC polymer thin films, XRD, electrical conductivity, optical properties.
PACS: 79.60.Fr, 78.70.Ck, 73.61.-r, 78.20.-e.

INTRODUCTION

Studies on the electrical and optical properties of polymers have attracted much attention in recent years in view of their application in electronic and optical devices. Electrical conduction in polymers has been studied aiming to understand the nature of the charge transport prevalent in these materials while the optical properties are aimed at achieving better reflection, antireflection, interference and polarization properties. [1-2].The electrical and optical properties of polymers can be suitably modified by the addition of dopants depending on their reactivity with host matrix[3,4].In the present work, the electrical and optical properties of pure and $KClO_3$ doped PVC polymer films were studied. Since K^+ is fast conducting ion in a number of crystalline and amorphous materials, its incorporation within a polymeric system may be expected to enhance its electrical and optical performance. A relatively small number of studies have been reported on the development of PVC based polymer electrolyte systems [5,6]. This paper presents the results of such investigation on the electrical and optical properties of $KClO_3$ doped PVC films.

CP1004, *Thermophysical Properties of Materials and Devices-NCTP '07*,
edited by P. Predeep, S. Prasanth, and A. S. Prasad

EXPERIMENTAL

Films (thickness~130μm) of pure PVC and $KClO_3$ doped PVC were prepared in the weight percent ratios 95:05, 90:10, 85:15 and 80:20 with tetrahydrofuran (THF) as solvent using solution cast technique.

The X-ray diffraction studies were performed by means of SEIFERT X-ray diffractometer in the range 10 - 65°. The d.c conductivity of pure and $KClO_3$ doped polymer electrolytes was studied at different temperatures ranging from 313–425K. The optical absorption spectra of these films were recorded at room temperature with the aid of JASCO spectrophotometer (model V-570) in the wavelength region 200 - 550 nm.

RESULTS AND DISCUSSION

XRD analysis

In order to investigate the influence of potassium salt on PVC structure XRD studies were performed. Fig.1 (a-f) shows the X-ray diffraction (XRD) patterns of pure PVC, $KClO_3$ salt and PVC complexed with $KClO_3$ salt. Comparison of the XRD patterns of complexed PVC films with those of pure PVC and $KClO_3$ reveals the following differences. Peaks observed for 2θ values around 23° are less intense in complexed PVC films compared to those in pure PVC films. This indicates that the addition of $KClO_3$ salt causes a decrease in the degree of crystallinity of the PVC complex. This is in agreement with the results reported by other researchers [7,8].

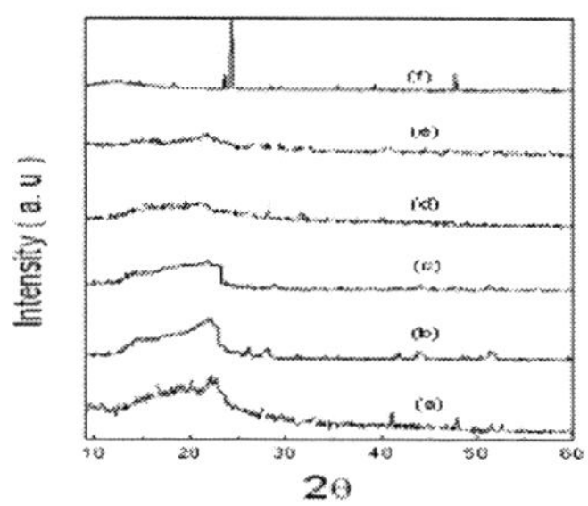

FIGURE 1. XRD patterns of (a). Pure PVC. (b). PVC+$KClO_3$(95:05),(c).PVC−$KClO_3$(90:10), (d). PVC+$KClO_3$ (85:15), (e). PVC+$KClO_3$ (80:20), (f). $KClO_3$ Salt

The crystalline peaks for 2θ values around 24, 27 and 49° corresponding to $KClO_3$ (Fig.1f) are absent in complexed PVC. This indicates the absence of any excess (uncomplexed) salt in the complexed polymer electrolyte films. Therefore, it may be confirmed that complexation has taken place in the amorphous phase.

Electrical properties

The conductivity increases with dopant concentration as well as temperature. The increase in conductivity with increasing concentration of $KClO_3$ is attributed to the decrease in the degree of crystallinity and increase in the degree of amorphousity of polymer matrix [3,4].

Fig.2 shows the variation of d.c conductivity as a function of temperature for pure and differently complexed PVC polymer electrolyte in the temperature range 313–425K. From the plots it was found that the conductivity increases with the temperature and follows the Arrhenius behavior throughout with two regions having different activation energies.

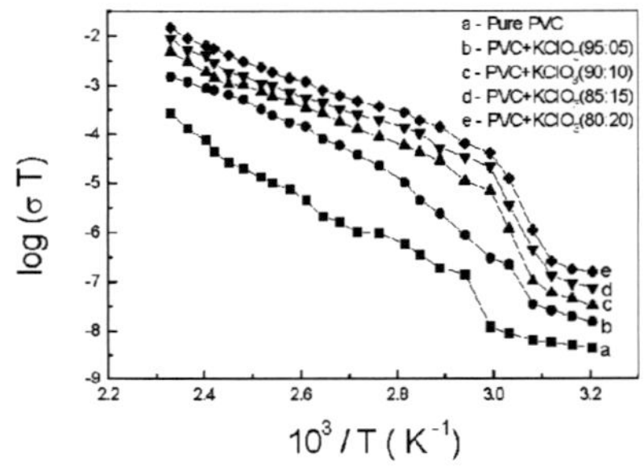

FIGURE 2. Temperature dependent conductivity of (a) pure PVC; (b) ($PVC+KClO_3$)(95:05);(c) ($PVC+KClO_3$) (90:10) ;(d) ($PVC+KClO_3$) (85:15) and (e) ($PVC+KClO_3$) (80:20).

Conductivity data fits the Arrhenius relation,

$$\sigma = \sigma_0 \exp(-E_a / kT)$$

where σ_0 is a proportionality constant, E_a, the activation energy, T, the absolute temperature and k, the Boltzman constant.

The activation energies in the two regions are tabulated in Table 1. The activation energies show a decreasing trend with increasing dopant concentration which may be explained in terms of formation of charge transfer complexes.

Optical studies

Optical absorption studies on pure and doped films were carried out to determine the optical constants such as optical band gap (E_g) and the position of the fundamental band edge. The absorption coefficient was determined from the spectra using the formula,

$$\alpha = A / d$$

where A is the absorbance and d, the thickness of the film.

When a direct band gap exists, the absorption coefficient has the following dependence on the energy of the incident photon [9,10],

$$\alpha h\upsilon = C (h\upsilon - E_g)^{1/2}$$

where E_g is the band gap, C, a constant dependent on specimen structure, υ, the frequency of incident light, and h, the Planck's constant.

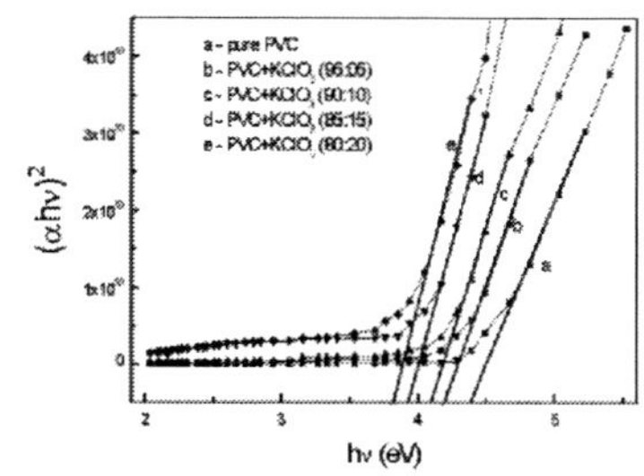

FIGURE 3. $(\alpha h\upsilon)^2$ vs $h\upsilon$ (photon energy) plots of $KClO_3$ doped PVC films for different dopant concentrations

Thus a plot of $(\alpha h\upsilon)^2$ vs $h\upsilon$ (photon energy) as shown in Fig 3 should be linear. The intercept on the energy axis on extrapolating the linear portion of the curves to zero absorption value may be interpreted as the value of the band gap.

For indirect transitions which require phonon assistance, the absorption coefficient has the following dependence on the photon energy [9,10],

$$\alpha h\upsilon = A\,(h\upsilon - E_g + E_p)^2 + B\,(h\upsilon - E_g - E_p)^2$$

where E_p is energy of the phonon associated with the transition and A and B are constants depending on the band structure.

The indirect band gaps were obtained from the plots of $(\alpha h\upsilon)^{1/2}$ vs $h\upsilon$ (photon energy) as shown in Fig 4.

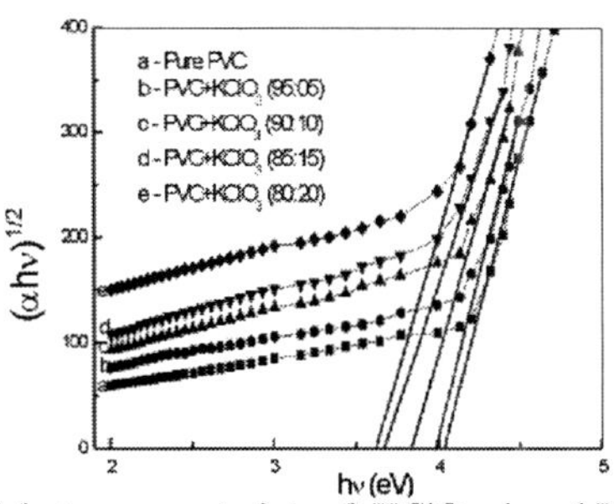

FIGURE 4. $(\alpha h\upsilon)^{1/2}$ vs $h\upsilon$ (photon energy) plots of $KClO_3$ doped PVC films for different dopant concentrations.

The position of absorption edge was estimated by extrapolating the linear portions of the α vs $h\upsilon$ plots (Fig 5) to zero absorption value.

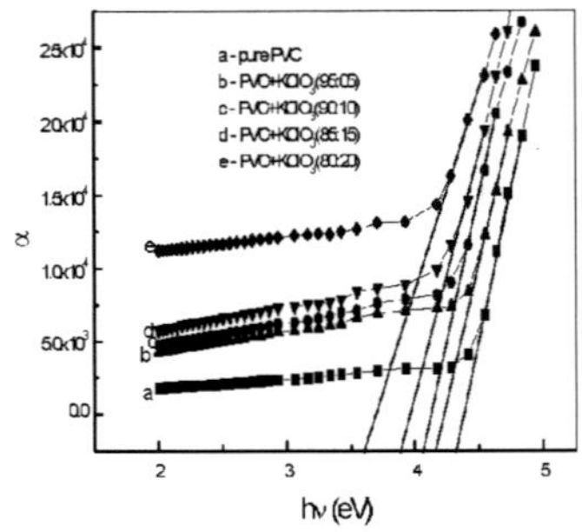

FIGURE 5. α vs hυ (photon energy) plots of $KClO_3$ doped PVC films for different dopant concentrations.

It is clear from Table.1 that the direct and indirect band gaps showed a decrease on doping with $KClO_3$. Similar behavior can also be seen in the activation energies obtained from the d.c conductivity data as shown in Table 1. This type of behavior has been observed in many polymer films [11, 12]. The magnitudes of the activation energies obtained from d.c conductivity data are small in comparison with optical band gap energies. This is due to the fact that their nature is different. While the activation energy corresponds to the energy required for conduction from one site to another, the optical band gaps corresponds to interband transition.

TABLE 1. Activation energies, absorption edge and band gaps of pure and ($PVC+KClO_3$) polymers.

Polymer electrolyte	Activation energy (E_a) (eV) Region – I	Region – II	Optical band gaps (eV) Direct	Indirect	Absorption edge (eV)
Pure PVC	0.68	1.03	4.37	4.05	4.32
$PVC+KClO_3$(95:05)	0.62	0.93	4.18	3.98	4.17
$PVC+KClO_3$(90:10)	0.56	0.70	4.07	3.84	4.08
$PVC+KClO_3$(85:15)	0.51	0.65	3.92	3.67	3.91
$PVC+KClO_3$(80:20)	0.47	0.56	3.81	3.59	3.63

The decrease in optical band gap and activation energy on doping may be explained on the basis of the fact that the incorporation of small amounts of dopant forms charge transfer complexes (CTCs) in the host lattice. These charge transfer complexes increase the electrical conductivity by providing additional charges in the lattice. This results in a decrease of activation energy.

CONCLUSIONS

The XRD study reveals the amorphous nature of the polymer electrolytes. The complex formation in polymer electrolyte system has been confirmed from FTIR studies. Electrical studies indicate that PVC can be effectively doped with $KClO_3$ to enhance its electrical conductivity and alter its conduction behavior. This enhancement has been explained on basic of charge transfer complex formation. Optical absorption edge and optical energy gap (both direct and indirect) showed a decreasing trend with increased dopant concentration. It was observed that these values shifted to lower energies on doping. This behavior is in agreement with the results on activation energies obtained from conductivity data.

REFERENCES

1. M.Tabata ,M. Satoh.K. Kaneto and K.Yoshino ,J.Physics Soc. Japan **5**,1305(1986).
2. M. Onoda,Y.Manada ,S.Morita and K.Yoshino J.Physics Soc.Japan **58,**1895(1989).
3. S.Sreepathi Rao S. Jaipal Reddy E. Laxminarasaiah and UV. Subba Rao Mater. Sci. Engg. B. **33**,173(1995).
4. Ch.V. Subba Reddy,A.K.Sharma and VVR.Narasimha Rao Journal of material Science Letters **21**,105-108(2002).
5. S. Ramesh and A.K. Arof Solid State Ionics **1197**,136-137(2000).
6. S. Rajendran,T. Uma and T. Mahalingam Ionics, **5,** 227(1999).
7. T. Uma ,T. Mahalingam and U. Stimming J. Materials Science **39**, 2901-2904(2004).
8. J.Li,X. Huang and K. Chen J. Electrochem. Soc.**147** (7) 2653(2000).
9. DS. Davis and JS Shalliday Phys, Rev. **118**,1020(1960).
10. GM.Thutupally and SG. Tomlin J.Phys, D. Appl. Phy. **9**,1639(1976).
11. A.K. Sharma and Ch.Ramu J. Mat. Sci. Letters **10**,1217(1991).
12. V. Raja,A.K. Sharma and VVR Narasimha Rao Mater. Lett. **57,** 4678 – 4683(2003).

*** Corresponding author : profaks_svu@yahoo.co.in**

Photoluminescence study of Cupric Telluride thin films

K.Neyvasagam[a*], G.Muralidharan[b], V.Vasu[a] and N.Soundararajan[a]

[a*]*Rajapalayam Rajus' College, Rajapalayam. 626 117,India*
[a]*Thin film lab, School of Physics, Madurai Kamaraj University, Madurai.625 021,India*
[b]*Department of Physics, Gandhigram Rural University, Gandhigram .624 302, India*
Email: nsrthinfilms@yahoo.com

Abstract. Thin films of Cupric Telluride (CuTe) of thickness from 50nm to 150nm have been prepared by thermal evaporation technique onto well-cleaned glass substrates kept at 300K under the vacuum better than $2X10^{-5}$ mbar. After annealing the deposited films at 375K for one-hour, the photoluminescence spectra of CuTe thin film was recorded. The X ray diffraction analysis confirmed the composition and polycrystalline nature of CuTe films. The grain size of CuTe thin film, estimated by Atomic Force Microscopy (AFM) measurements, is around 40nm. From Photoluminescence spectra, a strong emission peak at visible range was observed for all the films. Further a shift in the peak position was observed for the CuTe thin films of 150nm thickness. The observed PL spectra were attributed to confinement effect.

Keywords: Semiconductor, CuTe, thin films, Photoluminescence, grain size.
PACS: 78.20.-e, 78.55.Hx, 78.66.Li, 78.55.-m

INTRODUCTION

Semiconducting compounds such as I-VI Copper chalcogenides are widely used in the fabrication of photoconductive and photovoltaic devices [1-2]. Thin films of copper chalcogenide have been a subject of interest for many years mainly because of their wide range of applications in solar cells, superionic conductors, photodetectors, photothermal converters, electroconductive electrodes, microwave shielding etc [3-7]. The Cu-Te system exists in a few possible crystalline compounds such as CuTe, Cu_4Te_3, Cu_7Te_5, Cu_2Te, etc. [8]. Pashinkin and Pavlova [9] have analysed the thermodynamic stability of Cu-Te system. Brunner et al [10] have concluded that there is a metal to semiconductor transition in Cu-Te system for copper concentration of more than 48%at. by ultraviolet photon spectroscopy. Copper-Tellurium alloys exhibit interesting behaviours like the structural phase transition and the composition dependent band gap variation. Ishizaki et al [11] have carried out electro deposition of copper –Tellurium system under diffusion-limiting control.

Martha Symko-Davis [12] has reported that the optical band gap of CuTe thin films is around 1.5eV at room temperature. It is a direct band gap material. From the electron transport study of Newport et al [13], liquid CuTe exhibits the metallic

CP1004, *Thermophysical Properties of Materials and Devices-NCTP '07*,
edited by P. Predeep, S. Prasanth, and A. S. Prasad

behaviour and the sign of the thermoelectric power is positive. Baranova et al [14] have studied the electron diffraction analysis of CuTe thin films. The crystallographic structure of CuTe is orthorhombic. It is highly birefringent and pleochroic [15,16]. Bahl [17] has studied the K absorption studies on CuTe and estimated that the K absorption edge shift towards the high-energy side is due to the transfer of electrons from tellurium to copper. Optical anisotropy of laser synthesized CuTe thin films was studied by Hanus et al [18]. Hanus and Laude [19] suggested that the CuTe material is a convincing competitor in the optical storage WORM disc market.

Though Cu_2Te compound has five structural phase transitions between room temperature and 850K, [20,21] the phase transformation is absent in CuTe [22]. Though there are detailed studies on the optical and the electrical properties of Cu_2Te, a very limited work has been carried out on Cupric Telluride (CuTe) and also in the thin film state. Photoluminescence spectroscopy is a contactless, non-destructive method of probing the electronic structure of materials. The present work is the study of the photoluminescence spectra of cupric telluride thin films.

EXPERIMENTAL

Cupric Telluride (CuTe) alloy purchased from M/s Aldrich (India) company with 99.99% purity was used for preparing thin films by thermal evaporation. Known amount of CuTe material was taken and the entire charge from a molybdenum boat was evaporated under a vacuum better than $2x10^{-5}$ mbar on well cleaned glass substrates of $0.01x0.03m^2$, kept at room temperature. The glass substrates were cleaned with warm chromic acid and distilled water before mounting them in the vacuum chamber. Cupric Telluride thin films of thickness 50nm, 100nm and 150nm were deposited at the rate of 2.5nm/Sec. The thickness of films and the deposition rate were monitored with a digital quartz-crystal thickness monitor.
The X ray diffraction (XRD) measurements were performed on Phillips PANalytical 2KW X-ray generator using Cu-K_α line. Scanning Electron Microscopy (SEM) and Energy Dispersive Analysis of X-ray (EDAX) measurements were done on JEOL (JSM 5600) make set up to study the particle morphology, to determine the approximate grain size and abundance of atomic ratio of Cu and Te. Atomic Force Microscopy (AFM) micrographs were taken using Digital Instrument Nanoscope to estimate the grain size and analysed sample morphology. The Photoluminescence spectra of CuTe thin films were recorded in Shimadzu RF 5000 at excitation wavelength of 250nm. All the measurements were performed at room temperature.

RESULT AND DISCUSSION

Figure 1 shows the X ray diffraction patterns of CuTe thin films, which indicates CuTe composition formation and their polycrystalline nature. Indeed, while the XRD peaks are identified for orthorhombic CuTe compound. The composition of CuTe thin films as determined from EDAX was 1:1.

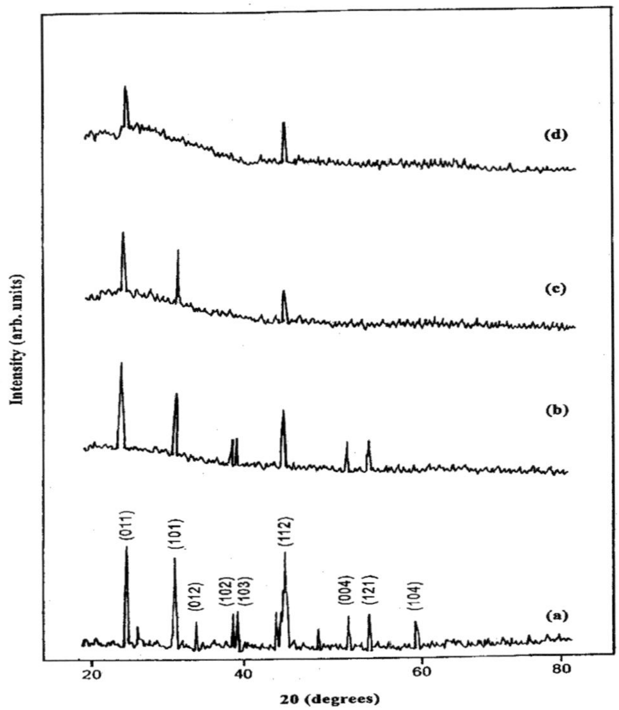

FIGURE 1. XRD pattern of CuTe (a)bulk, (b) thin films of thickness 150nm (c)100nm and (d)50nm

Figure 2(a) &2(b) shows the representative SEM and AFM picture of the CuTe film of thickness 100nm, which indicates rather uniform distribution of grains. The average grain size of all the CuTe thin film, estimated from Gaussian fit for the particle distribution, is approximately 40nm and surface roughness was estimated around 3.2nm. This was matched with that from XRD data using Debye-Scherrer's method.

FIGURE 2(a). SEM image of CuTe thin film

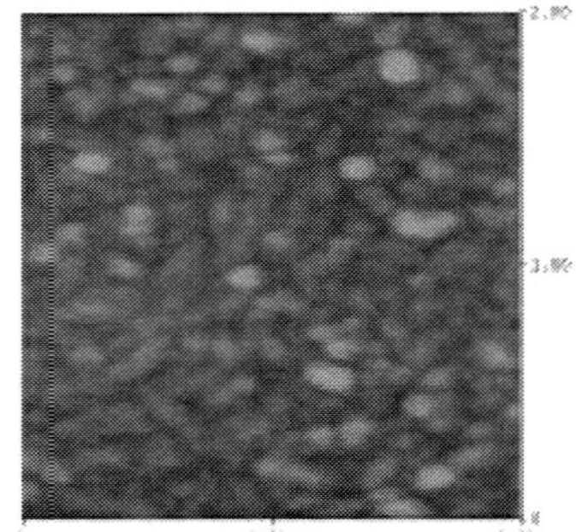

FIGURE 2(b). AFM image of CuTe thin film

The Photoluminescence spectra of CuTe thin films of thickness 50nm, 100nm and 150nm recorded in the room temperature are shown in Figure 3. A strong emission band in visible range at around 660nm could be observed for the CuTe films of thickness 50nm and 100nm. The observed luminescence peaks are above the bandgap energy of bulk CuTe. This may be due to the nanocrystalline effect. Further it may also be noted that there is a shift (blue shift) in the peak position for the CuTe films of thickness 150nm. This shift may be due to reduced crystalline size.

It is well known from the quantum mechanics and Uncertainty principle that the kinetic energy of the particle increases on confinement. The increase in energy $\Delta E_u \propto 1/d^2$, where d is the confining length and ΔE_u is also referred to the confinement energy.

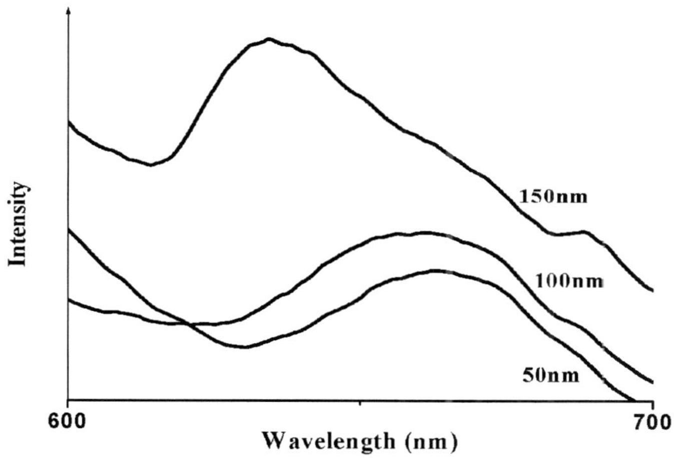

FIGURE 3. Photoluminescence spectra of CuTe thin films.

The peak Photoluminescence energy may be written as

$$h\omega_p = E_g + \Delta E_u - \Delta E_r$$

Where E_g is the optical band gap of bulk CuTe and ΔE_r is the reduction due to excitonic and photon related effect [23].

CONCLUSION

From XRD, SEM and AFM studies the surface of CuTe thin films found to be of relatively uniform grain distribution and polycrystalline. The average grain size was estimated around 40nm from AFM measurement. A strong emission band in visible range was observed at 660nm. Further it is observed that there is a shift in peak intensity of the CuTe film of thickness 150nm. The observed luminescence may be due to quantum size effect.

ACKNOWLEDGEMENT

K.Neyvasagam is thankful to UGC, for the FIP fellowship under the X plan grant. The authors are thankful to UGC-DRS support for the Phase II upgradation of the thin film laboratory. They are also thankful to Ragha Sudha for XRD at IISc Bangalore and Mohan Gongarade for the AFM measurement at UGC-DAE, CSR, Indore.

REFERENCES

1. O.Mahammad Hussain, B.Srinivasulu Naidu and P.Jeyarama Reddy, *Thin Solid films*, **193/194,** 777-781 (1990)
2. H.Okimura, M.Togasaki, M.Iwase,Y.Gekka. *Jap.J.Appl.Phys.*, **21**, 83 - 87 (1982)
3. B.Lio, Y.Xie, J.Huang, Y.Liu and Y.Qian, *J.Mater. Chem.* **12**, 2747 - 2748 (2002)
4. W.S. Chen, J.M.Stewart and R.A.Mickelsen., *Appl.Phys.lett.*, **46,** 1095 -1097 (1958)
5. C.Nascu, I.Pop,V.Ionscu, E.Indra, I.Bratu, *Mater.Lett.*, **32,** 73 -77 (1997)
6. H.Okimura,T.Matsumae, R.Makabe, *Thin Solid films* **71,** 53 -59 (1980)
7. M.A.Korzhuev, *Phys.Solid State*, **40**, 217 –219 (1998)
8. M.Hansen and K.Anderko, *Constitution of binary alloys*. (McGraw-Hill, New york, 1958), pp.638 - 641
9. A.S. Pashinkin and L.M. Pavlova, *Inorganic Materials*, **41**, 1050 -1054 (2005)
10. A.J.Brunner,H.Bretscher,R.Lapka P.Oelhafen, R.Schogl and H.J.Guntherodt *J.Phys. C:Solid State Phys*, **20**, 5233 -5239 (1987)
11. T.Ishizaki, D.Yata and A.Fuwa., *Mat.Trans.*, **44**, 1583 -1587 (2003)
12. Martha Symko-Davies Procedings of 19th *European Photovoltaic solar energy conference.* Paris,France **6** (2004).
13. R.J. Newport, R.A.Howe and J.E.Enderby, *J.Phys.C:Soild State Phys.*,**15**, 4635 - 4640 (1982)
14. R.V.Baranova,V.P.Aref'ev and S.S.S.R, *Neorg.Mater.*, **13**, 2157- 2159 (1977)
15. C.Paparoditis, C.Stella, D.Darmagna and J.Bernard, (Comm..coll.Int.Phys.couches Minces.Clausthal, Gottingen, 1965 pp. 732
16. F.Pertlik, *Mineralogy and Petrology*, **71**, 149 -154 (2001)
17. M.K.Bhal , *J.Phys.C:Solid StatePhys.*, **8**, 4107 -4116 (1975)
18. F.Hanus, M.Wautelet and L.D.Laude, *J.Appl.Phys*, **72**, 4901 -4906 (1992)
19. F.Hanus and L.D.Laude, *SPIE Proce.*, **1279**, 196 - 197 (1990)
20. N.Vouroutzis, N.Frangis and C.Manolikas, *Phys.Stat.Sol.*(a), **202**, 271-280 (2005)
21. A.L.Dawar, Anil Kumar, Pratap kumar and P.C.Mathur, *J.of less Com.Metals*, **91**,83-88 (1983)
22. K.Neyvasagam, C.Vijayan, M.Pandiaraman,and N.Soundararajan, *National seminar on Advances in Material Science* (2006) pp.22
23. G.C.John,V.A.Singh, *Phy Reports* **263,** 93-151 (1995)

Simple and Economical Method to Prepare $Mg_{1-x}Mn_xB_2$ Superconductor

Kiran Singh[a]*, Rajneesh Mohan[a], Ravi K. Shukla[a], N. K. Gaur[a] and R. K. Singh[b]

[a] *Superconductivity Research Laboratory, Department of Physics, Barkatullah University, Bhopal- 462026 (M. P.), India*

[b] *Institute of Professional Scientific Studies and Research, Chaudhary Devilal University, Sirsa -125055 (Haryana), India*

Abstract. The polycrystalline samples of $Mg_{1-x}Mn_xB_2$ ($0 \leq x \leq 0.03$) have been synthesized by using the solid state reaction method. The properties of the samples depend on the method of preparation. Generally, Tantalum foil is used for the preparation of MgB_2 superconductor to reduce the formation of MgO phase. Here, we have reported a very simple and economical method for the preparation of $Mg_{1-x}Mn_xB_2$ ($0 \leq x \leq 0.03$) samples by using the stainless steel sheet instead of Ta foil. These results are well in agreement as prepared by using the Ta foil. All the samples were sintered at 850 °C for 2 hrs in the presence of high purity argon flow. The structural and electrical properties were studied by using the X-Ray diffraction and four-probe resistivity measurement techniques, respectively. The XRD patterns revealed the formation of MgB_2 as the major phase. The superconducting transition temperature (T_c) decreases with increasing the Mn concentration. The Mn doping suppresses the transition temperature sharply as compared to the other doping like Al, Cu and Ag. This abrupt suppression in transition temperature is due to the pair breaking by the introduction of magnetic ions.

Keywords: MgB_2, Doping, Phase formation, Transition temperature.Enter Keywords here.
PACS: 74.70.Ad, 74.25.Fy, 74.25.62.Dh

INTRODUCTION

The discovery of superconductivity in MgB_2 [1] evoked a new enthusiasm among the experimentalists and theoreticians to understand the mechanism of superconductivity in this material. This discovery offers a new era in the field of superconductivity and as a guide to search the superconductivity in simple binary compounds. Unlike high temperature superconductors (HTS), MgB_2 has lower anisotropy, larger coherence length, high critical current density and transparency at grain boundaries to the current flow. This makes MgB_2, promising material for practical applications. This superconductor has the wide range of operating temperature, approximately twice the value of the present application of Nb_3Sn and Nb-Ti superconductors, which can be easily attained by cryogen/cryocoolers and more economical than the liquid helium. The density of this new intermetallic superconductor is very less (2.6 g/cm^3). Therefore, this superconductor is very useful in the applications where light material is preferable, like space science. The MgB_2

CP1004, *Thermophysical Properties of Materials and Devices-NCTP '07*,
edited by P. Predeep, S. Prasanth, and A. S. Prasad

wire has been used in X-ray spectrometer in 'Suzaku' satellite launched by Japan on July 10, 2005 [2]. Also MgB_2 has the prospects to make high field superconducting magnet [3].

The other groups prepared MgB_2 by different methods using the sealed tube method and vacuum furnace etc. The major problem to prepare MgB_2 is the formation of MgO phase. Generally, Tantalum foil is used for the preparation of MgB_2 superconductor to reduce the formation of MgO phase. We have already optimized the processing parameters for the preparation of Cu doped MgB_2 [4]. Our group has also studied the thermodynamical properties of this superconductor [5, 6]. Here, we have reported a very simple and economical method for the preparation of $Mg_{1-x}Mn_xB_2$ ($0 \leq x \leq 0.03$) samples by using the stainless steel sheet instead of Ta foil. The preparation, phase formation and electrical properties of $Mg_{1-x}Mn_xB_2$ ($0 \leq x \leq 0.03$) are discussed systematically in the following section.

EXPERIMENTAL PROCEDURE

We have used the standard solid-state reaction method for the synthesis of polycrystalline $Mg_{1-x}Mn_xB_2$ ($0 \leq x \leq 0.03$) samples. For this purpose, the stoichiometeric amount of Mg, B and Mn powders were taken and thoroughly ground in an agate mortar for 4 hrs and compacted into the circular pellets of 12 mm diameter by applying hydraulic pressure. We have made a boat of commercially available stainless steel sheet and put the pellets in the above said boat and heated the same in a quartz tube at 850°C for 2 hrs in the presence of high purity argon gas flow followed by the furnace cooling. These samples were characterized by X-Ray diffractometer (Cu K_α radiation) in the range $20° \leq 2\theta \leq 80°$ for identifying the phase formation. The low temperature electrical resistivity was measured by the standard four-probe method in the range of 30 K to 300 K by closed cycle refrigeration equipment (CTI Cryogenics).

RESULTS AND DISCUSSION

The XRD patterns of $Mg_{1-x}Mn_xB_2$ ($0 \leq x \leq 0.03$) samples are shown in fig.1. These patterns are indexed in P6/mmm space group by using PowderX program. In fig. 1, we have also introduced the XRD patterns of pure MgB_2 sample, which we have prepared by using the Tantalum foil for the reference purpose. It is clear from fig. 1 that all the peaks are indexed according the P6/mmm space group. The XRD patterns of pure MgB_2 by using the Tantalum foil and stainless steel sheet (fig. 1) are similar. In both the cases there is small impurity peak of MgO at $2\theta = 62.36$. The presence of this peak is also seen in other reports [7-10]. The small MgO peak is also observed near the major peak of MgB_2 in the case of 2% and 3% Mn concentration. There is no trace of Mn impurity.

The variation of normalized electrical resistivity of $Mg_{1-x}Mn_xB_2$ ($0 \leq x \leq 0.03$) in the temperature range 30 K to 300 K is shown in fig. 2. All the samples show metallic to superconducting transition. The superconducting transition temperature decreases as we increase the Mn concentration. The transition onset temperature of 1% Mn doped sample is 35 K, which is 3 K less than that of pure MgB_2. This decrease in transition

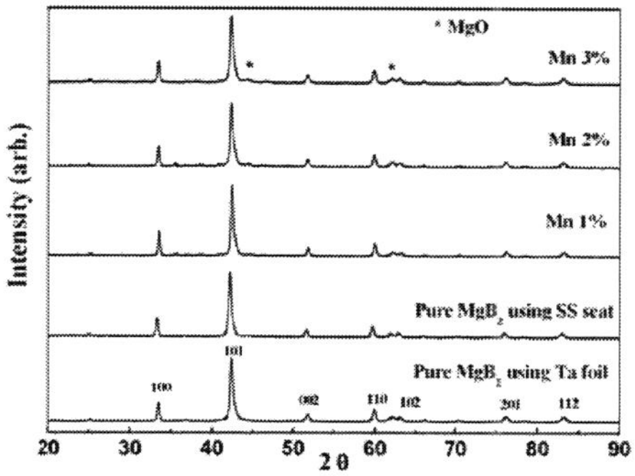

FIGURE 1 XRD patterns of $Mg_{1-x}Mn_xB_2$ $(0 \leq x \leq 0.03)$ superconductors

temperature is due to the substitution of magnetic Mn atoms at Mg site, which breaks the cooper pairing. Although the Mn doping suppresses the transition temperature but does not affects the transition width. The transition width in all the samples is very less

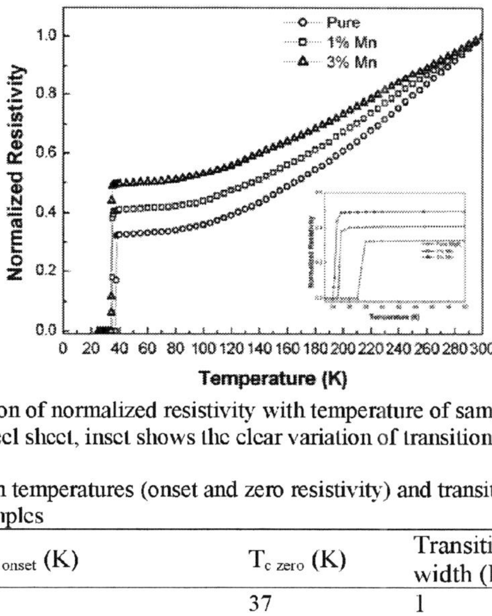

FIGURE 2. The variation of normalized resistivity with temperature of samples prepared by using the stainless steel sheet, inset shows the clear variation of transition temperature.

TABLE 1. Transition temperatures (onset and zero resistivity) and transition width of prepared samples

Sample	$T_{c\ onset}$ (K)	$T_{c\ zero}$ (K)	Transition width (K)
Pure MgB_2	38	37	1
1% Mn	35	34.50	0.50
3% Mn	35	34.10	0.90

(<1 K), which also support the monophasic nature of the prepared samples.

The variation of transition temperature is clearly seen from the inset of fig. 2. The normal state resistivity also increases with Mn doping. The variation of transition temperature with Mn doping is given in Table I.

CONCLUSION

We have synthesized $Mg_{1-x}Mn_xB_2$ ($0 \leq x \leq 0.03$) superconductors through solid state reaction method by using the stainless steel sheet instead of Tantalum foil. On the basis of above discussion, we can conclude that the prepared samples (using stainless steel sheet) show the same properties as we obtained by using the Tantalum foil. Thus, our method of preparation of MgB_2 is economical and simple because the Tantalum foil is very costly as compared to the stainless steel sheet. Although the Mn doping decreases the transition temperature abruptly but no impurity traces of Mn is observed in the MgB_2 matrix. This decrease in transition temperature is due to substitution of magnetic Mn atoms at Mg site, which breaks the cooper pairing.

REFERENCES

1. J. Nagamatsu, N. Nakagawa, T. Muranaka, Y. Zenitani, J Akimistu, *Nature* **410**, pp. 63-64 (2001).
2. S. I. Sclachter, W. Goldacker, A. Frank, B. Ringsdorf, H. Orshulko, *Cryogenics* **46,** pp. 201-207 (2006).
3. K. Vinod, R. G. Abhilash, U. Syamaprasad, *Supercond. Sci. Technol.* **20**, pp. R1-R13 (2007).
4. K. Singh, R. Mohan, N. Kaur, N. K. Gaur, M. Dixit, V. Shelke, R. K. Singh, *Physica C* **450**, pp. 124-128 (2006).
5. N. Kaur, R. Mohan, N. K. Gaur, R. K. Singh, *Physica C* **451**, pp. 24-30 (2007).
6. N. Kaur, R. Mohan, N. K. Gaur, R. K. Singh, *J. Phys. Chem. Solids* **(In press)** (2007).
7. V. P. S. Awana, M. Isobe. K. P. Singh, E. T. Muromachi, H. Kishan, *Supercond. Sci. Technol.* **19**, pp. 551-555 (2006).
8. R. G. Abhilash Kumar, K. Vinod, R. P. Aloysius, U. Syamaprasad, *Mater. Lett.* **60**, pp. 3328-3331 (2006).
9. D. Wang, Y. Ma, Z. Yu, Z. Gao, X. Zhang, K. Watanabe, E. Mossang, *Supercond. Sci. Technol.* **20**, pp. 574-578 (2007).
10. W. Du, H. Xu, H. Zhang, D. Xu, X. Wang, X. Hou, Y. Wu, F. Jiang, L. Qin, *J. Crystal Growth* **289**, pp. 626-629 (2006).

A photoacoustic study on Dimethoxymethane

G.Krishna Bama[a] and K.Ramachandran[b]
[a]Sri Meenakshi Government Arts College for women, Madurai 625002,India.
[b]School of Physics, Madurai Kamaraj University, Madurai, 625021, India.

Abstract. Thermal diffusivity of a highly inflammable liquid dimethoxymethane (DMM) is measured by photoacoustic technique. Thermal conductivity obtained from this technique is 0.11 $Wm^{-1}K^{-1}$ respectively at room temperature and the results are compared with the already reported value and discussed.

Keywords: Dimethoxymethane, photoacoustic, thermal conductivity , thermal diffusivity
PACS: 43.35.-c, 78.20Hp, 65

INTRODUCTION

Alternate clean fuels are the current burning issues, as petroleum resources become depleted and environmental pollution becomes more serious, especially in developing countries. Recently, methylal (CH_3-O-CH_2–O–CH_3) known as dimethoxymethane or DMM was found to be a fuel additive that can greatly improve diesel and petrol combustion properties [1, 2] as it has a large oxygen content (42% by mass); hence it is regarded as a good fuel additive and potential alternative fuel in the future. Recently Pan et al. [3] have measured the value of thermal conductivity of DMM by using transient hot wire technique for the temperatures from 240K to 362 K and this is the first ever result reported for this liquid. But, there are no experimental data for the thermal diffusivity of DMM reported so far by non-destructive technique. Here, photoacoustic method is used for the measurement of thermal diffusivity of DMM, which will be compared with the only reported value of Pan et al [3].

Photo Acoustics (PA)

Among the various photothermal techniques, the photoacoustic (PA) technique offers a powerful tool for determining thermal and optical properties of liquid samples [4,5].The principle of PA technique is that when a modulated light is absorbed by the sample located in a sealed PA cell, the non-radiative decay of the absorbed light produces a modulated transfer of heat to the surface of the sample which produces pressure waves in the gas inside the cell that can be detected by the microphone attached to the cell [6]. The present photoacoustic spectrometer consisted of a 400W Xe- lamp (Jobin Yvon), an electro - mechanical chopper (Model number PAR 650), a monochromator (Model Triax 180, Jobin Yvon), a digital storage oscilloscope (DSO) and a PA cell. The sample (DMM) is placed in the PA cell and the microphone is placed very close to the sample in a specially made PA cell for the liquid.

CP1004, *Thermophysical Properties of Materials and Devices-NCTP '07*,
edited by P. Predeep, S. Prasanth, and A. S. Prasad

Photoacoustic Cell

The PA cell here is unconventional in the sense that it is made up of glass looking like a needle shape with diameter of 1mm and length of about 5cm with one end sealed. The liquid is injected into this needle shaped cell (cuvette) and sealed at the other end.

A glass funnel of length 10 cm, diameter 5 cm is completely blackened inside by carbon black and the wider end is sealed with a concave watch glass. Along the surface of the funnel on the top conical side a small hole of diameter of 1.5 mm is drilled. The sample tube (cuvette) now is inserted into this hole and sealed with paste. A condenser microphone is inserted and placed close to the cuvette. The whole compartment, the PA cell, should be airtight and soundproof. It is known that the amplitude of PA signal will increase when the dimension of the PA cell is decreased, but cannot be decreased indefinitely as unwanted resonances will take place then. In general the amplitude of the PA signal will be very weak compared to solids and so the PA cell was designed carefully. Now the complete set up will be as shown in Fig.1.

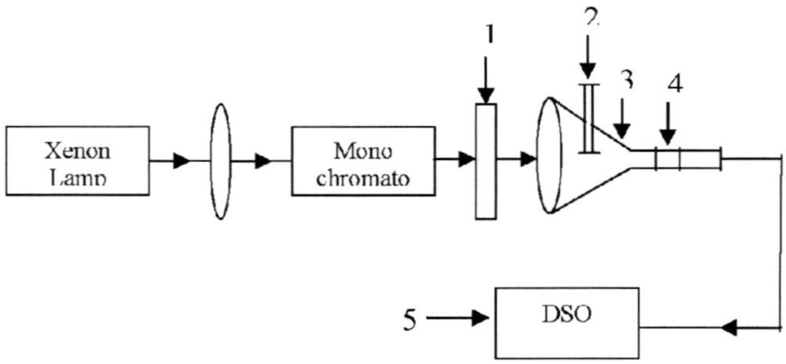

1. Mechanical chopper
2. Glass cuvette in the form of needle.
3. Glass funnel
4. Microphone
5. Digital storage oscilloscope

FIGURE.1. Schematic diagram of PA Setup

Thermal Diffusivity

In general thermal diffusivity is measured by studying the variation in the amplitude of the PA signal for various chopping frequencies for a fixed wavelength of the incident light. The dependence of PA signal on both the thermal and optical properties of the sample makes it a convenient method for the study of materials in different forms.

Thermal diffusion length in any system is a unique property, which means that the thermal energy from the surface can diffuse only to a particular distance in the sample and beyond which thermal energy cannot diffuse. Naturally this becomes a unique property and hence the thermal diffusivity. So, thermal diffusivity measurements were carried out for DMM using the present technique.

Following Rosencwaig's theory of photoacoustics, the intensity at the centre of the xenon lamp beam that is coming out of the PA cell can be expressed by [7]

$$S = \frac{I_t - I_\infty}{I_t} = S_{t=0} \frac{1}{\left(1 + 2n\frac{t}{t_c}\right)^2} \tag{1}$$

$$where \quad S_{t=0} = \frac{I_\infty - I_0}{I_0}$$

Where I_∞, I_0, and I_t are the steady state PA signal, the signal at time t=0, and the signal at time t, respectively. Here n is the number of photons absorbed. The variation of PA signal with chopping frequency for the liquid studied showed a linear dependence, suggesting one photon absorption (n=1).

The PA signal produced by the sample, DMM, was measured for different chopping frequencies. The variation of reduced PA signal $\sqrt{\frac{S_0}{S}}$ (where S_0 - signal at time t = 0 and S - signal at time t) as a function of time is shown in Fig.2. From the slope (of $\sqrt{\frac{S_0}{S}}$, as a function of time), of this graph ,the characteristic time constant (t_c) corresponding to the liquid DMM is found out. To begin with water and acetone are taken as the samples for calibration. The measured values of t_c and thermal diffusivity for water and acetone are in good agreement with those reported in the literature [7].

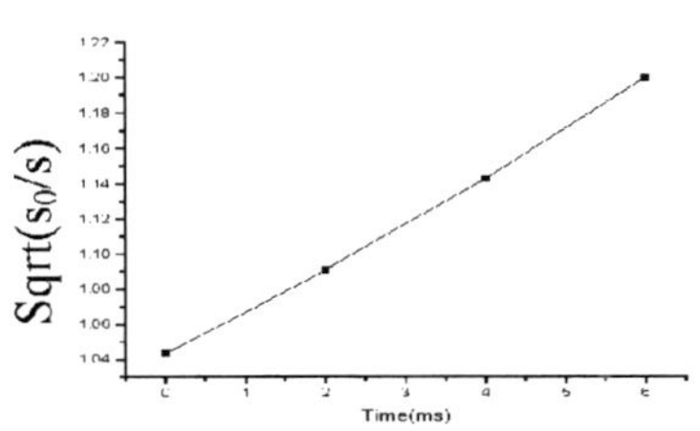

FIGURE 2. Reduced PA signal as a function of time in DMM .

The characteristic time constant t_c is related to the thermal diffusivity (D) and beam radius (ω) through the relation $t_c = \omega^2/4D$ [7]. A reference sample (water) with known thermal diffusivity is used to determine the thermal diffusivity of the unknown sample (to eliminate the uncertainty in the determination of beam radius). Thus,

$$D = \frac{D_{water} t_c^{water}}{t_c}$$

where D and t_c correspond to thermal diffusivity and characteristic time constant of DMM. From the measured value of thermal diffusivities, thermal conductivities can be calculated by using the relation $\alpha = \frac{\lambda}{c_p \rho}$, where λ is the thermal conductivity of DMM and ρ is the density and Cp is the isobaric specific heat capacity of DMM. These values of thermal conductivities are compared with the reported values of Pan etal [3].

RESULTS AND DISCUSSION

For the first time, photoacoustics measurements are reported here for the highly inflammable liquid DMM. The observed values of time constant (t_c) and thermal diffusivity (D) of liquid DMM are given in Table.1. Since photoacoustics (PA) is being a non destructive method, the results on thermal properties here will be more reliable than hot wire technique. Apart from that the present results are used to correlate the results of Pan et al. [3].Thermal conductivity at room temperature, from PA measurement is 0.11 $Wm^{-1}K^{-1}$ whereas from hot wire technique (Pan et al.) this is 0.14 $Wm^{-1}K^{-1}$. Since Pan et al [3] kept the saturated vapour pressure throughout the measurements; there is a slight increase in the thermal conductivity. The PA measurements can give only the thermal diffusivity and thermal conductivity is deduced from this using specific heat which is taken from literature.

The repeatability of the present PA measurements was verified and the error in thermal diffusivity is around 3 % and the accuracy claimed by Pan et al [3] is also reasonable and within these limits, the present PA results could explain the results of Pan et al. [3].

TABLE 1. The measured t_c ,D and λ for DMM at room temperature for various chopping frequencies using the PA technique.

Chopping frequency Hz	t_c in ms	D in 10^{-8} m^2s^{-1}	λ in $Wm^{-1}K^{-1}$	λ in $Wm^{-1}K^{-1}$
17	167	3.77	0.10	
18	164	3.84	0.11	0.14 [3]
20	163	3.86	0.11	
22	160	3.93	0.11	

(Error in measurements < 3%)

CONCLUSION

In conclusion, the photoacoustic method is effectively used to measure the thermal conductivity in this additive liquid. The measurements are comparable with that of hot wire technique [3]. Here the advantage is that the sample will not be contacted at any instant. So, minimum error is expected. It is concluded that such risky liquids can be well studied by photoacoustics.

ACKNOWLEDGEMENTS

One of the authors (G.Krishna Bama) acknowledges University Grants Commission, New Delhi for providing a fellowship under FIP scheme to carry out this work in the School of Physics, Madurai Kamaraj University, Madurai.

REFERENCES

1. J.C.Ball, C.Lapin,J.Buckingham,E. Frame,D.Yost,M.Gonzalez, E.Liney,M.Natarajan, and J.P.Wallace, SAE Paper No.2001-01-3627.
2. J.C.Ball, C.Lapin,J.Buckingham,E. Frame,D.Yost,M.Gonzalez, E. Liney,M.Natarajan, and J.P.Wallace, SAE Paper No.2001-01-3628.
3. J.Pan,J.T.Wu,Z.G. Liu and X.G.Jin, *Int.J.Thermophys.*, **25**, 701-708, (2004).
4. J.A.P.Lima,E.Marin,S.L.Carodoso,O.DelgadoVasallo,M.G.daSilva,M.S.Sthel, C.E.N.Gatts,A.Mariano,C.E.Rezende,A.R.C.Ovalle,M.S.Suzuki,and H.Vargas., *Analytical Sciences.*, **17**:,530-533 ,(2001).
5. Yaochun shen, Lu Zuhong,Stephen Spiers, Hugh A.Mackenzie,Helen S.Ashton,John Hannigan, ScottS. Freeborn, and John Lindberg,*Appl. Opt.* **39**, 4007-4011, (2000).
6. A.Rosencwaig, Photoacoustics and photoacoustic spectroscopy, New York. Wiley. 1980.
7. Bindhu C V, Harilal S, Nampoori VPN, Vallabhan CPG, *Opt. Eng.* 37(10),2791-94 ,(1998)

Analytical and Dielectric Studies of the Ferroelectric $Ba_5Ti_2O_7Cl_4$

N. M. Patil [a], N. V. Pradnyakar [b] and S. H. Shamkuwar [b]

[a]Lecturer & Head, Department of Applied Physics, Laxminarayan Institute of Technology, Rashtrasant Tukadoji Maharaj Nagpur University, Nagpur-440 033 (M.S.), India
Email: naresh_patil2006@yahoo.co.in
[b]Research Scholar, Department of Applied Physics, Laxminarayan Institute of Technology, Rashtrasant Tukadoji Maharaj Nagpur University, Nagpur-440 033 (M.S.), India

Abstract. The thermo analytic analyses are necessary for general characterization of the materials, i.e. to know if any important changes are occurring in the material. The important changes are basically of two types: 1) changes occurring as a result of chemical reaction, and 2) hanges occurring as a result of some physical change, e.g. change of crystal structure, change of some atomic groups within the given crystal structure, some physical change involving mission or absorption of thermal energy, as for example, change of electron structure or electronic interactions. The changes involving chemical reactions usually involve variation in mass content of the material system, and these changes are studied by thermo gravimetric analysis (TGA). The changes of the second type are studied by differential thermal analysis (DTA) and / or by differential scanning calorimetry (DSC). The present work reports the studies of DTA and DSC, while TGA studies showed no significant changes.

Keywords: DTA, DSC techniques.
PACS: 81.70.Pg, 65.40.b

INTRODUCTION

The thermal analysis has wide range of properties that may be studied as a function of temperature. These techniques are use a) to measure property of interest, such as specific heat, latent heat or heat of reaction, b) to assess the high-temperature stability of materials, c) to detect phase transformations, d) fundamental studies of reaction kinetics. The thermoanalytic studies are necessary for general characterization of the materials, i.e. to know if any important changes are occurring in the material. The important changes are basically of two types: 1) changes occurring as a result of chemical reaction, and 2) changes occurring as a result of some physical change. The changes involving chemical reactions usually involve variation in mass content of the material system, and these changes are studied by thermogravimetric analysis (TGA). The physical changes involving emission or absorption of thermal energy are studied by differential thermal analysis (DTA)

CP1004, *Thermophysical Properties of Materials and Devices-NCTP '07*,
edited by P. Predeep, S. Prasanth, and A. S. Prasad

and / or by differential scanning calorimetry (DSC). Both Differential Thermal Analysis (DTA) and Differential Scanning Calorimetry (DSC) are concerned with the measurement of energy changes in materials. Thus, they are most generally applicable of all thermal analysis methods, since every physical or chemical change involves a change in energy. The present work reports the studies of DTA and DSC, while TGA studies showed no significant changes.

ANALYTICAL STUDIES BY

a) DTA Technique

DTA involves heating or cooling a sample and an inert reference under identical conditions, and the temperature difference between the sample and reference is noted and recorded as a voltage signal. Changes in the sample which lead to the absorption or evolution of heat can be detected relative to the inert reference. The physical change may absorb the thermal energy (endothermic) or evolve thermal energy (exothermic). Correspondingly, the signal may be negative or positive. A significant physical change is indicated as a valley in the continuously recorded signal. The signal recorded in the form of a graph with temperature. The DTA curve recorded in a sample of $Ba_5Ti_2O_7Cl_4$ is shown in Fig.1.

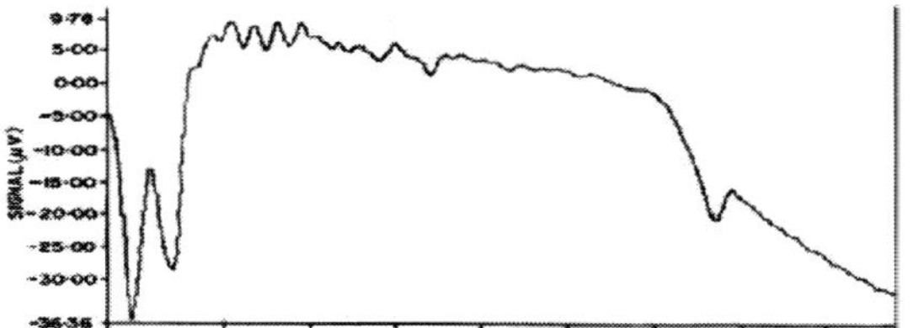

FIGURE 1. The DTA (differential thermal analysis) curve for the crystal.

The DTA curve can be used as a finger print for identification purposes. There are two dips or valleys found at temperatures of 90 and 150^{0}C which indicate absorption of energy and shows endothermic property. It is endothermic up to 200^{0}C, while above 200^{0}C, the behavior is exothermic with some small endothermic changes occurring as the temperature is increased up to 500^{0}C. After 500^{0}C the crystal indicates endothermic behavior again with a valley at about 775^{0}C. The physical changes occurred at 90, 150 and 775^{0}C are also reflected in the measurement of the dielectric constant measured as a function of temperature. A typical graph of dielectric constant versus temperature is shown in Fig.2.

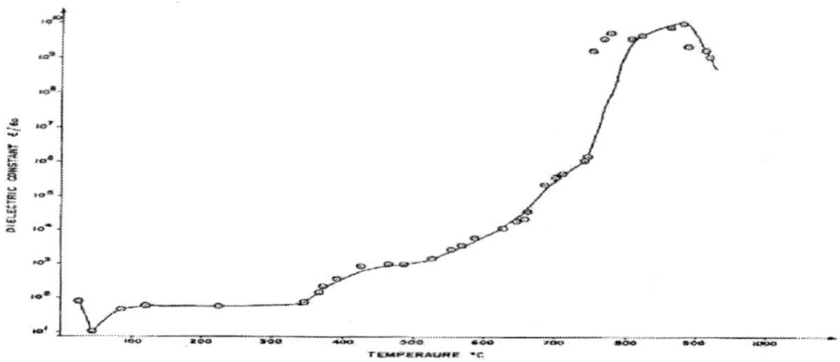

FIGURE 2. The variation of dielectric constant with temperature.

The measurement of the dielectric constant on the small samples formed by cutting and mechanical polishing showed that the dielectric constant at room temperature was of the order of 200. There is a little variation in dielectric constant values from sample to sample observed at room temperature but in every case, the dielectric constant values increase with increasing temperature, with a sudden change (spurt) at the Curie temperature, and then there is a decrease in the dielectric constant (Fig.2). The temperature at which the decrease in dielectric constant starts is the Curie temperature. The dielectric constant at this temperature lies in the range of 10^{10} to 10^{20} which shows that the crystal belongs to the category of induced dipole type ferroelectrics. It is also seen from the graph in Fig. 2 that two dips are observed on the curve. One dip occurs at about 90^0, and the other near 800^0C. Clearly, these two dips correspond to DTA dips at 90^0 and 775^0C. The dip at 800^0C is very easy to understand as it corresponds to ferroelectric transition, the fall indicating Curie-Weiss behavior. Thus, the DTA dip near 775-800^0C indicates a crystal structure change. The exact temperature of change changes from sample to sample. The reason for the valley at 90^0C, though also indicated by the dielectric constant curve, is not still not obvious. Hence, the spontaneous polarization measurements as a function of temperature were obtained through hysteresis (P-E) curves shown in the Fig. 3.

It is seen that at around 90^0C the Ps value is falling rapidly. It is known that the crystal has two ferroelectric mechanisms [1], and the fall in spontaneous polarization could be due to disappearance of one of the ferroelectric mechanism. The unit cell of $Ba_5Ti_2O_7Cl_4$ is shown in Fig. 4 for ready reference. This was confirmed by observing the domain structure as a function of temperature. It is known from our earlier work on domains in this crystal [2] that the crystal contains 180^0 and 90^0 domains associated with Cl^{1-}, and 79^0 domains associated with Ti^{4+}. It was found that 79^0 domains were removed from the structure at/near 90^0C. Thus, it was confirmed that the endothermic reaction indicated by the DTA at 90^0C corresponded to the disappearance of 79^0 domain structure at this temperature.

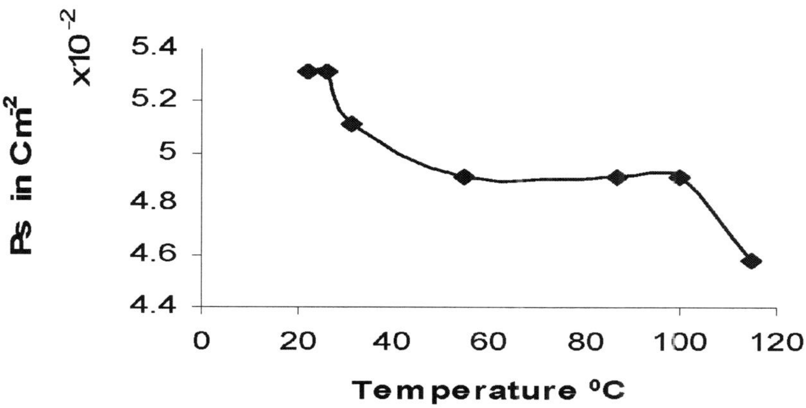

FIGURE 3. The variation of Ps with temperature.

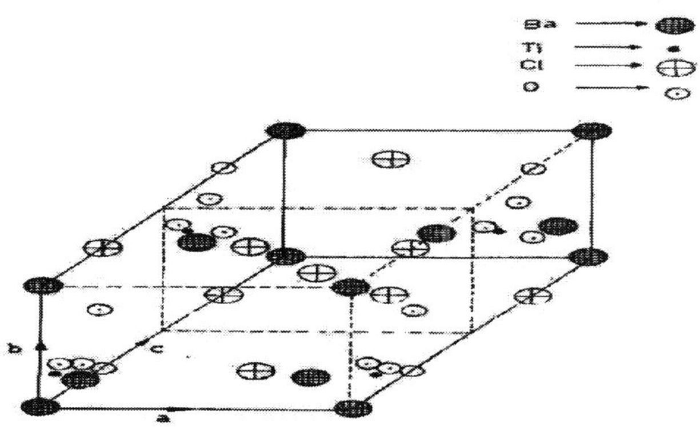

FIGURE 4. Unit cell of $Ba_5Ti_2O_7Cl_4$.

The disappearance of the domain structure clearly corresponds to the change in the position of the Ti^{4+} in the structure, i.e. a crystal structural change. This change of Ti^{4+} must cause a change in the neighboring oxygen ions, and hence the valley at 130^0C indicates such a structural change. This way, all the three important valleys shown by

DTA have been analyzed. The remaining valleys occurring over a temperature range 200 to 500^{0}C correspond to rearrangement of microdomain structures. It is known that the domain structures are formed around point defects, and the ordering of these point defects goes on changing with temperature. Thus, a significant change in the ordering of point defects leads to a significant change in the domain structure, and this physical process is indicated by a small dip or valley on the thermogram.

The general endothermic behavior after 500^{0}C shows gradually higher degree of ordering of the microdomains, when finally phase transition occurs at about 800^{0}C. Thus, making use of auxiliary physical measurements, the DTA gives comprehensive ideas of the changes that occur in the materials over a temperature range. The DTA data shown in Fig.1 was obtained at Regional Sophisticated Instrumentation Centre [RSIC] of R. T. M. Nagpur University, Nagpur (India). The sample was heated at the rate of 15^{0}C per minute. The mass of the sample was 38.8 mg.

b) DSC Technique

Differential scanning calorimetry (DSC) is a technique for measuring the energy necessary to establish a nearly zero temperature difference between a sample and an inert reference materials, as the two specimens are subjected to identical temperature regimes in an environment heated or cooled at a controlled rate. In these studies, the experimental sample and the reference sample are simultaneously heated in a furnace, and the difference in the power required to keep the two samples at the same temperature is noted as function of temperature. If the experimental sample is requiring more power than the reference sample to be maintained at a particular temperature, it means that the experimental sample is absorbing energy at that temperature. The excess energy required is shown in mW and thermogram is a excess energy in mW Vs temperature. A typical DSC curve is shown in Fig. 5. The heating rate was 10^{0}C per min. when the experimental sample weighed 3.875 mg.

It is seen that the thermogram shows two valleys, one at 80^{0}C and the other at 130^{0}C. The valley at 80^{0}C clearly corresponds to the valley in the DTA diagram obtained at 90^{0}C, and the valley at 130^{0}C corresponds to the valley at 150^{0}C. Thus, there is correspondence in valleys but with change of temperature. Since the changes are associated with shifts in ionic positions, it is obvious that the two types of thermograms must be referring to same changes. Thus, the temperature of ionic shift changes somewhat from specimen to specimen on account of local conditions. In some ferroelectric crystals such shift of temperature can be quite large [3]. Similar shifts of temperature are also seen to occur in phase transition temperature from sample to sample [4]. It is seen that both the type of ionic shifts are associated with latent heat as indicated by power absorption in the DSC curve. Thus, while DTA curve is more sensitive to specific heat changes, the DSC curve is more sensitive to latent heat. A comparative study between two types of curves is often useful.

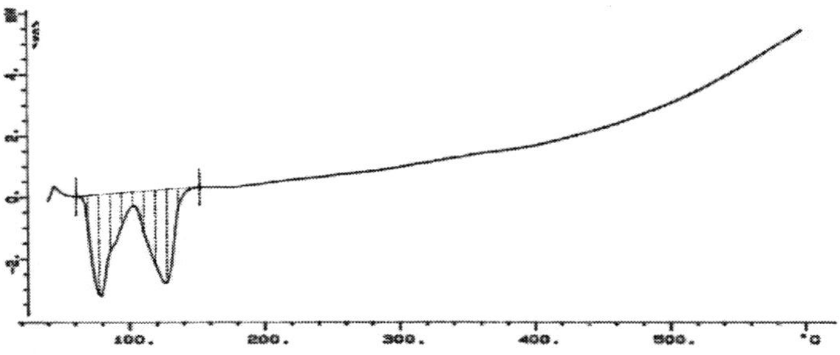

FIGURE 5. The DSC (differential scanning calorimetry) curve for the crystal.

CONCLUSION

The thermo analysis of the crystal indicates two structural changes, one at around 80-90^{0}C, and the other at 130-150^{0}C , a ferroelectric transition at about 800^{0}C, microdomain behavior in the temperature range 200 to 500^{0}C , and the vacancy generation with increased temperature. The technique is highly powerful when judiciously used in conjunction with other physical measurements that can corroborate the results obtained by thermograms (DTA and DSC).

REFERENCES

1. S. G.Ingle and N. M. Patil, Jpn. J. Appl. Phys. **39** , 2670, (2000).
2. S. G. Ingle and N. M. Patil, Ferroelectrics **274** , 143, (2002).
3. Baletskii, D.Yu., Bareha, D. M., Koperles, B. M., Turyanitsa, I. D., Fiz. Tverd. Tela **16,** 278 (1974)
4. Teng , M. K. , Massot , M. , Balkanski , M., and Ziolkiewicz , S., Phys. Rev. B. **17**, 3695 (1978).

Optical Constants of Cadmium Telluride Thin Film

P.Nithyakalyani*, M.Pandiaraman*, P.Pannir*, C.Sanjeeviraja+,

N.Soundararajan*

* *School of Physics, Madurai Kamaraj University, Madurai-625 021.*
+ *Department of Physics, Alagappa University, Karaikudi-630 003.*

Abstract: Cadmium Telluride (CdTe) is II-VI direct band gap semiconductor compound with potential application in Solar Energy conversion process. CdTe thin film of thickness 220 nm was prepared by thermal evaporation technique at a high vacuum better than 10^{-5} m.bar on well cleaned glass substrates of dimensions (1cm x 3cm). The transmittance spectrum and the reflectance spectrum of the prepared CdTe thin film was recorded using UV-Vis Spectrophotometer in the wavelength range between 300 nm and 900 nm. These spectral data were analyzed and the optical band and optical constants of CdTe Thin film have been determined by adopting suitable relations. The optical band gap of CdTe thin film is found to be 1.56 eV and this value is also agreeing with the published works of CdTe thin film prepared by various techniques. The absorption coefficient (α) has been higher than 10^6 cm^{-1}. The Refractive index (n) and the Extinction Coefficient (k) are found to be varying from 3.0 to 4.0 and 0.1 Cm^{-1} to 0.5 Cm^{-1} respectively by varying the energy from 1.0eV to 4.0 eV. These results are also compared with the literature.

Keywords: CdTe, thin films, reflectance, transmittance, optical contants
PACS: 78.40. -q, 78.40.Fy

INDRODUCTION

Cadmium Telluride is an important material in semiconductor technology due to increasing applications in various electronic devices, such as room temperature X-ray and gamma ray detectors, electro-optical and acousto-optical modulators, partial reflector, high efficient solar cells etc [1]. CdTe is most promising polycrystalline thin film material for producing photovoltaic solar cells because of its high absorption coefficient (> 10^4 cm^{-1}) and optimum band gap (1.5 eV) [2]. CdTe exhibit cubic structure with a = 0.637 nm. CdTe can also exist in the hexagonal form like other II-VI semiconductors but this phase is metastable [3]. From the electrical studies on CdTe, the activation energy of bulk CdTe is about 0.64eV [1]. CdTe exhibits piezoelectric effect [4]. From the electronic band structure studies by Gundel et al [5] have determined the band gap of CdTe is to be 2 eV. The as-deposited CdTe layer exhibits p-type conduction [6]. Franc et al have studied the temperature dependence of galvanomagnetic effects of CdTe about its melting point. They have observed that liquid CdTe is also semiconducting [7]. Photoluminescence studies on CdTe shows the maximum of the excitonic peak is localized at 1.57 eV [8]. The optical band gap of the sintered CdTe film is found to be 1.47eV [9]. The reflection and transmission have made it possible to determine the optical constants conveniently [10]. The present

CP1004, *Thermophysical Properties of Materials and Devices-NCTP '07*,
edited by P. Predeep, S. Prasanth, and A. S. Prasad

work is the determination of the optical band gap, Eg, of CdTe thin film and its optical constants like refractive index (n), extinction coefficient (k), absorption coefficient (α) by analyzing the transmittance and reflectance spectra.

EXPERIMENTAL

CdTe films of thickness 220 nm were prepared by thermal evaporation technique in a vacuum better than 10^{-5} mbar on well-cleaned glass substrates of dimension (1cm x 3cm) kept at 303 K at the deposition rate of 0.2nm/sec. A quartz crystal thickness monitor was used to monitor the film thickness and the deposition rate during the film preparation. The transmittance and reflection spectra of these samples were recorded in the wavelength range between 300nm and 800nm using UV-Vis single beam spectrophotometer.

RESULT AND DISCUSSION

From the powder X-ray diffraction, it is confirmed that CdTe has cubic structure with (111) orientation. The transmittance spectrum of the CdTe film of thickness 220 nm is shown in fig (1).

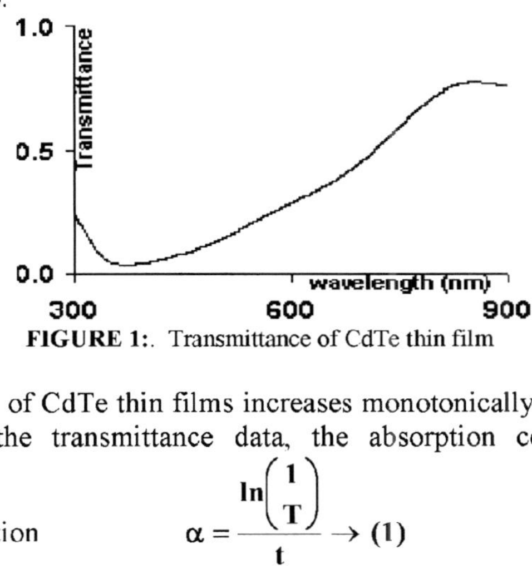

FIGURE 1:. Transmittance of CdTe thin film

The transmittance of CdTe thin films increases monotonically with the increase in wavelength. Using the transmittance data, the absorption coefficient has been calculated by the relation

$$\alpha = \frac{\ln\left(\frac{1}{T}\right)}{t} \rightarrow (1)$$

The spectral dependence of the absorption coefficient is shown in fig (2). The absorption is very strong above 1.5 eV. The value of the absorption coefficient is higher than 10^5 cm^{-1}, which confirms the direct transition in CdTe thin films.

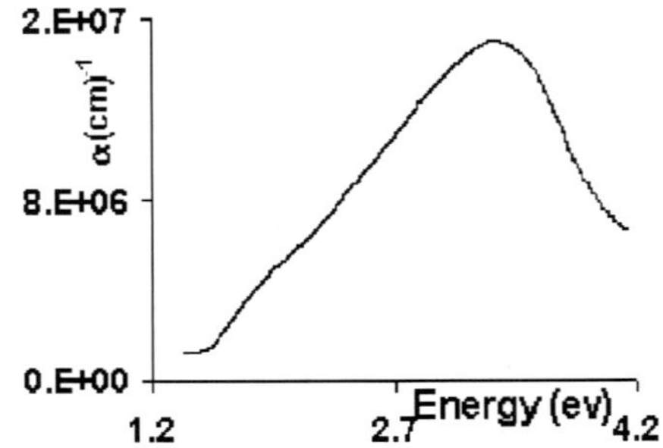

FIGURE 2:. Dependence of absorption coefficient (α) on Energy (hν)

The energy band gap of CdTe thin films is determined using the relation [11].

$$\alpha h\nu = A(h\nu - E_g)^n \rightarrow (2)$$

In this relation, the value of n depends on the type of transition. For direct transition, n takes the value ½. For indirect transition, n is equal to 2.The optical band gap of CdTe thin films is determined by carrying out a linear fitting on the plot $(\alpha h\nu)^2$ against the energy (hν) [10,12,13]. The optical band gap of the CdTe thin film is 1.56 eV. This optical band gap of CdTe thin film is in agreement with the value reported earlier by several authors [10,12,13].

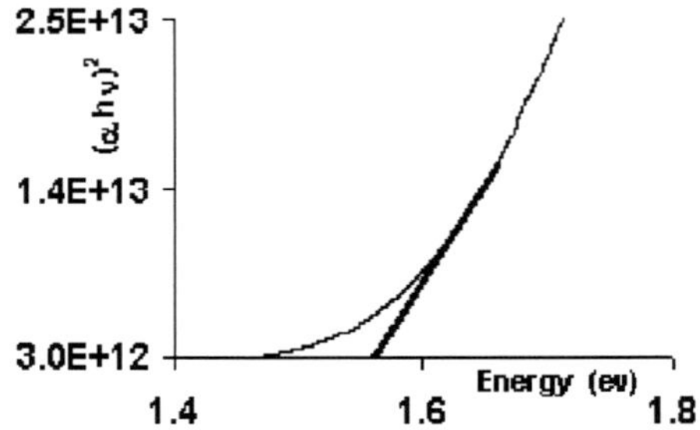

FIGURE 3 Dependence of $(\alpha h\nu)^2$ on the Energy of CdTe thin film of thickness 220 nm

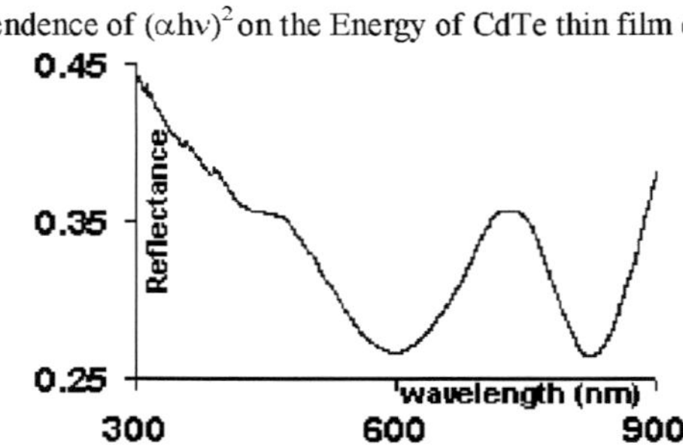

FIGURE 4. Reflectance of 220 nm CdTe thin film

Using the spectral dependence of the absorption coefficient, the extinction coefficient of CdTe thin film had been determined using the relation (3). [14]

$$k = \left(\frac{\alpha\lambda}{4\pi}\right) \rightarrow (3)$$

The energy dependence of the extinction coefficient (k) is shown in fig (5).

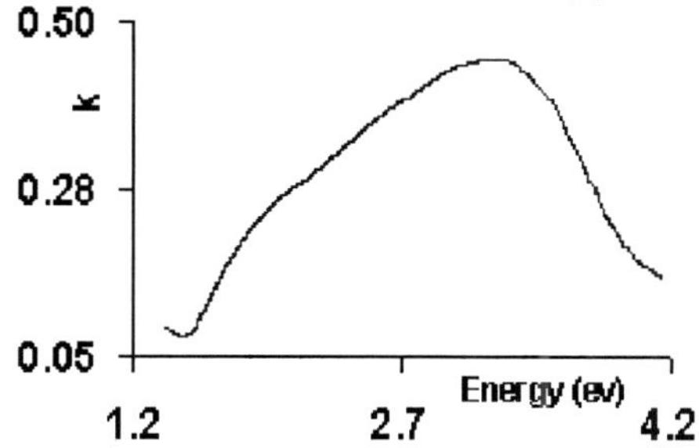

FIGURE 5:. Dependence of Extinction coefficient (k) on Energy (hν)

The extinction coefficient shows a maximum at 3.26 eV. The extinction coefficient is increasing with incident photon energy [14].
Using the transmittance and reflectance spectrum the refractive index of the CdTe thin film is determined by the relation

$$\mathbf{R = \frac{(n-1)+k^2}{(n+1)+k^2} \rightarrow (4)}$$

The refractive index of the CdTe thin film is determined and plotted against the energy (eV) in the fig (6).

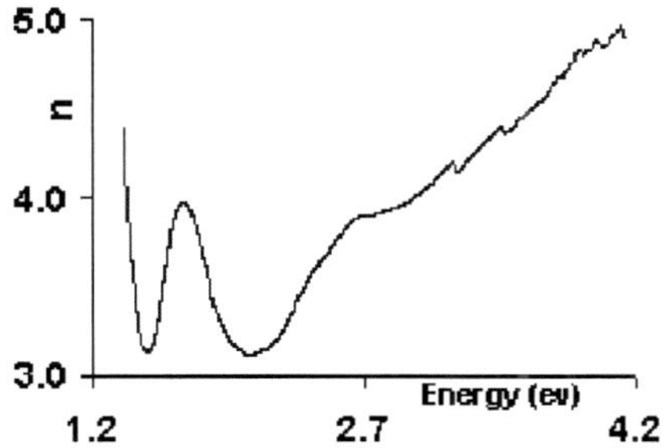

FIGURE 6: Dependence of Refractive index (n) of 220 nm CdTe thin film on Energy (hν)

The energy dependence of refractive index shown in fig (6) is similar to the one observed by Lalitha et al [10]. The refractive index is very low at the fundamental band gap region. The variation may be due to the surface effect and volume imperfection on the micro scale.

CONCLUSION

The CdTe thin film of 220 nm was prepared and optically studied by UV-Vis single beam spectrophotometer. The transmittance and reflectance spectra of the CdTe thin film were recorded and analyzed. The absorption coefficient of the film is found to be greater than 10^5 cm^{-1} conforming direct band gap. The optical band gap of CdTe thin film was determined to be 1.56 eV. Some of the optical constants like extinction coefficient and refractive index of CdTe thin film are determined and their spectra's are analyzed. The extinction coefficient decreases with the increase in the incident photon wavelength. The refractive index and extinction coefficients are oscillatory in nature. This may be due to multiple reflectances on the two surfaces of films.

ACKNOWLEDGMENTS

We kindly Acknowledge Mr.K.Neyvasagam, senior scholar in our lab, for his kind help during the preparation of CdTe Thin films and Mr.Deepan, research scholar, Department of Physics, Alagappa University, for his help in recording the optical spectrums of CdTe thin films.

REFERENCES

1. A.Nasar, M.Shamsuddin, *Journal Material Science* **35** 1465-1468 (2000)
2. Wagner Anacleto Pinheiro, Vivienne Densie Falco, Leila Rosa dde Olivera cuz, Carlos Luiz Ferreira, *Materials Research* **9** 47-49 (2006)
3. Pierre Lefebvre, Tristan Richard, Jacqics Allegre and Henry Maticu *Physical Review B* **53** 15440-15442 (1996)
4. Andrea Dal Corso,, Raffaele Resta and Stefano Baroni, *Physical Review B* **47** 16252-16256 (1993)
5. S. Gundel, A. Fleszar, W.Faschinger, W.Hanke, *Physical Review B* **59** 15261- 15269 (1999)
6. Kentaro Arai, Souichi Hagiwara, Kuniaki Murase, Tetsuji Hirato, and Yasuhiro Awakura, *J. Electrochem Soc.*, **152** 588-592 (2005)
7. Franc J.; Höschl P.; Grill R.; Turjanska L.; Belas E.; Moravec P. *Journal of Electronic Materials*, **30**, No. 6, 939-943 (2001)
8. I.Caraman, S.Vatavu, G.Rusu, P. Gasin, *Chalcogenide Letters* **3** 1-7 (2006)
9. Seema Sirohi, Vipin Kumar, T.P.Sharma, Optical Materials **12** 121-125 (1999)
10. S.Lalitha,S.Zh.Karazhanov, P.Ravindran, Physica B, *Condensed matter*, **38** 227-238 (2007).
11. X.Mathew, *Journal of Material Science Letters*, **21** 529-531 (2002).
12. R.Chakrabarti, S.Ghosh, S.Chaudhuri and A.K.Pal, *J.Phys.D. Appl Phys* **32** 1258-1262 (1999)
13. N.El.Hardy, M.F.Ahmed. K.Abdul Hady, *Thin solid films* **274** 120-127 (1996).
14. "Thin Film Fundamentals" A.Goswami, New age international (p) Ltd. 1976

Structural and Elastic Properties of Strontium Selenide under High Pressure

Purvee Bhardwaj[a] Sadhna Singh[b] and N.K. Gaur[b]

[a] Centre for Basic Sciences M. P. Bhoj (Open) University, Bhopal-462016, India.

[b] *Department of Physics Barkatullah University, Bhopal-462026, India.*

Abstract. In the present paper we have investigated the high-pressure structural phase transition of strontium chalcogenides using the modified three-body potential (MTBP) incorporated by covalency effect. The phase transition pressures and associated volume collapses obtained from TBP show a reasonably good agreement with experimental data. It is observerd that Further more, the elastic constants and their combinations as a function of with pressure are also reported. It is found that MTBP incorporated by covalency effect has a promise to predict the phase transition pressure, the elastic constants and their pressure derivatives of other chalcogenides as well.

Keywords: Chalcogenides, phase transition, three-body potential, Gibbs free energy and volume collapse.
PACS: 64.70.Kb.,62.20.Dc., 62.50.+p

purveebhardwaj@gmail.com

INTRODUCTION

High pressure research on structural phase transformations and behavior of materials under compression based on their calculations or measurements have become the subject of increasing past few years as it provides insight into the nature of the solid state theories, and determine the values of fundamental parameters. An example can be found in the III-V semi conducting compounds, where the structural optical and electrical properties at high pressure have been extensively studied [1]. The alkaline earth chalcogenides (AX: A= Ca, Sr, Ba; X= O, S, Se, Te) form a very important closed shell ionic system crystallizing in NaCl-type structure at room conditions. Among these compounds strontium selenide is found to exhibit a structural phase transition even with the application of moderate pressure 14 GPa. These compounds under normal condition crystallize in rock salt structure and show a different phase at high pressures. The new phase often shows different physical and chemical properties. At elevated pressure these compounds undergo pressure induced structural phase transition from six fold coordinated NaCl to more closed eight fold coordinated CsCl structure [2-6]. Recently, many efforts were made to interpret the experimental results by using a variety of theoretical models [7-8].

CP1004, *Thermophysical Properties of Materials and Devices-NCTP '07*,
edited by P. Predeep, S. Prasanth, and A. S. Prasad

We have studied phase transition phenomena in ionic compounds using modified three-body potential model (MTBP), which includes long range Coulombic, three body interactions, vander Waals and short-range overlap repulsive interactions, and covency effect. Inclusion of covalency effects makes the present approach realistic and the present model takes care of the covalent interaction. To fulfill this need, we have included the effect of covalency in three-body potential on the lines of Motida [9] and applied it to study phase transition and elastic properties of strontium selenide.

Interaction Potential and Method of Calculation

Application of pressure directly results in compression leading to the increased charge transfer (or three body interaction effect [10]) due to the deformation of the overlapping electron shell of the adjacent ions (or non-rigidity of ions) in solids. These effects have been incorporated in the Gibbs free energy (G = U+PV-TS) as a function of pressure and three body interactions (TBI), which are the most dominant among the many body interactions. Here, U is the internal energy of the system equivalent to the lattice energy at temperature near zero and S is the entropy. At temperature T=0K and pressure (P) the Gibbs free energies for rock salt (B1, real) and CsCl (B2, hypothetical) structures are given by:

$$G_{B1}(r) = U_{B1}(r) + PV_{B1}(r) \qquad (1)$$

$$G_{B2}(r') = U_{B2}(r') + PV_{B2}(r') \qquad (2)$$

With V_{B1} (=2.00r^3) and V_{B2} (=1.54r'^3) as unit cell volumes for B_1 and B_2 phases respectively. The first terms in (1) and (2) are lattice energies for B_1 and B_2 structures and they are expressed as:

$$U_{B_1}(r) = \frac{-\alpha_m z^2 e^2}{r} - \frac{(12\alpha_m z e^2 f_m(r))}{r} - [\frac{C}{r^6} + \frac{D}{r^8}] + 6b\beta_{ij}\exp[(r_i + r_j - r)/\rho] \qquad (3)$$
$$+ 6b\beta_{ii}\exp[(2r_i - 1.414r)/\rho] + 6b\beta_{jj}\exp[(2r_j - 1.414r)/\rho]$$

$$U_{B2}(r') = \frac{-\alpha'_m z^2 e^2}{r'} - \frac{(16\alpha'_m z e^2 f_m(r'))}{r'} - [\frac{C'}{r'^6} + \frac{D'}{r'^8}] + 8b\beta_{ij}\exp[(r_i + r_j - r')/\rho] \qquad (4)$$
$$+ 3b\beta_{ii}\exp[(2r_i - 1.154r')/\rho] + 3b\beta_{jj}\exp[(2r_j - 1.154r')/\rho]$$

With α_m and α'_m as the Madelung constants for NaCl and CsCl structure respectively. C (C') and D (D') are the overall vander Waal coefficients of B1 (B2) phases, β_{ij} (i,j=1,2) are the Pauling coefficients. Ze is the ionic charge and b (ρ) are the range (hardness) parameters, r(r') are the nearest neighbour separations for NaCl

(CsCl) structure $f_m(r)$ is the modified three body force parameter which includes the covalency effect with three body interaction r_i (r_j) are the ionic radii of ions i (j).

The expression of elastic constants, vander Waals coefficients and pressure derivatives are directly used from ref [11-12].The values of A_i, B_i, and C_i have been evaluated from the knowledge of b and ρ and the vdw coefficients. We have added the effect of covalency in three-body interaction on the lines of Motida [9]. Now modified three body parameter $f_m(r)$ becomes

$$f_m(r) = f_{TBI}(r) + f_{cov}(r)$$

$$f_{cov}(r) = \frac{4V^2_{sp\sigma}e^2}{rE_g^3}$$

$$\frac{V^2_{sp\sigma}}{E_g^2} = \frac{1-e^*_s}{12},\ E_g = E - I\frac{(2\alpha-1)e^2}{r}$$

With $V_{sp\sigma}$ is the transfer matrix between the outermost P orbital of anion and lowest excited state of cation E_g is the transfer energy of electron from anion to cation. Denoting the static and optical dielectric constant ε_0 and ε respectively and the transverse optical phonon frequency at zone centre by ω_t, e_s represented as

$$(e^*_s)^2 = \frac{9\mu\omega_t^2(\varepsilon_0-\varepsilon_\infty)}{4\pi N(\varepsilon_\infty+2)^2}$$

and (5)

$$\frac{(e^*_s)^2}{(e)^2} = \frac{9v\mu\omega_0^2(\varepsilon_0-\varepsilon_\infty)}{4\pi e^2(\varepsilon_\infty+2)^2}$$

Where ν denotes the unit cell volume: $2r^3$, r equilibrium value of the separation of the nearest neighbouring ions, ε_0 the static dielectric constants, μ reduced mass of the ions, and ω_0 the infrared dispersion frequency.
The values of derivatives of $f_m(r)$ obtained as

$$f_m(r) = f_0 e^{-r/\rho} + f_{cov}(r)$$

RESULT AND DISCUSSIONS

The Modified Three Body potential for NaCl (B_1) and CsCl (B_2) structures contain three model parameters [b ρ, $f_m(r)$], namely range, hardness and modified three-body interaction parameter. These parameters are calculated using the lattice energy and equilibrium condition. The input crystal data and model parameters for strontium chalcogenides. are given in Table 1.Using the measured values of the equilibrium lattice constant (r_0) and cohesive energy for B_1 phase the model parameters thus calculated are also given in Table 1.

We have followed the technique of minimization of U_{B1} (r) and U_{B2} (r') at different pressures in order to obtain inter ionic separation r and r' for B_1 and B_2 phases respectively associated with minimum energies.

TABLE 1. Input data and model parameters of SrSe.

Input data and Model parameters	Strontium Selenide
r_i (Å)	1.27[a]
r_j (Å)	1.91[a]
r (Å)	3.115[a]
U (KJ/mol)	2736[a]
b (10^{-12} ergs)	0.411
ρ (10^{-8})	0.155
f (r)	-0.627
ε_0	4.9[b]
ε_∞	4.33[b]
ω_t (10^{13} S^{-1})	4.68[b]

a-ref [16], b- ref [17].

TABLE 2. A Phase transition Pressure and volumecollapse of SrSe.

Crystal	Phase transition (GPa)	Volume Collapse
SrSe(Present)	13.7	9.86%
(Others)	14.5a	-
(Expt.)	14b	10.7%

a-ref [18], b-ref [15]

We have evaluated the corresponding G_{B1}(r) and G_{B2} (r') and their respective differences $\Delta G = G_{B1}(r) - G_{B2}(r')$ which we have plotted with pressure (P) as shown in Fig. 1(a) for SrSe.

TABLE 3. elastic constants and pressure derivatives of elastic modulii of SrSe

Strontium Selenide	Present	Exp[t.
C_{11}(GPa)	99.8	97.2-13.1a
C_{12}(GPa)	13.8	15.6-17.6a
C_{44}(GPa	34.3	28.6-30.4[a]
dk/dp	6.02	-
ds/dp	4.65	-
dC_{44}/dp	-0.46	-

a-ref [18]

The phase transition pressure (P_t) is the pressure at which ΔG approaches zero. Fig. 1 (a) and Table 2 shows that the present phase transition from NaCl-type [B_1] to CsCl-type [B_2] structures in SrSe at 13.7GPa. Relative volume changes $V_{(p)}/V_{(0)}$ have been plotted with pressure and depicted in Fig. 1(b) SrSe. It is clear from Table 3 and Fig: 1(b) that our calculated volume collapses $-\Delta V_{(p)}/V_{(0)}$ from our modified model for SrSe is 9.86% which are closer to the experimental data 11-14%. The second order elastic constants (SOEC's) and pressure derivatives of SrSe are given in Table 3 The volumes change and % change in volume along with the experimental data [15] for B1-B2 phase are given in Fig: 1.

The success achieved in the present investigation can be ascribed to the inclusion of the charge transfer (or three body) and covalency effect seems to be of great importance at high pressure when the inter-ionic potential reduces considerably and the coordination number increases.

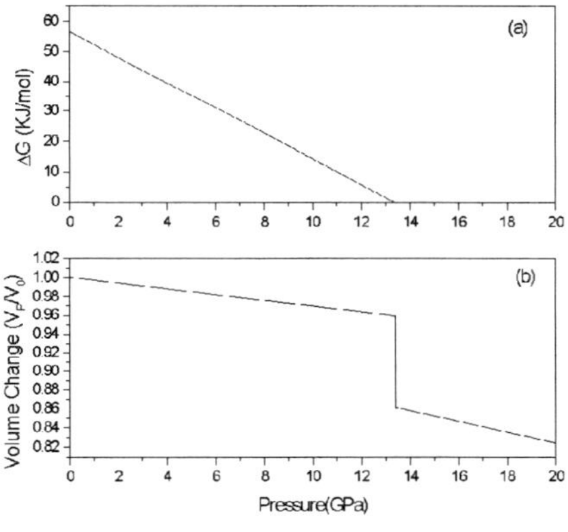

FIGURE 1: (a). Variation of ΔG (KJ/mol) with pressure for SrSe.**1 (b).** Variation of volume change V_P/V_0 with pressure for SrSe.

Finally, we can conclude that during the phase transition from NaCl to CsCl, the volume discontinuity in pressure volume phase diagram identifies the same trend as obtained the experimentally. On the basis of above it may be concluded that the present modified three body potential model is adequately suitable for describing the phase transition phenomena and elastic properties of strontium selenide.

ACKNOWLEDGMENTS

The authors are grateful to the Madhya Pradesh Council of Science and Technology (MPCST), Bhopal for the financial support to this work.

REFERENCES

1. M. P. Brown and K. Austin, *The New Physique*, Publisher City: Publisher Name, 2005, pp. 25-30.
2. M. P. Brown and K. Austin, *Appl. Phys. Letters* **85**, 2503-2504 (2004).
3. R. T. Wang, "Title of Chapter," in *Classic Physiques,* edited by R. B. Hamil, Publisher City: Publisher Name, 1999, pp. 212-213.
4. C. D. Smith and E. F. Jones, "Load-Cycling in Cubic Press" in *Shock Compression of Condensed Matter-2001*, edited by M. D. Furnish et al., AIP Conference Proceedings 620, American Institute of Physics, Melville, NY, 2002, pp. 651-654.
5. B. R. Jackson and T. Pitman, U.S. Patent No. 6,345,224 (8 July 2004)
6. D. L. Davids, "Recovery Effects in Binary Aluminum Alloys", Ph.D. Thesis, Harvard University, 1998.

Optical and Structural Characterizations of Tin Phthalocyanine Thin Films

Regimol C. Cherian and C.S.Menon

School of Pure and Applied Physics, Mahatma Gandhi University
Priyadarshini Hills P.O, Kottayam-686560, Kerala, India
regijose2004@yahoo.co.in

Abstract. Phthalocyanines are today regarded as optical materials, which applies to organic dye lasers. The analysis of the optical properties of these thin films enforces the application in the field of thin film optics. Tin phthalocyanine (SnPc) thin films used for the characterization studies are prepared by thermal evaporation technique. The variation of optical band gap with irradiation of heat radiation and post deposition heat treatment are studied from the absorption spectra. Structural properties have been analyzed using the X-ray diffractogram of SnPc powder and thin films. The structure is identified as monoclinic with a=12.132 Å, b=8.712 Å, c=10.806 Å and β=108.85°. The grain size increases with increase of annealing temperature. The SEM images show a rough corrugated surface. Due to heat treatment, crystallites grow into bigger size

Keywords: Phthalocyanines; Thin films; Optical Properties; Structural Studies
PACS: 71.20Rv; 78.20Ci; 61.80Ed; 68.37.Hk

INTRODUCTION

Phthalocyanines are small organic molecules characterized by their high symmetry, planarity and electron delocalization and are used as an active layer for electronic devices [1-3]. Metal phthalocyanines (MPc's) have the potential to be used as solar cells, liquid crystals, photovoltaic cells, gas sensors and organic thin film transistors [4-6]. Recently metal phthalocyanine has attracted the researchers for the possibility of using these materials for radiation measurement. The electrical, optical, structural and surface morphological studies of several phthalocyanine thin films have been studied in recent years [7-10], To the author's knowledge, no such studies have been reported for tin phthalocyanine (SnPc) thin films. The object of this paper is to probe the optical and structural properties of Tin Phthalocyanine thin films.

EXPERIMENTAL DETAILS

Tin Phthalocyanine (SnPc) powder of purity 97% supplied by the Sigma-Aldrich company (USA) is used as the source material. The purity of the material is crosschecked by the CHN analysis (C=61.06%, H=2.471%, N=18.0%). TGA-DTA techniques are employed to find the physicochemical changes of the material during the heat treatment. The decomposition temperature of the material is found as about 300°C. Thin film samples of suitable thickness for the investigations are

CP1004, *Thermophysical Properties of Materials and Devices-NCTP '07*,
edited by P. Predeep, S. Prasanth, and A. S. Prasad

deposited on to purely cleaned micro-glass substrates by vacuum sublimation of the powder from a resistively heated molybdenum boat using a "Hind Hi-Vac 12A" vacuum coating unit at a pressure of 10^{-5} Torr by thermal evaporation technique. Thin films prepared at room temperature is annealed in air for one hour in a temperature controllable furnace. The thickness of the film is measured by interference technique.

Absorption and reflection spectra are recorded using the Shimadzu 160A UV-Visible Spectro-photometer. SnPc thin films are irradiated with gamma rays by keeping the sample in a gamma-ray chamber where ^{60}Co is used as the gamma ray source. Irradiations are performed at radiation dose of 0.280Mrad for different exposure times. From the X-ray diffractogram of SnPc thin films annealed in air, the variation of average grain size with annealing temperature is studied. The scanning electron micrographs are analyzed to study the surface morphology of the films.

RESULTS AND CONCLUSIONS

The wavelength dependence of the absorbance spectra of SnPc thin films for different annealing temperatures and gamma ray irradiated samples for different exposure times are shown in Fig. 1 (a) and 1(b) respectively. The spectra originate from the molecular orbitals within the 18π electron system and from the overlapping orbitals on the central metal atom [11]. The electronic transitions a_{1u} (π) to eg(π*) and b_{2u}(π) to eg(π*) are responsible for the intense visible absorption Q and ultraviolet B (Soret) bands, positioned at about 705 nm and 345 nm respectively. To obtain information about interband transitions, the fundamental absorption edge of the absorption spectra is analyzed. The absorption coefficient α is related to the energy band gap Eg by the relation

$$\alpha = \alpha_0 \left(h\nu - E_g\right)^n \quad (1)$$

where the value of n is 3/2 for direct forbidden transitions, 1/2 for direct allowed transitions, 2 for indirect allowed transitions and 3 for indirect forbidden transitions. The direct band gap determined by extrapolating the graph α^2 versus photon energy hν at α^2=0 is 3.06 eV±0.01 eV.

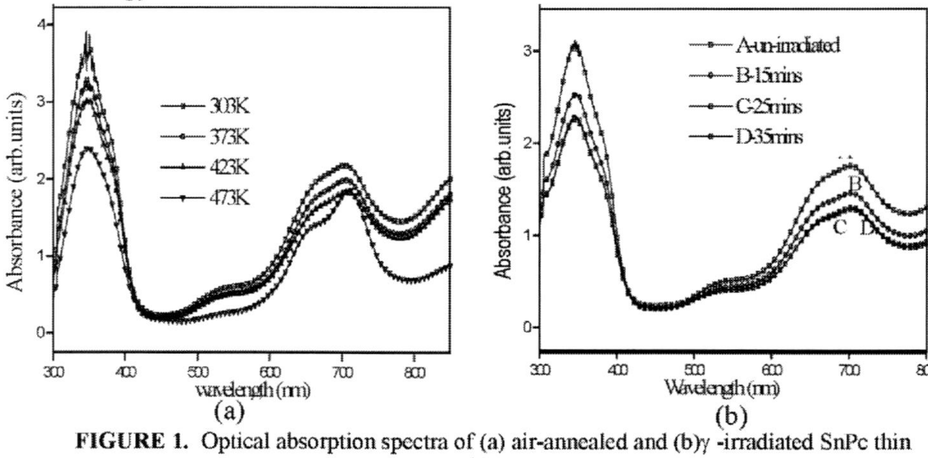

FIGURE 1. Optical absorption spectra of (a) air-annealed and (b)γ -irradiated SnPc thin films.

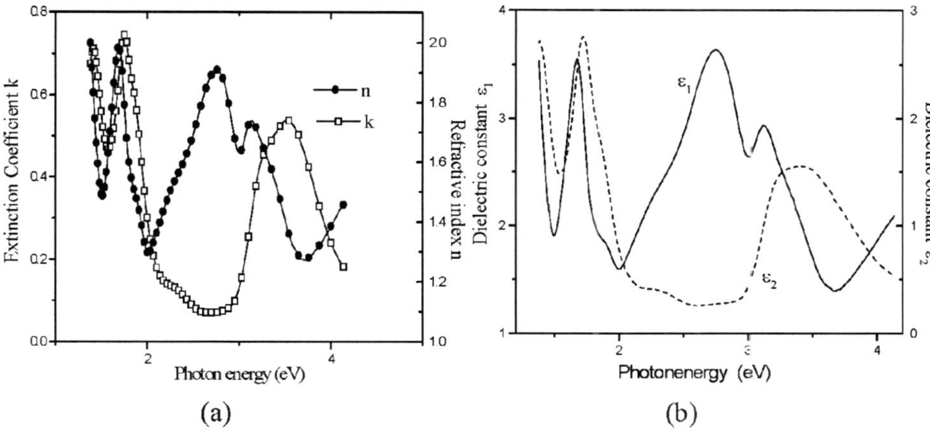

FIGURE 2. Variation of optical parameters with photon energy for SnPc thin film

Due to air annealing and γ-ray irradiation, there is no variation of fundamental energy band gap, which shows the stability of the material in this range of annealing temperature and γ-ray exposure times .The reflectivity of the absorbing medium of indices n and k in air for normal incidence is given by

$$R = \frac{(n-1)^2 + k^2}{(n+1)^2 + k^2} \qquad (2)$$

By knowing R and k, n is calculated using Eq.2. The optical properties of the films are characterized by the complex refractive index (N=n-ik) and complex dielectric constant ($\varepsilon_1 = \varepsilon_1 - i\varepsilon_2$). Figure 2 shows the variation of optical parameters n, k, ε_1 and ε_2 with photon energy hν for the as-deposited SnPc thin film. The real part ε_1 related to dispersion and the imaginary part ε_2 related to the dissipation rate of the wave in the material are calculated using the expressions

$$\varepsilon_1 = n^2 - k^2 \;\&\; \varepsilon_2 = 2nk \qquad (3)$$

MPc's exists in different polymorphic forms. X-ray diffractogram of SnPc powder and thermally evaporated films air annealed at different temperatures are analyzed to study the structure related properties. Figure 3 shows X-ray diffractogram of SnPc powder. The peaks are identified and the observed and calculated lattice spacing d_{hkl} are given in table 1. The unit cell parameters are determined and the structure is identified as monoclinic with a=12.132 Å, b=8.712 Å, c=10.806 Å, α=γ=90° and β=108.54°. X-ray diffractogram of SnPc thin films annealed to different temperatures given in Fig. 4 are analyzed. It is found that films grow with a preferential orientation along (1 1 0) direction.

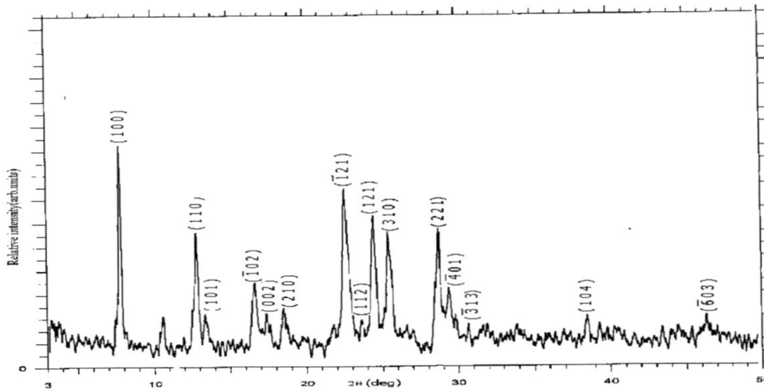

FIGURE 3. X-ray diffraction pattern of SnPc

TABLE 1. Observed and calculated interplanar distances (d_{hkl}) with the corresponding miller indices and 2θ values for SnPc powder.

2θ (deg)	d_{hkl} observed (Å)	d_{hkl} calculated (Å)	Δd_{hkl} (Å)	(h k l)
7.694	11.480	11.480	0	1 0 0
12.745	6.940	6.940	0	1 1 0
13.311	6.646	6.644	0.002	1 0 1
16.529	5.359	5.359	0	¯1 0 2
17.321	5.110	5.113	-0.003	0 0 2
18.436	4.808	4.793	0.015	2 1 0
22.530	3.943	3.943	0	¯1 2 1
23.578	3.770	3.779	-0.009	1 1 2
24.415	3.643	3.643	0	1 2 1
25.391	3.505	3.504	0.001	3 1 0
28.683	3.109	3.107	0.002	2 2 1
29.362	3.039	3.030	0.009	¯4 0 1
30.668	2.912	2.912	0	¯3 1 3

The crystallite size is calculated using the Scherrer's formula [12]

$$L = \frac{K\lambda}{\eta_{1/2} \cos\theta} \quad (4)$$

Where λ=1.5406Å, η½ is the full width at half maximum in radians and θ is the diffraction angle of the most intense peak. Table 2 gives the value of average grain size for different annealing temperature. As the annealing temperature increases, grain size increases, which is also evident from the SEM micrographs given in Fig.5 Grain size is maximum for the film annealed at 423K. Since charge mobility is higher in crystalline materials than in amorphous ones, the existence of dense small crystalline grains would favour the fabrication of TFTs with higher field effect mobility [13].

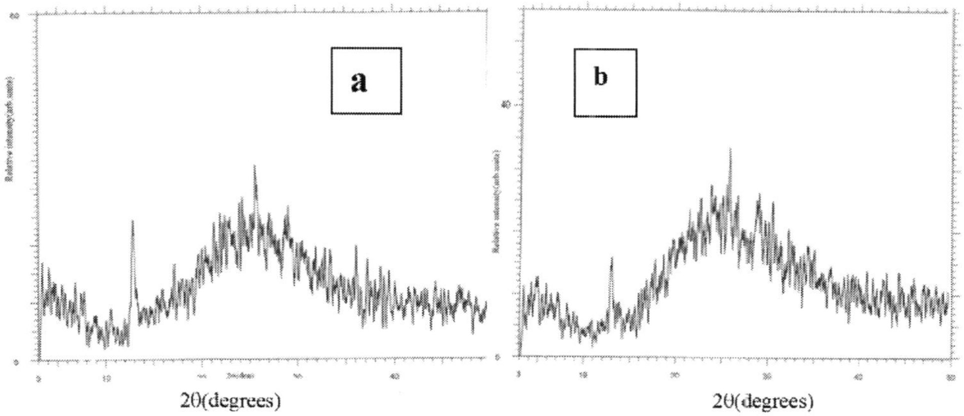

FIGURE 4. X-ray diffractogram of SnPc thin films (a) As-deposited (b) annealed 473K

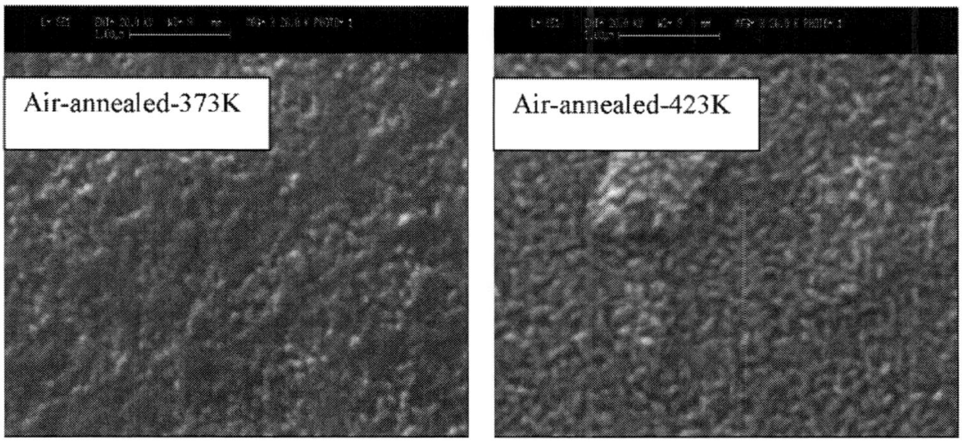

FIGURE 5. SEM images of air annealed SnPc thin

TABLE 2. Variation of grain size with annealing temperature

Annealing Temperature (K)	Grain Size (nm)
As-deposited	18.8
373	23.3
423	26.6
473	24.5

The basic optical properties of SnPc thin films have been studied. Two absorption peaks are identified at high-energy region (345nm) and low energy region (705nm) results to B (Soret) band and Q-band respectively. A direct allowed transition at 3.06eV is observed. Fundamental optical parameters are determined from both absorption and reflection data. The optical band gap of SnPc thin films irradiated by gamma radiations is found unchanged. The grain size increases with annealing temperatures. Surface morphological study shows that annealing improves the crystallinity.

REFERENCES

1. K.Kudo, D.X.Wang, M.Lizuka, S.Kuriyoshi, K.Tanaka – Synth.Mate., **111**, 11(2002).
2. F.Schauer, I.Zhikov S. Nespurek – J.Non-Cryst.Solids, **266**, 999 (2000).
3. N.R.Amstrong– J. Porphyrins Phthalocyanine, **4,** 417(2000).
4. S.R. Forest –Chem. Rev.,**97**, 1793 (1997).
5. T.D.Anthopoulos, T.S.Shaffai –Thin Solid Films,**441**, 207 (2003).
6. C.D. Dimitrakopoulos, D.J.Mascaro –IBM J.Res&Dev.,**45**, 11 (2001).
7. R.Rajesh, C.S. Menon – Eur.Phys.J.B, **47**, 171(2005).
8. M.M.El-Nahass,K.F.Abd-l_Rahman,A.A .Al-Ghamdi, A.M.Asiri –Physica B, **344**,398 (2004).
9. A.M. Saleh, A.O. Abu-Hilal, R.D.Gould –Current Applied Phys., **3**, 345 (2003).
10. M.E. Azim-Araghi, A.Krier – Pure. Appl .Opt..**6**, 443(1997).
11. E.A.Ough, J.M.Stillman –Can.J.Chem.,**71**, 1891(1993).
12. G.Gordillo, J.M.Florez, L.C.Hemandez – Sol.Energy.Mater.Sol.Cells, **37**, 273 (1995).
13. J.Puigdollers, C.Voz, M.Fonrodona, B. Cheylan, M.Stella, J.Andreu, M.Vetter, Alcubilla–J. Non-Cry. Solids, **352,** 1778(2006).

Nonlinear Wave Modulation and Stability Criteria for Strongly Dispersive Ion-Acoustic Waves

Runmoni Gogoi, Latika Kalita and Nirupama Devi

Department of Mathematics, Cotton College, Guwahati-781001, Assam, INDIA.

Abstract. We have derived the Nonlinear Schrödinger Equation (NLSE) for strongly dispersive ion-acoustic plasma waves taking into consideration the electron inertia effect through the parameter Q (= m_e/m_i, the ratio of electron to ion mass). In the present studies, the NLS equation is obtained by the modified reductive perturbation method and the derived equation is studied for the nonlinear wave modulation and the stability criteria. The parameter Q leads to higher dispersion compared to the case of electrons considered as Maxwellian. Consequently stronger nonlinearity comes into play for the studies of solitary wave propagation. In short it can be said that electron inertia effect modifies the nonlinear and dispersive coefficients of the NLS equation. The coefficient of the nonlinear term in the NLS equation turns to be complicated owing to the presence of electron mobility. A parametric range is obtained for angular frequency (ω) and the wave number k in order to get the stable region. It is found that instability occurs for small values of k due to electron inertia effect. A progressive wave type of solution of the NLS equation is obtained and plotted numerically to observe the modulation of nonlinear wave structure. Also a steady state solution of the NLS equation is derived.

Keywords NLSE, stability, wave modulation, nonlinear structure.

PACS: 43.25-x, 43.25Ba,43.25 Fe.

INTRODUCTION

Nonlinear Schrödinger Equation [1] has been obtained heuristically in the field of nonlinear optics [2], problem of the heat pulse in solids [3] etc. Optical solitons in fibre are pulses that propagate without any change in pulse shape or intensity. Due to their remarkable stability properties, optical solitons are now at the center of an active research field of nonlinear wave propagation in optical fibres. Results of experimental observation of modulational instability (MI) in optical fibres [4] are in good agreement with theory. The problems of nonlinear wave modulation and instability may be discussed on the basis of NLSE. The corresponding envelope soliton solutions of NLSE can exist in an electrical transmission line, and can be created by an input with a pulse shape. The remarkable phenomenon, which is known as MI was discovered by Lighthill [5]. Experimentally the MI was observed by Watanabe [6]. Various procedures for treating the nonlinear problems were suggested by several authors [7, 8]. Among these the significant one appears to be the Krylov-Bogoliubov-Mitropolsky (KBM) [9] method that was developed for the study of the theory of nonlinear oscillations. Kakutani and Sugimoto [1] applied the KBM method to consider the

CP1004, *Thermophysical Properties of Materials and Devices-NCTP '07*,
edited by P. Predeep, S. Prasanth, and A. S. Prasad

wave number shifts as well as long time slow modulation of the amplitude of plasma waves by deriving the NLSE. From the NLSE with KBM method, Salahuddin *et al.* [10] observed that the critical wave number k_c is highly dependent upon the ion streaming velocity V_0. In the absence of ion streaming velocity, the system is unstable for $k_c \geq 1.47$, which reconfirms the results of Kakutani and Sugimoto. Our present work differs from these earlier works due to the consideration of electron inertia through the parameter Q $(=m_e/m_i)$ in a collisionless, unmagnetized plasma.

Recently, Demiray [11] has presented a method so called "modified reductive perturbation method" to examine the contribution of higher order terms in reductive perturbation expansion and applied the method to ion-acoustic plasma waves for weakly dispersive case. Later on Demiray [12] extended this work to strongly dispersive case. Here, we used their method to derive the NLSE for strongly dispersive ion-acoustic plasma waves to study the nonlinear wave modulation and stability. To the best of our knowledge there is no theoretical evidence of such work considering the electron inertia effect for strongly dispersive ion-acoustic plasma waves. Observed that electron inertia affect the nonlinear and dispersive coefficient of the NLSE and become more complicated as compared to Maxwellian electrons. Also parameter Q leads to higher dispersion and higher nonlinearity that maintains the appropriate balance between them. We have derived progressive, steady state solution of the NLSE to observe the wave modulation of the nonlinear structures. Also a Duffing-like equation from the NLSE is derived and obtained a series solution for it. MI may be used to generate a soliton train at high repetition rate in optical fibre communication.

Basic Equations and Derivation of the NLSE

Considering the continuity and momentum equations for both ions and electrons together with the Poisson equation, the basic equations are as follows

For ions,
$$\frac{\partial n_i}{\partial t}+\frac{\partial (n_i u_i)}{\partial x}=0 \qquad (1),$$

$$\frac{\partial u_i}{\partial t}+u_i\frac{\partial u_i}{\partial x}+\frac{\partial \psi}{\partial x}=0 \qquad (2)$$

For electrons,
$$\frac{\partial n_e}{\partial t}+\frac{\partial (n_e u_e)}{\partial x}=0 \qquad (3),$$

$$\frac{\partial u_e}{\partial t}+u_e\frac{\partial u_e}{\partial x}=\frac{1}{Q}\left(\frac{\partial \psi}{\partial x}-\frac{1}{n_e}\frac{\partial n_e}{\partial x}\right) \qquad (4)$$

Poisson,
$$\frac{\partial^2 \psi}{\partial x^2}+n_i-n_e=0 \qquad (5)$$

where u_i, u_e - flow variables of ions and electrons, n_i, n_e - ion and electron density , ψ the electric potential. We denote the fluctuation of the ion and electron density from its equilibrium value by n_i=1+n_i, n_e=1+n_e. Introducing the slow variables $\xi = \varepsilon(x - \lambda t)$, $\tau = \varepsilon^2 g t$ where ε is small parameter measuring the weakness of certain physical entities; $\lambda \geq 0$ and g are some scale parameters, the set of equation (1) gets modified as below:

$$\frac{\partial n_i}{\partial t}+\frac{\partial u_i}{\partial x}+\frac{\partial(n_i u_i)}{\partial x}+\varepsilon^2 g\frac{\partial n_i}{\partial \tau}+\varepsilon[\frac{\partial u_i}{\partial \xi}-\lambda\frac{\partial n_i}{\partial \xi}+\frac{\partial(n_i u_i)}{\partial \xi}]=0 \quad (6)$$

$$\frac{\partial u_i}{\partial t}+u_i\frac{\partial u_i}{\partial x}+\frac{\partial \psi}{\partial x}+\varepsilon^2 g\frac{\partial u_i}{\partial \tau}+\varepsilon[u_i\frac{\partial u_i}{\partial \xi}-\lambda\frac{\partial u_i}{\partial \xi}+\frac{\partial \psi}{\partial \xi}]=0 \quad (7)$$

$$\frac{\partial n_e}{\partial t}+\frac{\partial u_e}{\partial x}+\frac{\partial(n_e u_e)}{\partial x}+\varepsilon^2 g\frac{\partial n_e}{\partial \tau}+\varepsilon[\frac{\partial u_e}{\partial \xi}-\lambda\frac{\partial n_e}{\partial \xi}+\frac{\partial(n_e u_e)}{\partial \xi}]=0 \quad (8)$$

$$\frac{\partial u_e}{\partial t}+u_e\frac{\partial u_e}{\partial x}-\frac{1}{Q}(\frac{\partial \psi}{\partial x}-\frac{1}{1+n_e}\frac{\partial n_e}{\partial x})+\varepsilon^2 g\frac{\partial u_e}{\partial \tau}+\varepsilon[u_e\frac{\partial u_e}{\partial \xi}-\frac{\lambda\partial u_e}{\partial \xi}-\frac{1}{Q}(\frac{\partial \psi}{\partial \xi}-\frac{1}{1-n_e}\frac{\partial n_e}{\partial \xi})]=0 \quad (9)$$

$$\frac{\partial^2 \psi}{\partial x^2}+n_i-n_e+\varepsilon^2\frac{\partial^2 \psi}{\partial \xi^2}+2\varepsilon\frac{\partial^2 \psi}{\partial \xi \partial x}=0 \quad (10)$$

In this work it is assumed that the field variables are functions of slow variables (ξ, τ) as well as the fast variables (z, t). We further assume that the variables and the scale parameter g may be expressed as asymptotic series of ε as given below

$$n_i=\sum_{j=1}^{\infty}\varepsilon^j n_{ij}(\mathrm{z,t},\xi,\tau)\ ,\quad n_e=\sum_{j=1}^{\infty}\varepsilon^j n_{ej}(\mathrm{z,t},\xi,\tau)\ ,\quad u_i=\sum_{j=1}^{\infty}\varepsilon^j u_{ij}(\mathrm{z,t},\xi,\tau)\ ,$$
$$u_e=\sum_{j=1}^{\infty}\varepsilon^j u_{ej}(\mathrm{z,t},\xi,\tau)\ ,\quad g=1+\sum_{j=1}^{\infty}\varepsilon^j g_j\ ,\quad \psi=\sum_{j=1}^{\infty}\varepsilon^j \psi_j(\mathrm{z,t},\xi,\tau) \quad (11)$$

where the coefficients functions n_{ij}, n_{ej}, u_{ij}, u_{ej}, ψ_j and the constants are to be determined from the solution of the field equations. The form of equations of different orders in ε suggests us to seek the following type of solution to these field equations:

$$n_{i1}=N_{i1}(\xi,\tau)\mathrm{e}^{\mathrm{i}(\omega t-kx)}+\mathrm{c.c}\ ,\quad u_{i1}=U_{i1}(\xi,\tau)\mathrm{e}^{\mathrm{i}(\omega t-kx)}+\mathrm{c.c}\quad n_{e1}=N_{e1}(\xi,\tau)\mathrm{e}^{\mathrm{i}(\omega t-kx)}+\mathrm{c.c}$$
$$n_{e1}=N_{e1}(\xi,\tau)\mathrm{e}^{\mathrm{i}(\omega t-kx)}+\mathrm{c.c}\ ,\quad u_{e1}=U_{e1}(\xi,\tau)\mathrm{e}^{\mathrm{i}(\omega t-kx)}+\mathrm{c.c}\ ,\quad \psi_1=\phi_1(\xi,\tau)\mathrm{e}^{\mathrm{i}(\omega t-kx)}+\mathrm{c.c}$$
$$n_{i2}=N_{i2}^{(0)}+(N_{i2}^{(1)}\mathrm{e}^{\mathrm{i}\phi}+N_{i2}^{(2)}\mathrm{e}^{2\mathrm{i}\phi}+\mathrm{c.c})\ ,\quad u_{i2}=U_{i2}^{(0)}+(U_{i2}^{(1)}\mathrm{e}^{\mathrm{i}\phi}+U_{i2}^{(2)}\mathrm{e}^{2\mathrm{i}\phi}+\mathrm{c.c}) \quad (12)$$
$$n_{e2}=N_{e2}^{(0)}+(N_{e2}^{(1)}\mathrm{e}^{\mathrm{i}\phi}+N_{e2}^{(2)}\mathrm{e}^{2\mathrm{i}\phi}+\mathrm{c.c})\ ,\quad U_{e2}=U_{e2}^{(0)}+(U_{e2}^{(1)}\mathrm{e}^{\mathrm{i}\phi}+U_{e2}^{(2)}\mathrm{e}^{2\mathrm{i}\phi}+\mathrm{c.c})$$
$$\psi_2=\phi_2^{(0)}+(\phi_2^{(1)}\mathrm{e}^{\mathrm{i}\phi}+\phi_2^{(2)}\mathrm{e}^{2\mathrm{i}\phi}+\mathrm{c.c})\ \text{etc.}$$

with ω the angular frequency, k the wave number, U_1, N_1 and ϕ_1 are the complex amplitude functions, c.c. complex conjugate of the corresponding quantities, $\phi=\omega t-kx$. Charge neutrality condition, $N_i = N_e = N$ and some mathematical simplifications, gives

$$i\frac{\partial N}{\partial \tau}+p\frac{\partial^2 N}{\partial \xi^2}+q|N|^2 N=0 \quad (13)$$

which is the nonlinear Schrödinger equation with p and q as the dispersive and nonlinear coefficients respectively and are defined as

$$p=[\frac{1}{\omega^2}+\frac{Q}{4k^2}+\frac{1-4\omega^2}{4k^2(1-\omega^2)}]/[2(\frac{\mathrm{k}}{1-\omega^2}-\frac{1}{\omega k}+\frac{Q}{\omega})] \quad (14)$$

$$q=[\frac{1}{\omega^2(1+Q)-4k^2}\{24Q\omega^2+4\omega^2-6\omega Q+8\omega^2Q^2+\frac{\omega^4}{1-\omega^2}(8+Q)\}+\frac{9}{2}+2Q-\frac{1}{2\omega^2}$$
$$-4Q(1-\omega^2)+\frac{2(1-\omega^2)^2}{3\omega^2}Q(Q-2)+\frac{(1-\omega^2)}{\omega^2}(\frac{3}{2}+Q)-4Qk^2]/[2(\frac{\mathrm{k}}{1-\omega^2}-\frac{1}{\omega k}+\frac{Q}{\omega})] \quad (15)$$

Solution sets of the Nonlinear Schrödinger Equation (NLSE)

Progressive wave solution:

Using the transformation $N(\xi,\tau)=f(\zeta)\exp[i(K\xi-\Omega\tau)],\ \zeta=\xi-c\tau$ where K, Ω, c - constants and $f(\zeta)$- unknown real function to be determined through the solution, the NLSE (13) is

$$f''(\zeta)+(\frac{\Omega}{p}-K^2)f(\zeta)+\frac{q}{p}f^3(\zeta)=0,\ \text{where}\ c=2kp \quad (16)$$

Introducing $f(\zeta)=A\tanh\beta\zeta$, where A is the amplitude and β is constant in (16), the solution is obtained as $N(\xi,\tau)=A\tanh\{\sqrt{\frac{-q}{2p}}A\}\varsigma\exp[i\{K\xi-(K^2p\text{-}qA^2)\tau\}]$ (17)

Steady state solution:

For steady state, the transformation $N(\xi,\tau)=f(\xi)\exp(-i\alpha\tau)$ (with α - constant, $f(\xi)$ is an unknown function) (13) is modified as $\alpha f(\xi)+pf''(\xi)+qf^3(\xi)=0$ (18)

Using $f(\xi)=A\tanh\beta\xi$, in (18) we get $N(\xi,\tau)=A\tanh\{\sqrt{\frac{-q}{2p}}A\}\xi\exp[(iqA^2)\tau]$ (19)

Here (18) is a Duffing–like equation, which gets modified by using the transformation, $f=\nu\phi$ with $\nu=1,\ z=\tanh(\xi)$, and $\phi(\xi)=\mathrm{W}(z)$ as,

$$p(1-z^2)\frac{d^2W}{dz^2}-2zp(1-z^2)\frac{dW}{dz}+\alpha W+qW^3=0 \quad (20)$$

Balance of power order gives $N=1$, that seeks a series solution $W(z)=\sum_{r=0}^{\infty}a_r z^{p-r}$ and get,

$$N(\xi,\tau)=\{\sqrt{\frac{2p-\alpha}{3q}}+\sqrt{\frac{-2p}{q}}\tanh\xi\}\exp[(iqA^2)\tau] \quad (21)$$

RESULTS AND DISCUSSIONS

Our present work is related to the studies of nonlinear wave modulation and MI in plasma through the NLSE and some solution sets for the NLSE. The value of the parameter Q=0.00054 (fixed) for all numerical results throughout this work.

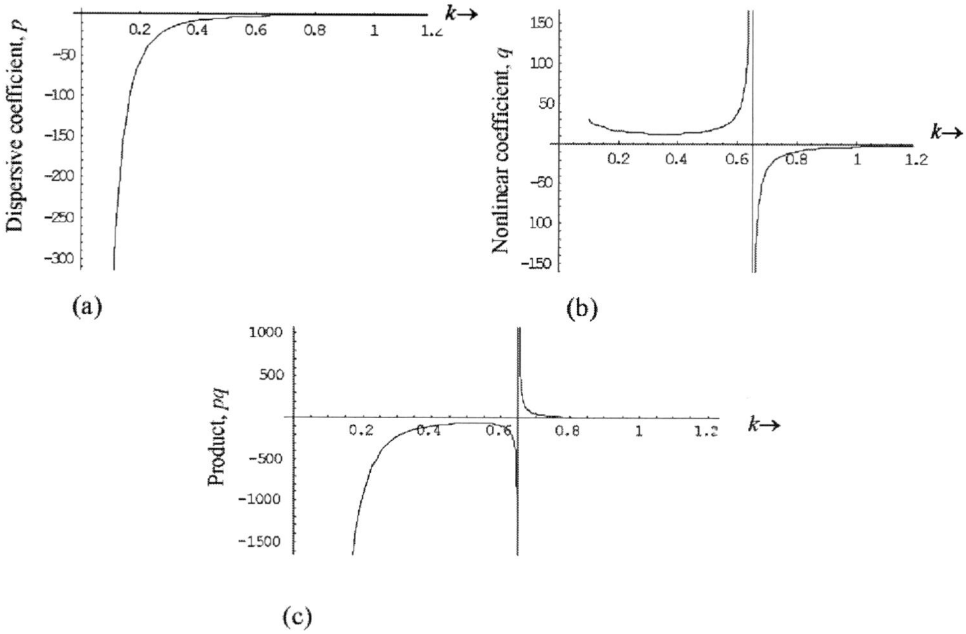

FIGURE 1. Nature of the dispersive coefficient p, nonlinear coefficient q and their product pq versus wave number k for angular frequency ω=1.3.

The NLSE gives comprehensive information about the MI, which we highlight first. As shown by earlier works [13, 14], the plane wave solution of the NLSE is modulationally unstable if $pq<0$ and stable if $pq>0$, depending on the wave number k. From Equation (14) and (15), it is seen that the coefficients p and q both depend on ω and k. Also observed that for ω =1.3, MI occurs when $k\leq0.65$ and stability is found for $k\geq0.66$. Here the two measured values of p and q confirm the well-known Lighthill criterion, i.e. the product pq must be positive for the self-contraction of the wave towards stability. p is always negative for all ks as shown in Fig. 1.(a) which agrees with Bailung and Nakamura [15]. Hence, for the existence of stability $q<0$, the region can be seen from Fig. 1.(b).

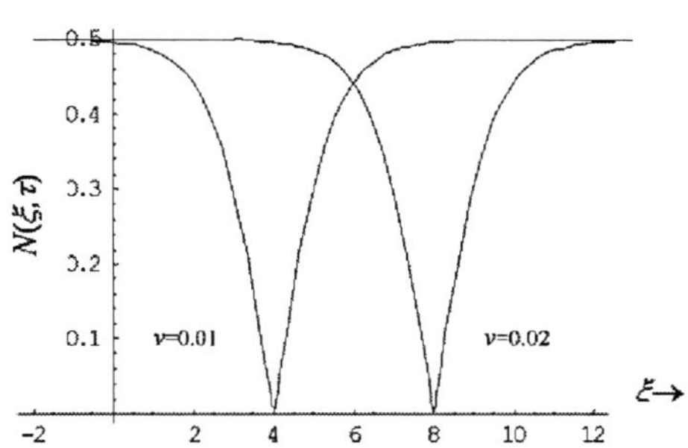

FIGURE 2. Progressive wave solution (17) versus ξ is plotted for k=0.5, w=1.3, τ=400 and different values of v.

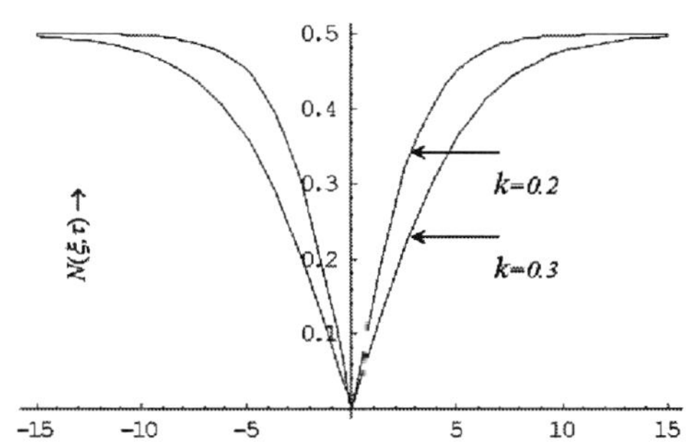

FIGURE 3. Steady state solution (19) versus ξ is plotted for w=1.3, τ=400 and different values of k.

Again Fig.1.(c) shows that pq undergoes a change in sign from negative to positive, from which the critical wave number k_c (between 0.65-0.66) can be observed, where pq=0. We noticed that electron inertia could generate stability earlier.

Progressive, steady state solutions of the NLSE are derived and plotted numerically to study the nonlinear wave modulation. The parameter v adds energy to the progressive wave, which is depicted in Fig. 2. For the steady state solution (19), the orientation is regular for different wavelengths, which is shown in Fig. 3. In case of progressive wave modulation, the wavelength is reduced in comparison with that of the steady state, the self-focusing being originated with similar angular frequency (Fig. 2, Fig. 3).

In NLSE the linear dispersive term and the nonlinear terms, which are of the same order, can balance to yield envelop and hole soliton solutions. Plotting of N (ξ, τ) vs. ξ is depicted in Fig. 3 by using (19) for steady state solution to get the dark soliton. Such a soliton which is known as a hole soliton (a dark soliton in optics) was originally found by Zakharov and Shabat [16]. Hole soliton corresponds to a hole in the amplitude or envelop function. Hence there is a relationship between the envelope pulses and the envelope soliton solutions.

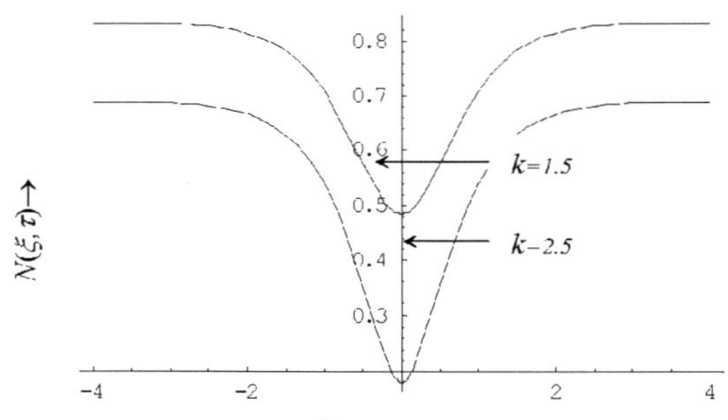

FIGURE 4. Solution (21) versus ξ is plotted for w=1.3, τ=400 and different values of k.

Fig. 4 shows the plotting of solution (21) of the Duffing–like equation (18) and found that modulation depth is decreasing with increase in k. Through the Duffing equation the basic development of the ordered chaotic structures of the plasma wave modulation described by NLSE can be observed.

ACKNOWLEDGMENTS

The authors (ND and RG) are grateful to the DST, Govt. of India for providing a Research Project grant. Thanks are also to the HOD of Mathematics, Cotton College, Guwahati for providing necessary facilities to carry out research in the Department.

REFERENCES

1. T. Kakutani, N. Sugimoto, *Phys. Fluids*, **17**, 1617-1625 (1974).
2. S. A. Akhrnanov, A. P. Sukhorukov, R. V. Khokhlov – Usp. Flz. Nauk, **93**, 19, (1967).
3. F. D. Tappert, C. M. Varma – Phys. Rev. Lett. **25**, 1108, (1970).
4. K. Tai, A. Haskuawa, A. Tomita - Phys. Rev. Lett. **56**, 135, (1986).
5. M. J. Lighthill – J. Inst. Math. Appl. **1**, 269, (1965).
6. S. Watanabe – J. Plasma Phys. **17**, 487, (1977).
7. A. H. Nayfeh – Perturbation Methods, Wiley, New York, (1973).
8. G. B. Whitham – Linear and Nonlinear Waves, Wiley, New York, (1974).
9. N. N. Bogoliubov, Y. A. Mitropolsky – Asymptotic Methods in the Theory of Nonlinear Oscillations (Hindustan Publishing Company, Delhi), Chap. **I**., (1961).
10. M. Salahuddin, I. R. Durrani, H. Saleem – Phys. Fluids, **B 2**, 741, (1990).
11. H. Demiray - J. Phys. Soc. Jpn., **68**, 1833, (1999).
12. H. Demiray - J. Phys. Soc. Jpn., **71**, 1921-1930, (2002).
13. T. Taniuti, N. Yajima - J. Math. Phys. **10**, 1369, (1969).
14. H. Hasimoto, H. Ono - J. Phys. Soc. Jpn. **33**, 805, (1972).
15. H. Bailung, Y. Nakamura, *J. Plasma Phys.* **50**, 231-242 (1993).
16. V. E. Zhakarov, A. B. Shabat – Interaction Between Solitons in a Stable Medium, Sov. Phys. JETP. **37**, 823-828, (1973).

Organic Semiconducting Materials in Film and Powder Forms from a Co-polymeric Elastomer-Styrene Butadiene Rubber

Santhamma G and P.Predeep

Condensed Matter Physics Laboratory, Sree Narayana College, Kollam, Kerala, India.

Abstract. Semiconducting materials in both film and powder forms are prepared by Antimony Pentachloride ($SBCl_5$) doping in Styrene Butadiene Rubber (SBR). SBR is a synthetic co-polymeric elastomer, insulating in undoped state, is mainly used for manufacturing tires, tubes etc. Synthesized conducting materials are proposed to have tremendous application potentials in optoelectronic, electronic and electrical industries. For example conducting films can be used as active elements for fabrication of organic light emitting diodes, photovoltaic cells etc. Electrical and optical properties of prepared samples are studied by measuring electrical conductivity and analyzing spectroscopic data. Electrical conductivity of samples lies in the range of that of semi-conducting materials. Presence of conjugated sequences in the back bone of prepared conducting materials, which is regarded as pre-requisite condition for a polymer to conductive, is confirmed by studying UV/Vis spectra.

Keywords: Polymers, Doping, Conducting
PACS:72.80.Le.

INTRODUCTION

The discovery by Heeger and MacDiarmid [1-2] that the polyacetylene would undergo an increase in conductivity of 12 orders of magnitude by oxidative doping quickly reverberated around the polymer and electrochemistry communities. Following this discovery, some other conducting polymers were also discovered. These polymers are assumed to have tremendous applications in different fields, especially in the electronic industry. But the major concern in this area is the processability of these polymers as most of them are not stable at processing conditions. Prime work in this area is the attainment of processability in conducting polymers. This fact prompted the researchers to search for the possibility of introducing conductivity in non-conjugated polymers like elastomers [3-4]. In this study, intrinsically conducting powders and films of one of the most popular artificial elastomer, Styrene Butadiene Rubber (SBR) are prepared by doping with Antimony Pentachloride ($SbCl_5$)) and the important results obtained are discussed.

CP1004, *Thermophysical Properties of Materials and Devices-NCTP '07*,
edited by P. Predeep, S. Prasanth, and A. S. Prasad

EXPERIMENTAL

0.43 $\times 10^{-3}$ kg of SBR is dissolved in $50 \times 10^{-6} m^3$ of CCl_4. 3.59$\times 10^{-3}$ kg of $SbCl_5$, dissolved in $5 \times 10^{-6} m^3$ of CCl_4 is added drop wise to the above solution. Precipitate formed is filtered off after stirring for 5 hrs. It is washed 3 times with plenty of CCl_4 .Yield of the product is measured using an electronic balance. Conductivity is measured using Keithly 6514 electrometer. The processes of preparation of conducting powder and measurements of yield and conductivity are repeated by varying the amount of dopant. Similarly conducting films are prepared by doping SBR films with solutions of $SbCl_5$ in CCl_4 of different concentrations for a time periods of 5 hrs. Optical absorption spectra of samples are taken using UV/Vis/NIR Spectrophotometer (GASCO V 550).

RESULTS

TABLE 1. Yield and conductivity of powders prepared by doping 0.43$\times 10^{-3}$ kg of SBR with different amount of $SbCl_5$, doping for 5 hrs.

Amount of $SbCl_5$ (10^{-3} kg)	Doping period (h)	Yield (10^{-3} kg)	Conductivity (S/m)
3.59	**5**	**1.424**	**5 x 10^{-2}**
2.99	**5**	**1.23**	**2.5 x 10^{-2}**
2.392	**5**	**1.20**	**2.5 10^{-3}**
1.795	**5**	**1.143**	**1 x 10^{-3}**

Prepared conducting powders are dark brown in color. Similarly, on doping colorless films become reddish brown. These colors of the materials prepared are indication of the introduction of conjugated sequences of alternate single and double carbon bonds on their backbone, on doping. Table 1 displays yield and conductivity of powder samples prepared by doping 0.43x10^{-3} kg of SBR with different amount of $SbCl_5$, doping for 5 hrs. It can be seen that conductivity and yield of products decrease with decrease in amount of dopant. Maximum conductivity obtained is 5 x 10^{-2}S/m, 10 orders of magnitude more than that of pristine SBR. Figures 1 shows the variation of electrical conductivity of SBR films as a function of dopant concentration.. Conductivity of samples is found to increases up to 10^{-2} S/m, 9 orders of magnitude more than that of pristine state.

Optical absorption spectra of the SBR films corresponding to different degree of doping are shown in Fig. 2.

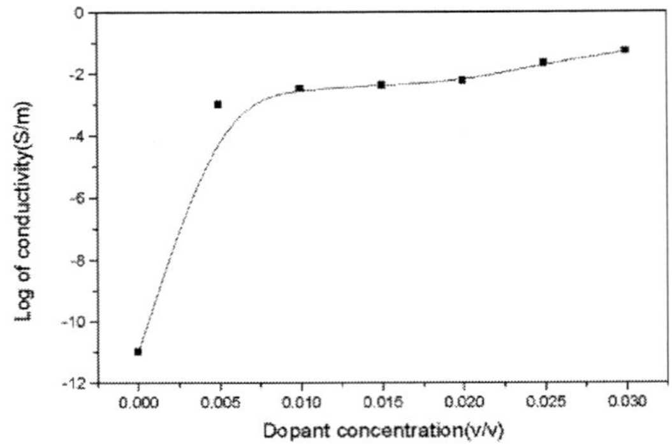

FIGURE 1. Conductivity of $SbCl_5$ doped SBR film as a function of dopant concentration.

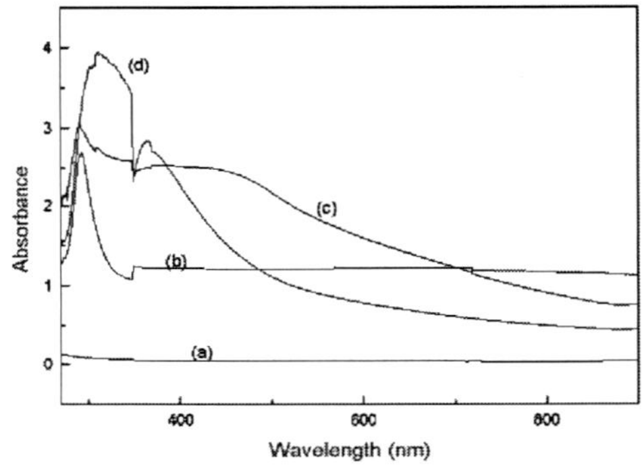

FIGURE 2. UV/Vis spectra of $SbCl_5$ doped SBR films corresponding to different degree of doping. (a). undoped film, (b) lightly doped (0.02 v/v), (b). intermediately doped (0.03 v/v) and (c). heavily doped (0.04v/v).

UV/Vis spectra of pristine SBR, conducting powders prepared by doping 0.43^{-3} kg of SBR using 2.392×10^{-3}kg of $SbCl_5$ for different time periods are shown in Fig 3.

No absorption is observed for pristine SBR. In this state polymer is colorless because of the isolated double bond structure. Conducting materials absorbs strongly in UV/Vis region. These results also indicate the formation of conjugated sequences of $(- CH = CH -)_n$ of various lengths [4-6] along the backbone of the conducting materials.

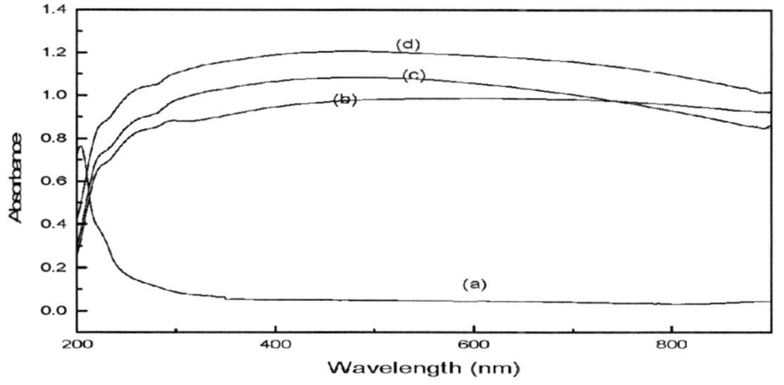

FIGURE 3. UV/ Vis spectra (a) pristine SBR (b) conducting powder prepared by doping 0.43×10^{-3}kg of SBR using 2.392×10^{-3}kg of $SbCl_5$ doping for 5 hrs (c) doping for 8 hrs(c) doping for 12 hrs.

ACKNOWLEDGEMENTS

The authors are thankful to all members of Condensed Matter Physics Laboratory, Sree Narayana College, Kollam, Kerala, India

REFERENCES

1. Alan Heeger & Alan MacDiarmid, *Physics Today* (December 2000).
2. A.J.Heeger, (Nobel Lecture) Angrew, *Chem. Int. Ed.* **40,** 2591-2611 (2001).
3. L.Dai, J.Phys.chem., **06**, 6469 (1992).
4. L.Dai, J.W.White, Polymer, **32** , 2120 (1001).
5. S. Najitha, N.S. Saxena, R. Sreeja, C.H. Unnithan, P. Pradeep, *Material letters,* **59**, 3431-3436 (2005).
6. P. Pradeep, R. Sreeja, R. Alex and P.D. Sharma, *Progress in Rubber Plastics and Recycling Technology,* Vol.**20.**,No.3 (2004).
7. R.Sreeja, S.Najitha, S.Remya Jayan, P.Predeep, Maciej Mazur, P.D.Sharma, *Polymer,* **47**, 617-623 (2006).

Photo Response in Zinc Oxide Doped Alizarin Thin Film

Sreekala C.O[a], P. Balraju[b], Y. S. Deol[c], P.Pradeep[a] and M.S.Roy[c]

aDepartment of Science & Humanities National Institute of Technology Calicut, Kerala, - 673 601

bJNV University, Jodhpur – 342 005

cDefence Laboratory, Jodhpur, Rajasthan-342 011

Abstract. Hybrid structure based on organic and inorganic materials are emerging as most promising materials for photovoltaic applications. We have designed a hybrid structure based on Alizarin (1, 2 dihydroxy anthraquinone) and a wide band gap zinc oxide. Alizarin acts as donor and zinc oxide acts as an acceptor material. The device having configuration of ITO/Aln:ZnO/Ag was characterized by analyzing its optical and electrical characteristics. It is found that the doping with zinc oxide imparts an accelerated improvement in device performance particularly in terms of its photo response. Impedance analysis of the device was carried out by recording Cole-Cole plot in between real and imaginary impedance under biasing varying from 0 to 5V.

Key words: Hybrid structure, photo response, impedance, cole-cole plot.
PACS: 01.10.Fv

INTRODUCTION

Organic materials have attracted interest as potential candidates for the use in optoelectronic devices [1-5], since these materials exhibit unique features such as flexibility, large-scale area, low cost fabrication and possibility to produce flexible devices on plastic substrates.In the present communication, we have studied the temperature dependent current – voltage (J-V) characteristics and impedance behavior of Alizarin chemically known as 1, 2 Dihydroxyanthraquinone (Aln) thin film sandwiched between indium tin oxide (ITO) and aluminum (Al) electrodes, having ITO/Aln/Al structure. Further, to get information about the charge injection mechanism, we have investigated J-V characteristics in dark and impedance spectroscopy of the device. We have also analyzed the influence of ZnO doping over the photo voltaic response.

EXPERIMENTAL DETAILS

Alizarin dye was chemically purified and the molecular structure of the Aln is shown in diagram 1(a). The thin film of the Aln:ZnO was deposited by spin casting technique over cleaned ITO coated glass substrate. The aluminum (Al) electrode was evaporated on the top of thin film and used as counter electrode. The structure of the ITO/Aln/Al device is as shown in diagram 1(b). The electrical charge transport in the

CP1004, *Thermophysical Properties of Materials and Devices-NCTP '07*,
edited by P. Predeep, S. Prasanth, and A. S. Prasad

device is studied via current –voltage (J-V) characteristics using Keithley electrometer (Model 6517A), in dark and under illumination using 100 W halogen lamp and impedance spectra were measured employing HP Impedance analyzer.

FIGURE 1. (a) Molecular Structure of Alizarin. (b) Structure of the Device.

RESULTS AND DISCUSSION

Current – Voltage Characteristics

Figure 2(a). shows, the current – voltage (J -V) characteristics in dark for ITO/Aln/Al device at different temperature ranging from 240 K to 340 K, in linear scale. The asymmetrical nature of the curves is the resultant of the difference in the work function of electrodes. Under forward bias, when ITO is positive with respect to Al the device clearly shows rectifying behavior, in dark, which can be explained in terms of small work function of Al electrode, which forms the schottky barrier with Aln, and where the HOMO and LUMO edges of Aln bends. As the work function of ITO electrode is high as compared to Al, it forms ohmic contact with Aln.

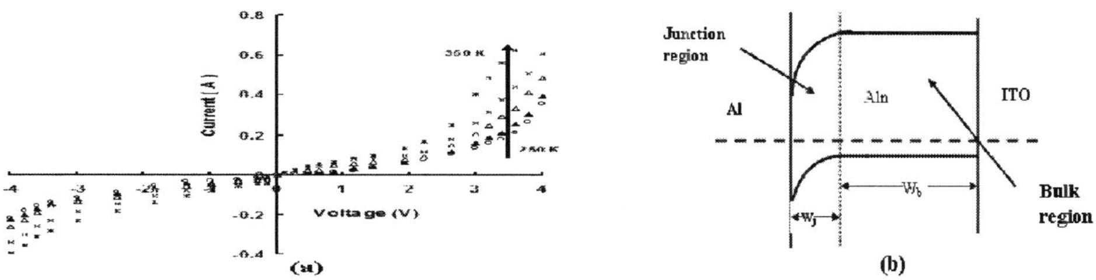

FIGURE 2. (a)Current-Voltage Chracteristics of ITO/Aln/Al Device in Dark from temperature range 240K to 340K (b) Energy Level Diagram of the Device under Equilibrium conditions

The proposed energy band diagram for the device is shown in Fig 2(b). According to the theory of Schottky barrier, for appearance of a rectifying barrier at interface, the work function of metal must be lower than the Fermi level of p type semiconductor. Since the work function of ITO and Al are 4.8 eV and 4.2 eV, respectively, rectifying behavior exists at Al- Aln interface and ohmic contact at ITO-Aln interface, respectively. The formation of schottky barrier and nearly ohmic contact at Al- Aln and ITO-Aln respectively, suggest that Aln behaves as p type organic semiconductor.

The current – voltage relationship for such device can be described by the standard diode equation:

$$J = J_0[e^{(qV/nkt)} - 1] \quad (i)$$

Where $J_0 = AT^2 e^{(-\phi_b/kT)}$ is the reverse saturation current.

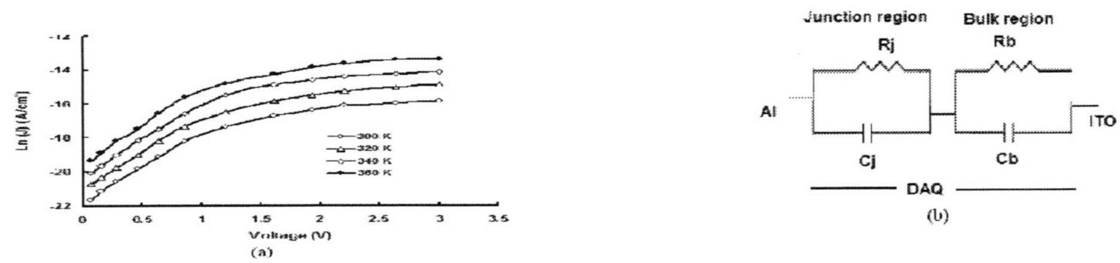

FIGURE 3. (a) Semi log Plot of I-V Characteristics of Al/Aln/ITO Device. (b) Equivalent Circuit of the Device

The schottky barrier device can be divided into two regions: (i) the barrier region of a depletion layer thickness W_j having capacitance C_j in parallel with R_j and (ii) a neutral bulk with dielectric capacitance C_b and resistance R_b [6] as shown in Fig.3(b). The ln (J) versus voltage (V) plot is linear up to 0.86 V at all temperatures. This depicts that the depletion region resistance is dominant in this voltage region, wherein the variation of junction and bulk voltages as a function of forward applied voltage, govern by following equations:

$$V_j = (nkT/q)\ln(J/J_0) \quad (2a)$$

$$V_b = V_{appl} - V_j \quad (2b)$$

where V_j and V_b are the voltage drops at junction and bulk, respectively, and V_{appl} is the voltage applied to device.

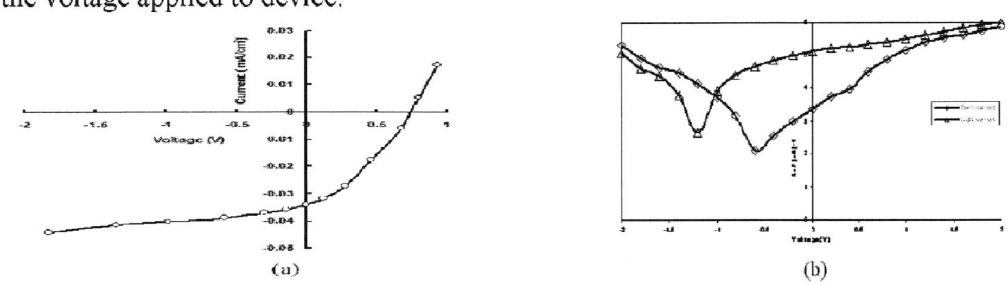

FIGURE 4. I-V Characteristics of (a) under illumination for Al/Aln/ITO Device (b) Ag/Aln:ZnO/ITO Device

Figure 4(a) shows the J-V characteristics of ITO/Aln/Al device under illumination through ITO. The various photovoltaic parameters such as short circuit photocurrent (J_{sc}), open circuit voltage (V_{oc}), Fill factor and power conversion efficiency (η %) have been derived from the analysis of J-V characteristics and are as follows : J_{sc} = 0.034 mA/cm^2, V_{oc} = 1.2 V, FF = 0.56 and η = 1.2 % . Fig.4(b) shows the variation of short circuit photocurrent (J_{sc}) and open circuit voltage (V_{oc}) with ZnO doping.

The relation between J_{sc} and V_{oc} is given by

$$V_{oc} = (nkT/q)\ln[(J_{sc}/J_0)+1] = (nkT/q)\ln(J_{sc}/J_0) \quad (3)$$

when Jsc >> Jo

In the present device, the open circuit voltage (V_{oc}) increases logarithmically with light intensity which is in agreement with the Eq. (3).

Impedance Characteristics

Figure 5(a) depicts the Cole–Cole plots at different bias voltages which shows the strong voltage dependence of the device impedance in the low frequency range. The impedance curves in high frequency region are independent of the bias voltage showing a small semicircle close to origin [7]. Fig.3(b) represents the equivalent circuit of the device. In ideal case, the proposed equivalent circuits for the device corresponds to two perfect semicircles in the complex impedance plane ($Z_{img} - Z_{real}$ plot) and can be explained on the basis of their frequency dependence capacitance. At low frequencies, the capacitance C_b and C_j lead to high impedance and hence the response of the circuit is dominated by R_b and R_j, where $R_j = 1/2\pi f_j C_j$ and $R_b = 1/2\pi f_b C_b$.

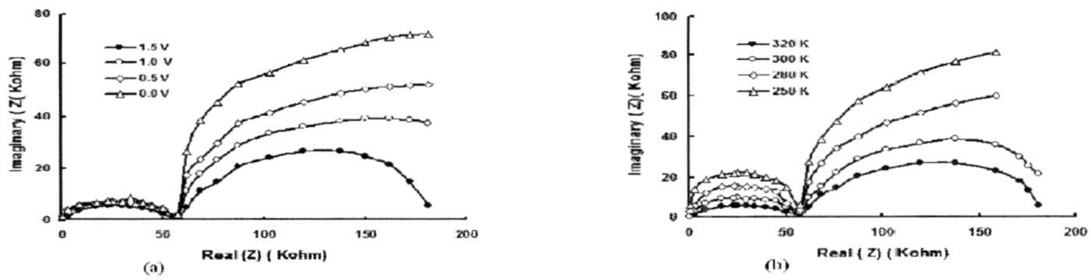

FIGURE 5. Cole-Cole Plots for AL/Aln/ITO Device (a) at different biasing voltage (b) at different temperatures

The experimental observations as revealed in the Fig.5(a), yields $C_j = 15.3$ nF, $C_b = 34$ nF, $R_j = 2.8 \times 10^6\ \Omega$ and $R_b = 1.6 \times 10^5\ \Omega$ and further from these values, the thickness of the bulk and junction region can be evaluated by employing $C_{j/b} = \varepsilon\, a/ d_{j/b}$. Assuming the value of $\varepsilon = 3.0 \times 10^{-11}$ F/m which is almost frequency independent, we obtained the thickness of the junction region and bulk region as 35 nm and 165 nm, respectively.

Figure.5 (b). demonstrates that the centers of the semicircles fitted through both junction and bulk impedance are depressed below x - axis, indicating the appearance of more than one relaxation time (τ) in the Aln which justifies the possibility of several hopping mechanisms such as inter-soliton hopping or hopping between unspecified localized states. Additionally, the subsequent increase in temperature and frequency may lead to interference in conduction mechanism with each other. The activation energies for the bulk and junction region are found to be 0.22 eV and 0.34 eV, respectively.

The variation of Im (Z) with Re (Z) as a function of biasing voltage for Ag/Aln:ZnO/ITO device is shown in the Fig.6(a). The curvature decreases with increase in biasing voltage. Therefore the equivalent circuit (Fig.6(b)) can be designed as a single parallel resistor R_b and a capacitor C_b net work with a series resistance R_s. R_s probably corresponds to the ohmic contact at the hole injecting ITO/Aln:ZnO interface. These data therefore demonstrates that a large fraction of volume in

Aln:ZnO composite is effectively at the D/A interface. i.e Aln:ZnO composite behaves as a bulk D/A hetero junction material sandwiched between Ag and ITO.

FIGURE 6. (a) Cole-Cole Plots for Ag/Aln:ZnO/ITO Device (b) Equivalent circuit of Ag/Aln:ZnO/ITO Device

CONCLUSIONS

We have studied the current–voltage characteristics in dark and under illumination, bias voltage and temperature dependence of impedance and frequency of ITO/Aln/Al device. We found that the Aln behaves as *p*-type organic semiconductor which forms the Schottky barrier with Al and ohmic contact with ITO. It is explained that current is limited by hole injection at Al/Aln interface when the device is in reverse bias and become space charge limited at high voltages wherein tunneling contributes significantly to the charge injection. In forward bias, current becomes space charge limited in the whole voltage and temperature range verifying the formation of ohmic contact for hole injection at ITO-Aln interface.

From the impedance data, it was concluded that the ITO/ Aln/Al device perfectly follows the Schottky model and can be modeled by the simple equivalent circuit. The higher activation energy and longer relaxation time associated with the junction region indicates that the charge transport across the junction is the limiting factor in controlling the conduction process. Doping with ZnO imparts a dramatic increase in V_{oc} and that leads to improvement in conversion efficiency.

ACKNOWLEDGMENTS

One of the authors (C.O.S) thanks Bipin Nair (Dean, Amrita School of Biotechnology) whose valuable advice and support helped for the successful completion of the work. Acknowledgements also go to Y.Janu, R.Mishra and N.Prasad (Defence Lab Jodhpur) for technical support.

REFERENCES

1. Y. Yang and A.J. Heeger, *Appl. Phys. Letters,* **64,** 1245 (1994)
2. J.B. Whitlock, P.Panayotatos, G.D.Sharma and G.R. Bird, *Opt. Eng.* **32,** 1921 (1993)
3. J.H. Schon, S. Berg, Ch. Kloc and B. Batlogg., *Science* 287 1022 (2000)
4. L. Sicot, B. Geffroy, A. Lorin, P. Raimond and C. Sentein, *J. Appl. Phys.* **90**, 1047 (2001)
5. G.D.Sharma, Dhiraj Saxena and M.S. Roy, *Syn.Met.* **123**, 189-196 (2001)
6. M.A. Lampert, and P. Mark, Current Injection into Solids *Academic Press*, New York (1970).
7. M. S. Roy, Manmeeta, Pratibha Jaiswal and G. D. Sharma., *J. Appl. Phys.*94, (2003)

Miscibility Studies of Blends of PPG and PEG using Viscosity Technique

K.Venkatramanan [a] and V.Arumugam [b]

[a] *Department of Physics, Sri Chandrasekharendra Saraswathi Viswa Mahavidyalaya, Enathur, Kanchipuram – 631 561, India. e mail: venkat_kannan @ rediffmail.com*
[b] *Biophysics Laboratory, Central Leather Research Institute, Adyar, Chennai – 600 020, India.*

Abstract. Viscosity method is simple and it offers very useful information about the relationship between dilute solution properties and bulk structure of the polymer blend. In the present study, an attempt has been made to blend PPG 500 with PPG 4000 and PEG 4000 in Toluene at 303K at 1% concentration. The blend compositions were formed in the ratio 0:1, 0.2:0.8, 0.4:0.6, 0.5:0.5, 0.6:0.4, 0.8:0.2, 1:0. Relative viscosity of the blend compositions was determined and the values are nonlinear which showed incompatibility. Miscibility nature of the blends were determined further using additive law, log additive law and free volume additive law and the experimental values in all the cases showed negative deviation, which is a characteristic of immiscible blend. The miscibility nature of the blend was further confirmed through ultrasonic velocity and optical microscopic techniques. Both the blends showed immiscibility.

Keywords: Blend, Miscibility, Polypropylene glycol, Polyethylene glycol, Viscosity.

PACS: N 61.25.hk

INTRODUCTION

Viscometry is one of the simplest methods for studying the interactions and properties of polymer-polymer systems in solutions. When two or more polymers are intimately mixed in a single continuous solid product, the composition is generally referred to as a polymer blend or polyblend. Depending upon the degree of miscibility or immiscibility, such a polyblend may resemble a random copolymer or a block or graft copolymer or it may show micro or macro heterogeneous separation into a multiphase system. Colin et al [1] measured the equation of state for the surface of the blend (PPG)+ poly(ethylene glycol) (PEG). Viscosity studies on polystyrene and High Impact Polystyrene [HIPS] in various solvents have been reported by Al-Jarrah et al [2]. Sun et al [3] has suggested viscometric method to study polymer-polymer miscibility. The miscibility studies of polymers were determined by Chee et al [4]. Arumugam et al [5] performed ultrasonic studies on PU/PMMA blends and analysed the miscibility of these blends. Varadharajulu et al [6] studied the miscibility of PVC/PMMA blends by ultrasonic and refractive index

CP1004, *Thermophysical Properties of Materials and Devices-NCTP '07*, edited by P. Predeep, S. Prasanth, and A. S. Prasad

method. Somanathan et al [7] performed compatibility studies on PS / Poly-n-butul methacrylate. Guerrica - Eche Varria et al [8] studied interfacial tension as a parameter to characterize the miscibility level of polymer blends. Viscosity studies on blends on Polypropylene Glycol were reported by Venkatramanan et al [9]. Viscometric study of PVC/PVAc blends in various solvents has been reported by Zhang et al [10]. In the present study, an attempt has been made to blend PPG 500 with PPG 4000 and PEG 4000 in Toluene at 303K at 1% concentration.

MATERIALS AND METHODS

Polypropylene glycol (PPG) of molecular weight 500, 4000 and PEG 4000 were obtained as gift from Manali Petro products, Chennai. In the present study PPG 500 is blended with PPG 4000 and PEG 4000. 1% concentration of PPG 4000, PEG 4000 and PPG 500 are separately prepared by dissolving 2.5 gms of each in 250 ml toluene (AR). Then various compositions of the blend PPG 4000 : 500 and PEG 4000 : PPG 500 in the following ratios 1:0, 0.8:0.2 , 0.6:0.4, 0.5:0.5 , 0.4:0.6 , 0.2:0.8 , 0:1 were separately prepared. The relative viscosity of the blend was studied at 30 ± 0.5^0C using an Ubbelhode suspended level Viscometer. The ultrasonic velocity measurements were performed using a Mittel multifrequency ultrasonic interferometer at 3 MHz. The temperature is maintained at 30^0C by circulating water from a thermostat with a thermal stability of $\pm 0.05^0$C through the double wall jacket of the cell. Carl Zeiss (JENAVAL) optical microscope was used for optical studies in Phase contrast mode.

RESULTS AND DISCUSSION

(A) Viscosity Method For Blends

The basic idea of using viscosity as a parameter for compatibility determination of polymer blends lies in the fact that in solution the repulsive interaction may cause shrinkage of polymer coils resulting in a viscosity of the polymer mixture that is lower than the value calculated from viscosities of the pure components on the assumption of additive law. Das and Banerjee [11] have used a few empirical and semi empirical equations for predicting the miscibility of polymer blends based on viscosity viz., the additive rule, log additive rule, and free volume additive rule given by Eq. 1, Eq. 2 and Eq. 3.

$$\eta_b = W_1 * \eta_1 + W_2 * \eta_2 \quad (1)$$

$$\log \eta_b = W_1 * \log \eta_1 + W_2 * \log \eta_2 \quad (2)$$

$$1 / (\log \eta_b) = W_1 * (1/\log \eta_1) + W_2 * (1/ \log \eta_2) \quad (3)$$

where η_b is the viscosity of the blend , η_1 , η_2 that of the components and W_1, W_2 the weight fractions of the components.

In the case of completely miscible blends, experimental values will be the same as the values calculated from the additive rule of the mixture. When the blend is partially miscible, the values predicted from log additive rule of the mixture will be close to the experimental value. Utrachi and Kamal [12] have described the complexity of the viscosity composition behaviour of polymer blends. Using rheological data, they are divided into three categories depending on the deviation from log additive rule.

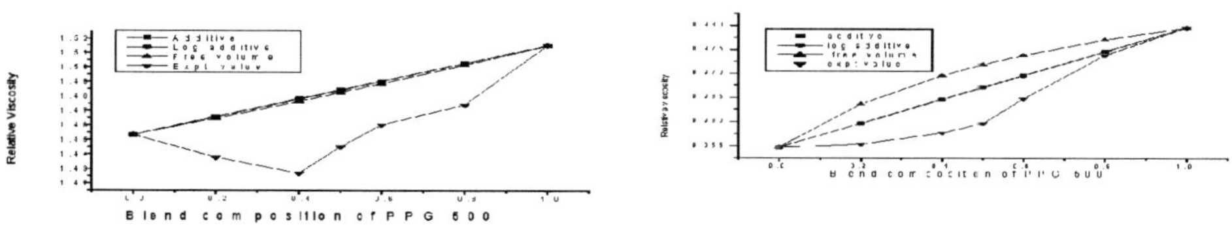

(a) **(b)**

FIGURE 1. Viscosity against blend composition PEG 4000:PPG 500 and PPG 4000:PPG 500 using additive law

i.e.) 1) positively deviating 2) negatively deviating 3) positively - negatively deviating. The negative deviation of the log additive rule in the case of immiscible blends was reported by Polockocki [13]. From Fig.1 (a&b), we could observe that the experimental values are negatively deviating and they are closer to log additive values, which refers negative interaction, which in turn causes the macromolecules to shrink. This shows that the blend PPG 4000:PPG 500 and PEG 4000 : PPG 500 is immiscible.

Sun et al [3] proposed a parameter α for the miscibility studies. The proposed interaction parameter α, is based on the long range hydrodynamic interactions for pairs of single molecules, the interaction due to double molecules and their inter molecular attraction, viscosities of double and single molecule species at molecular level, indicating miscibility, depending on attractive or repulsive forces in dilute solution. They have reported that Huggin's coefficient [14] K_m can be defined by two types of interaction,
[i] a long range hydrodynamic interaction of pairs of single molecules K_{m1} as in the Eq. 4 given below,

$$K_{m1} = \frac{K_a [\eta_1]^2 C_1 + K_b [\eta_2]^2 C_2 + 2 (K_a K_b)^{0.5} [\eta_1][\eta_2] C_1 C_2}{([\eta_1] C_1 + [\eta_2] C_2)^2} \quad (4)$$

where C_1 and C_2 correspond to the weight fraction of the components 1 and K_a and K_b are the Huggin's coefficient for the pure components.

[iii] the intermolecular attraction or repulsion represented by $K_{m3} = \alpha$.

In this condition K_m can be defined by $K_m = K_{m1} + K_{m3}$ where

$$K_m = \frac{K_a [\eta_1]^2 C_1 + K_b [\eta_2]^2 C_2 + 2 (K_a K_b)^{0.5} [\eta_1][\eta_2] C_1 C_2}{([\eta_1] C_1 + [\eta_2] C_2)^2} + \alpha \qquad (5)$$

TABLE 1. α Values with Blend Compositions

Blend Composition	PEG 4000 : PPG 500	PPG 4000 : PPG 500
1 : 0	0	0
0.8 : 0.2	-0.04583	-0.03952
0.6 : 0.4	-0.04186	-0.03246
0.5 : 0.5	-0.03848	-0.02715
0.4 : 0.6	-0.03141	-0.02142
0.2 : 0.8	-0.01726	-0.01424
0 : 1	0	0

Thus, $\alpha = K_m - K_{m1}$. Here, K_m is Huggin's constant for the blend, whereas K_A , K_b are that of pure components and were determined by Equation (6).

$$\eta_{sp} / c = [\eta] + K [\eta]^2 c \qquad (6)$$

If $\alpha >= 0$, then the system is miscible and it is immiscible when $\alpha < 0$. In the present study, both the blends PPG 4000: PPG 500 and PEG 4000 : PPG 500 show negative α values [Table 1.] which is the characteristic property of an immiscible blend.

(B) Ultrasonic Method

It has been reported by Sidkey et al [15] that [i] if the variation of ultrasonic velocity with blend composition is of S -type, then the blend is immiscible, [ii] if the ultrasound velocity shows a linear relation with composition, then the blend is compatible or miscible and [iii] for semi compatible system, the nature of the curve is in between straight line and s- shape. In the present blend system, the change in the ultrasonic velocity (Fig.2 a&b) with blend composition is of S -type and this variation shows that the blend is incompatible.

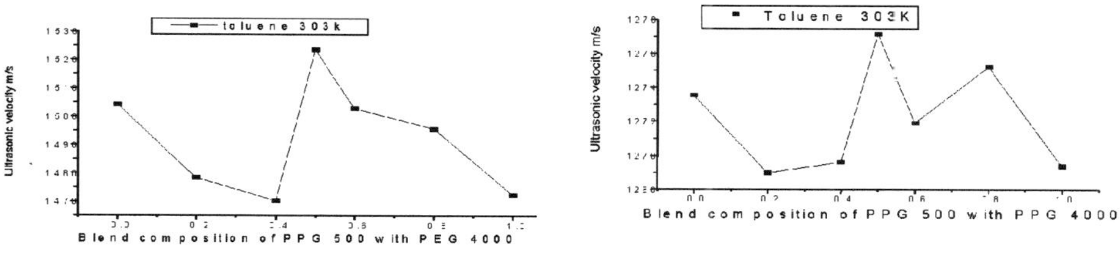

(a) **(b)**

FIGURE 2. Variation of Ultrasonic velocity against blend composition PPG 500 with PPG 4000 and PEG 4000

(C) Optical Studies

From optical microscopic studies (Fig 3. a,b,c), we find that the polymer solution in its pure state shows different character. During blending, domain formation occurs and they spread out in 50:50. Each polymer shows separate domains. The higher concentration of a polymer shows larger domains and it dominates as seen in the optical micrograph. This shows that the blend system is immiscible in nature.

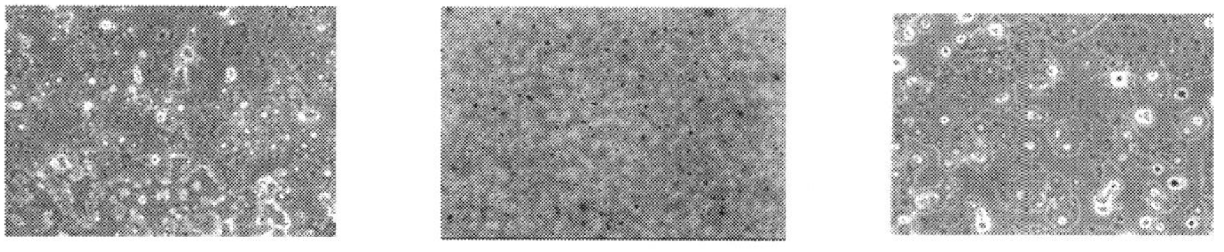

(a) 0:1 PPG 500 alone **(b)** 1 : 0 PPG 4000 alone **(c)** 0.5 : 0.5 PPG 4000:500

FIGURE3. Optical microscopic studies for blend composition PPG 500 with PPG 4000

CONCLUSION

PPG 500 is blended with PPG 4000 and PEG 4000 in Toluene at 303K at 1% concentration and the compatibility nature of the blend has been studied using viscosity technique, additive rule, ultrasonic studies and optical microscopic studies. The blend showed immiscibility.

REFERENCES

1. A. C. Colín, R.G. Rubio and A. Compositzo, *Polymer J*, **41(20),** 7407 (2000).
2. M. M. F. Al-Jarrah, A. M. Ai-Dujaili, S. K. Hassun and K. Subhi, *J. Polym. Sci*, **41,** 1339 (1990).
3. Z. Sun, W. Wang and Z. Fung, *Eur.Polym.J.* **28**, 259-1261(1992).
4. K. K. Chee, *Eur. Polym.J.*, **26(4)**, 423 (1990).
5. V. Arumugam, N. Somanathan, R.Sanjeevi and R.Balakrishnan, *J.Pure Appl. Ultrasonics*, **19**, 36 (1997).
6. A. V. Rajulu, R. L. N. Reddy, S. M. Raghavendra and S. A. Ahmed, *Eur.Polym.J.*, **35 (6),** 1183 (1999).
7. N. Somanathan, A. Sanjay, V. Arumugam, *J. Appl. Polym. Sci.*, **83**, 2322 (2002).
8. G. Guerrica - E chevarria, J. I. Eguiazabal and J. Nazabal, *PolymerTesing,* **19(7),** 849 (2000).
9. K. Venkatramanan and V. Arumugam, *Int. J. Thermophysics*, **27(1),** 66-78 (2006).
10. Y. Zhang, J. Qian, L. Ke, X. Zhu and H. Bi. K. Nien, *Eur. Polym. J.*, **38**, 333 (2002).
11. G. Das and A. N. Banerjee, *J. App. Polym Sci.*, **69**, 2577 (1998).
12. L. A.Uttrachi and M. R. Kamal, *Polym. Eng. Sci.*, **22,** 96 (1982).
13. A. P. Polockocki, *Polym. Eng. Sci.*, **22**, 1153 (1982).
14. M. L. Huggins, *J. Amer. Chem. Soc.*, **64**, 2716 (1942).
15. M. A. Sidkey, A. M. Abd El.Attah, A. A. Yehio and N. S. Abdel All, *J. Appl. Polym. Sci.*, **44,** 1441 (1991).

Structural and Electrical Characterization of Protonic Acid Doped Polyaniline

Vinodini Shaktawat*, Narendra S. Saxena*, Kananbala Sharma and Thaneshwar P. Sharma

Semi-conductor & Polymer Science Laboratory, 5, Vigyan Bhawan, Deptt. of Physics, University of Rajasthan, Jaipur- 302004, India.
**E-mail n_s_saxena@rediffmail.com, s_vinodini@rediffmail.com, Fax –91-141-2704056*

Abstract. Polyaniline doped with different protonic acids were chemically synthesized using ammonium persulfate (APS) as an oxidant. These samples were characterized through X- ray diffraction (XRD) and Fourier transform infrared (FTIR) spectroscopy, which confirms the amorphous nature and acid doping, respectively. Electrical conduction in these samples has been studied through the measurement of I-V characteristics at room temperature as well as in the temperature range from 313 K to 413 K. So obtained characteristic curves were found to be non-linear. The conductivity of phosphoric acid doped polyaniline sample is higher as compared to HCl doped polyaniline and pure polyaniline. Temperature dependence of conductivity suggests a semiconducting nature with increase in temperature. Activation energies have been found to be 50.86, 25.74 and 21.05 meV for pure polyaniline (base), polyaniline doped with hydrochloric, phosphoric acid, respectively.

Keywords: Conducting polymer, Structural characterization, Electrical conductivity, Temperature dependence.
PACS: 72.80.Lc

INTRODUCTION

Intrinsically conducting polymers such as polyaniline and polypyrrole have become very popular in field of material science due to their promising and novel electrical properties such as energy storage devices [1], gas sensors [2], electromagnetic shielding, electrostatic charge dissipation [3], light emitting diodes and flexible display devices [4], anticorrosive material [5], electrochromic materials [6] and electronic conducting fabrics [7]. Chemical doping and undoping can reversibly control the electrical conductivity of the polymers, which results into extensive use in sensor material for toxic gases and hazardous vapours. They are preferred over other conducting polymers because of their high chemical, atmospheric and electrical stability at ambient temperature [8,9]. It may offer viable alternative of inorganic semiconductors because it is possible to control the conductivity of the polymer by altering the organic molecular structure by attaching substituents in the polymer backbone.

CP1004, *Thermophysical Properties of Materials and Devices-NCTP '07*,
edited by P. Predeep, S. Prasanth, and A. S. Prasad

PANI can achieve its highly conductive state through the protonation of the imine nitrogen (=NH-) in its Emeraldine (base) state (Fig. 1a). In the first step of acid base reaction (Fig. 1b) imine groups of PANI are protonated by (H^+) of an acid and simultaneously anions (X^-) that neutralise the positive charge of the chain are incorporated to the polymeric system. This results in the formation of spinless dicationic species with quinoid-type sequence of bonds and is treated as a bipolaron. The positive charges in the form of dication are not stable for emeraldine PANI and further interactions of positive charges with π electrons lead to a more uniform charge distribution and is called "polaron lattice" with strong resonant stabilisation energy (Fig. 1c). Such acid-base doping of emeraldine lead to p-type conductors in which charge carriers are positive [10].

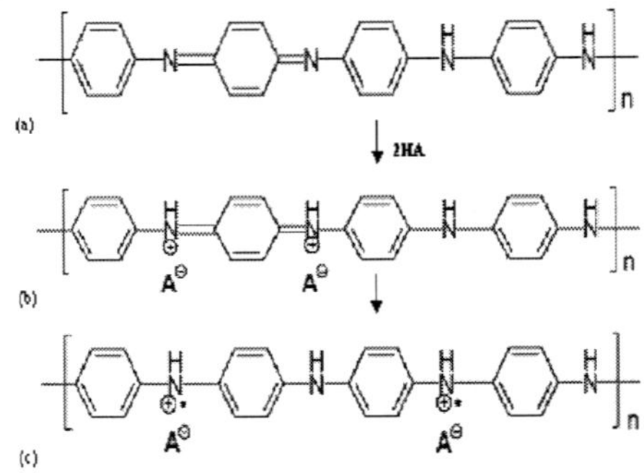

FIGURE 1. Principal of polyaniline doping (a) emeraldine base (b) intermediate dicationic species and (c) emeraldine salt

In the present work, electrical conductivity of pure polyaniline and polyaniline (doped with protonic acids HCl and H_3PO_4) has been studied. Electrical conductivity strongly depend on reaction parameter, oxidizing agent, synthesis method, and molecular weight of polymer.

MATERIAL PREPARATION AND CHARACTERIZATION

PANI is prepared by redox polymerization of aniline using ammonium peroxydisulphate [$(NH_4)_2S_2O_8$] as an oxidant as discussed by other authors [11]. This PANI salt was first neutralized with NaOH, filtered, washed thoroughly (till ph was 7), to obtain PANI base. This Emeraldine base form of polyaniline is redoped with HCl (1 M) and phosphoric acid and then filtered, washed thoroughly, dried under vacuum for 8 hrs at 60^0 C temperature to obtain doped PANI.

X-ray diffraction (XRD) pattern of pure and acid doped PANI (HCl, H_3PO_4) as shown in fig. 2 at room temperature in order to study the structure of the material. The materials were found to be amorphous in nature.

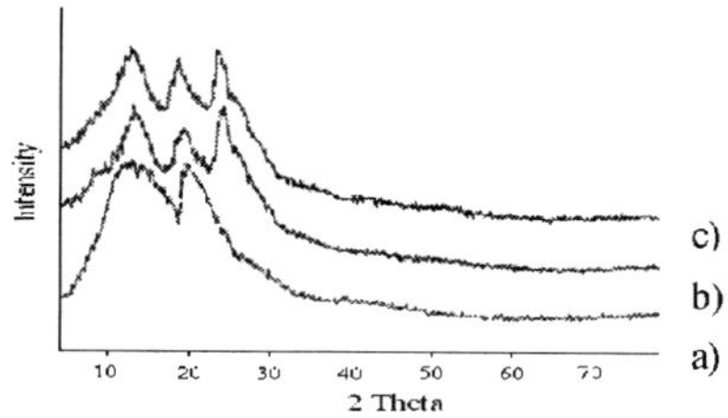

FIGURE 2. X ray diffraction pattern of pani samples a) Pure PANI, b) PANI doped with HCl and c) H_3PO_4doped PANI.

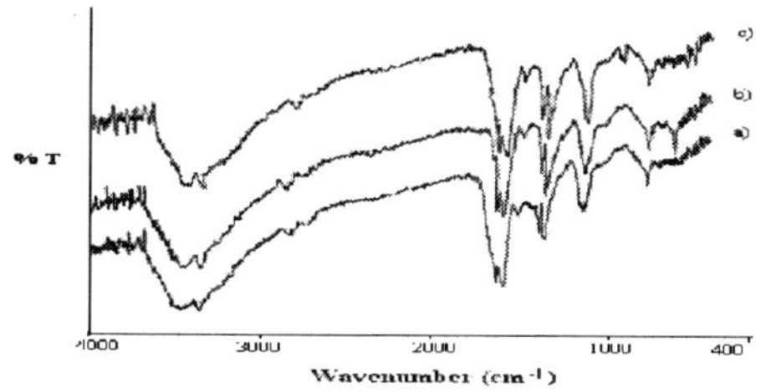

FIGURE 3. FTIR spectra of PANI samples a) Pure PANI, b) PANI doped with HCl and c) H_3PO_4 doped PANI.

The FTIR spectra for pure and doped polyaniline has been presented in fig. 3. The

presence of characteristic IR peaks of polyaniline in all the samples confirms the synthesis of PANI [12, 13]. The C–Cl stretching peak at 650 cm^{-1} in HCl doped PANI (S_1) is due to C-Cl stretching [14]. The peak observed at 1008 cm^{-1} in H_3PO_4doped PANI (S_3) is due to stretching of phosphoric group [15].

EXPERIMENTAL TECHNIQUE

The I-V characteristics were recorded in the temperature range from room to 473 K using Keithley Electrometer 6517A. The details about electrical measurements of I-V characteristics have been discussed elsewhere [16].

RESULT AND DISCUSSIONS

The I-V characteristics of all polyaniline samples are shown in fig. 4, where current increases non-linearly with increasing voltage. In the case of PANI, when nitrogen atoms in imine groups are protonated, nitrogen and its neighboring quinoid ring become a semiquinoid radical cation and the carriers generated in this process of doping are known to be self-trapped by the conjugated polymeric chains in the form of polarons or bipolarons [17]. The charged species are formed during protonation of the polymer, which are subsequently responsible for the increase in conductivity. Application of an external electric field makes both polaron and bipolaron mobile via the rearrangement of conjugation. As the applied field increases the formation of polarons and bipolarons increases. This contributes to the rapid increase in current with respect to applied voltage and is reflected in non-linear I-V curve. Elevations in temperature lead to an increase in lattice vibration, which results in chain stretching and leads to intra and inter chain hopping of charge carriers which inturn increase the conduction.

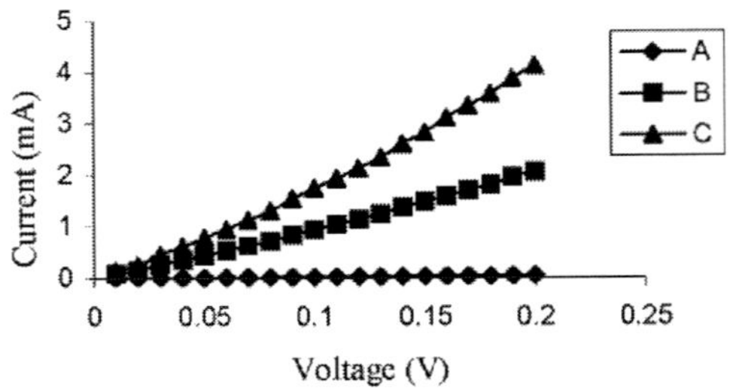

FIGURE 4. I-V Characteristics of A) Pure PPy B) PPy[Cl^-] and C) PPy[PO_4^{3-}]

Conductivity of pure polyaniline is minimum among all the samples i.e. 1.87E-05 S/cm. Conductivity (σ) H_3PO_4 doped PANI (9.18E-04) is higher than PANI[Cl^-] (9.52E-04) sample at 40^{0}C. The conductivity decreases in the order σ (H_3PO_4)> σ (HCl)> σ (Pure). This can be attributed to the strength of dopant acid and extent of doping. During solution polymerization, dissociation of ions increases with increase in acidic strength of protonic acid, which leads to creation of more and more quinoid structures along the polymer backbone. This results into formation of more and more polarons and bipolarons due to larger distorted lattice structure along the chain. Hence, conductivity increases with acidic strength except for PANI doped with H_3PO_4, which shows maximum conductivity with lower value of acidic strength.

In case of PANI doped with phosphoric acid, extent of doping of anions overweighs the effect of acidic strength. This is due to presence of lone pair and its triprotic nature [18]. Electronic configuration of N is $1s^2\ 2s^2\ 2p^3$ i.e. it is trivalent in nature. In case of PANI doped with H_3PO_4, more and more anions are introduced through out the matrix due to bond formation between N and PO_4^{3-} through local resonance and no lone pair is available on N. This leads to increased extent of doping, which initiates the formation of polarons and bipolarons, which are responsible for conduction. Hence the conductivity increases due to high extent of doping.

Temperature Dependence of Conductivity

The electrical conductivity of all the samples increases with the increase in temperature from 313 to 413K, showing the semiconducting nature of the samples (Fig. 5). In the temperature region from room to 333 K, structural relaxation of polymer chain takes place, which is not sufficient for any lattice vibrations. Beyond 333 K, lattice vibration increases which results into inter and intra chain hopping and thereby enhances the conductivity. The temperature dependence of electrical conductivity can be explained using Arrhenius model. Electrical conductivity data for all the PANI samples were found to follow the Arrhenius model [19]:

$$\sigma = \sigma_0 \exp\left[-E_{ac} / kT\right] \qquad (1)$$

where σ_0 is the conductivity at room temperature, E_{ac} is the carrier activation energy and k is the Boltzmann constant. Plots of ln(σ) vs 1000/T have been shown in fig. 6.

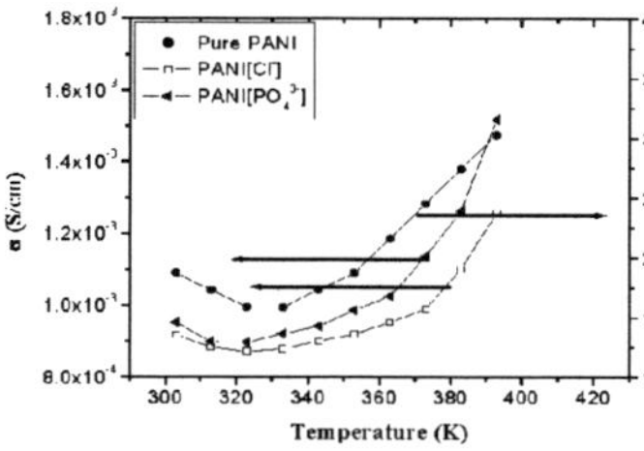

FIGURE 5. Plot of Conductivity vs. Temperature.

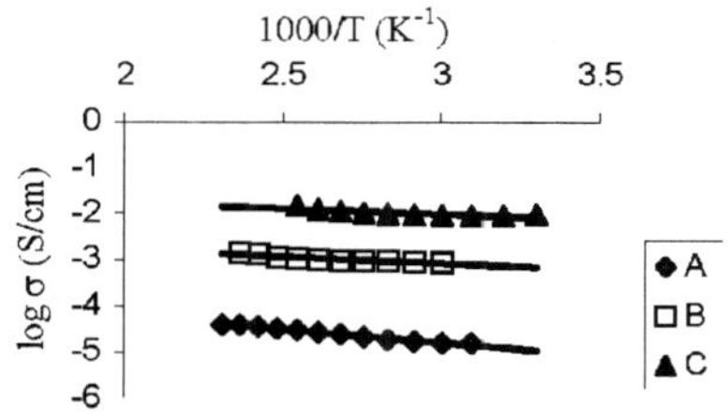

FIGURE 6. Plot of log (σ) vs. 1000/T A) Pure PPy B) PPy[Cl⁻] and C) PPy[PO_4^{3-}].

The slope [equal to E_{ac} /1000k] of these plots provides information about activation energies E_{ac}. It was found that activation energy is of the order of millielectron volt. Activation energies were found to be 50.86, 25.75, 21.05 meV for pure and PANI doped with hydrochloric and phosphoric, respectively. It is an interesting fact to note that the activation energy of PANI doped with phosphoric acid is the lowest among other doped ones, but the conductivity is largest. This is due to the fact that these conducting polymeric samples have extended states (Lowest unoccupied molecular orbital) as well as localized (or trapping) states. Both these extended and localized states are involved in charge carrier transport as they do in case of inorganic amorphous materials [20]. The increase in current in this temperature region is attributed to the increase in the number of polarons and bipolarons created in the sample i.e. increase in carrier density. This increase in carrier density is due to excitation of carriers from localized states to the extended states. The current is through the drift of carriers in the extended states with intermittent trapping-detrapping in the localized states, which enhance the conductivity of the samples [21]. This is consistent with the trapping model, which relies on the presence of both the extended states and localized state (traps). Conduction through these extended and localized states is responsible for lower value of activation energy and higher value of conductivity.

CONCLUSION

Conductivity of PANI doped with phosphoric acid is higher inspite of its low acidic strength. Temperature dependence of conductivity of samples suggests an enhancement in semiconducting nature with the increase of temperature. With doping of phosphoric acid, a large number of polaron and bipolaron states are formed with in the gap, which in turn decrease the activation energy.

ACKNOWLEDGEMENTS

The authors are indebted to DRDO, New Delhi for providing financial assistance for this work. One of the authors Vinodini Shaktawat is also thankful to Mr. Neeraj Jain, Mr. Vibhav K. Saraswat and Mrs. Rashmi Saxena for various help during this work.

REFERENCES

1. G. Kumar, A. Sivashanmugam, N. Muniyandi, S. K. Dhawan and D. C. Trivedi, *Synth. Met.* **80**, 279-282 (1996).
2. S. Koul, R. Chandra and S. K. Dhawan, *Sensors & Actuators B* **75,** 151-159 (2001).
3. R. Wycisk, R. Pozniak and A. Pasternak, *J. Electrostatics* **56**, 55-66 (2002).
4. L. Kumar, S. K. Dhawan, M. N. Kamalsannan and S. Chandra, *Thin Solid Films* **441,** 243-247 (2003).
5. C. K. Tan and D. J. Blackwood, *Corrosion Science* **45**, 545-557(2003).
6. Colin Pratt, Application of conducting polymers (2003).
7. K. Hwa, H. K. Wha Oh and T. J. Kang, *J. App. Polym. Sci.* **97,** 1326-1332 (2005).
8. K. Anuar, S. Murali, A. Fariz and H. N. M. MahmudEkramul, *Mater. Sci.* **10**, 255-259 (2004).
9. A. Kassim, Z.B. Basar and H. N. M. E. Mahmud, *Chem.Sci.* **114**, 155-162 (2002).
10. F. Zuo, M. Angelopoulos and A. G. MacDiarmid, *Phys. Rev. B* **36,** 3475-3478 (1987).
11. G.P. Joshi, N.S. Saxena, T.P. Sharma, V. Dixit and S.C.K. Mishra, *J. Phys. Chem. Solids* **64,** 2391-2396 (2003).
12. R. Murigesan and E. Subrammian, *Bull. Mater. Sci.* **25**, 613-618 (2002).
13. Y. R. Sharma, *Elementary organic spectroscopy*, S Chand and Comp Ltd, New Delhi (2002).
14. J. L. Koenig, *Spectroscopy of polymer*, 2nd Ed, Amsterdam, Elsevier (1999).
15. J. A. Asensio, S. Borrós and P. G. Romero, *Elect. Comm.* **5**, 967 (2003).
16. Vibhav K. Saraswat, K. Singh, N. S. Saxena, V. Kishore, T. P. Sharma and P. K. Saraswat, *Curr. App. Phys.* **6**, 14-18 (2006).
17. N. Jain, D. Patidar, N. S. Saxena and K. Sharma, *Ind. J. Pure & Appl. Phys.* **44**, 767-770 (2006).
18. Wikipedia: The Free Encyclopedia, http://en.wikipedia.org/wiki/phosphoricacid.
19. M. Gosh, A. Barman, A. K. Meikap, S. K. De and S. Chatterjee, *Phys. Lett. A* **260,** 138-140 (1999).
20. P. G. Le Comber and W. E. Spear, *Phys. Rev. Lett.* **25,** 509-511 (1970).
21. A. K. Kapoor, S. C. Jain, J. Poortmans, V. Kumar and R. Mertens, *J. Appl. Phys.* **92**, 3835-3838 (2002).

Low Dose Gamma Radiation Monitoring Through TiO_2 Doped Lead Phthalocyanine (Pb-Pc) Based Schottky Device

Yojana Janu, Anil Gautam, Manish Kumar, Narottam Prasad, Y.S.Deol and M.S. Roy

Radiation Processing and Devices Group, Defence Laboratory, Jodhpur, Rajasthan – 342 011, India

Abstract. The concept of Organic thin film based solid-state dosimeters is relatively new and more effective. The organic conductor based solid- state dosimeter provides a mean for low cost, ease to fabricate and sensitive radiation sensor which can be employed as pocket dosimeter for army personals getting exposed to nuclear radiation while working in the radioactive environment This concept is being utilized here for monitoring the effect of nuclear radiation on our organic material based sandwiched devices. In the present communication, lead Phthalocyanine (PbPc) doped with TiO_2 (5% by weight) is developed into the form of thin film structure. The developed ITO/PbPc: TiO2/Ag Schottky device structure was characterized in terms of change in its electrical and optical properties before and after exposure to radiation Exposure to radiation imparts an accelerated decrease in forward bias current and capacitance characteristics reveal a linear relationship between dose v/s current behavior which supports its suitability as pocket dosimeter for the dose ranging from 50 mR to 800 mR.

Keywords: Metal-Phthalocyanines (M-Pc); Schottky Device; Thin film; Gamma Radiation; Optical Band gap; Radiation Dosimeter

PACS: 61.82.-d Radiation effects on specific materials

INTRODUCTION

The metal-phthalocyanines are well-known p-type organic semi-conducting materials and have been extensively used in the form of thin-film based devices for solar cell and gas sensor [1-2]. Recently, these materials have attracted its attention as sensing element towards fabrication of radiation dosimeter [3-5]. The most interesting property of the phthalocyanines is their similarity in structure with biological molecules i.e. Chlorophyll and hemoglobin [6] and therefore such dosimeter are considered as tissue equivalent. . In the present communication, some metal-phthalocyanines which are p-type material, thermally very stable and can be sublimed into thin film structure were synthesized and taken as a semi-conducting sensing element for the fabrication of thin film Schottky and capacitive devices. The developed ITO/M-Pcs/Ag Schottky device structure was characterized in terms of change in its electrical and optical properties before and after exposure to radiation.

CP1004, *Thermophysical Properties of Materials and Devices-NCTP '07*,
edited by P. Predeep, S. Prasanth, and A. S. Prasad

EXPERIMENTAL

Synthesis of Lead-Phthalocyanine

The material was synthesized through microwave-assisted synthesis process using focused microwave based SYNTHEWAVE-402 (make Prolabo, France). In the procedure, the fine grounded phthalic anhydride (18 mmol), urea (92 mmol), ammonium chloride (17 mmol), ammonium molybdate (5 mg as catalyst) and lead chloride (0.6 mmol) were taken in a Quartz cylindrical reaction vessel (cap. 250 ml) and then irradiated with microwave (Power 300W) under the given temperature profile. The product obtained within 15 minutes reaction time was washed with hot water and then dried in vacuum oven for one hour. The re-precipitation through concentrated H_2SO_4 leads to the purest form of the material. Further, purification was carried out by soxlet extraction with acetone for 3 hour and then pure Pb-Pc was obtained approximately in 70% yield. The absorption spectrum of M-Pc, recorded with UV-Vis spectrophotometer reflects two absorption bands as reported for other metal-phthalocyanines. The structure of Pb-Pc is given in fig. (1).

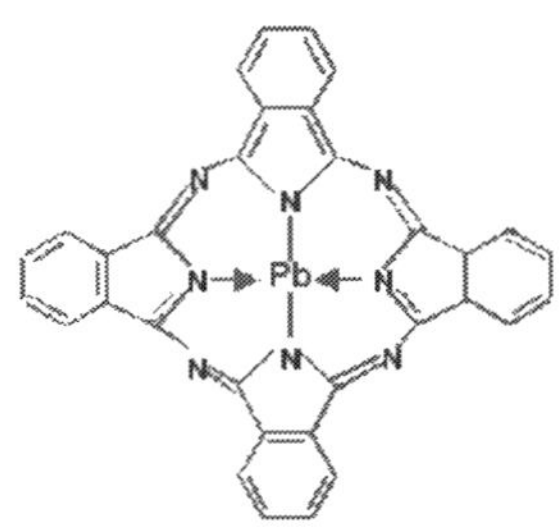

FIGURE 1: Structure of Metal Phthalocyanine

Fabrication of Schottky device

Schottky device was fabricated by developing the thin-film of Pb-Pc blended with TiO_2 employing spin coating technique on metallic substrate. The Silver (Ag) contact was then made over the top of the organic film covering 1cm^2 active area of the device using the vacuum evaporation method. The fabricated device configuration is shown in Fig. 2.

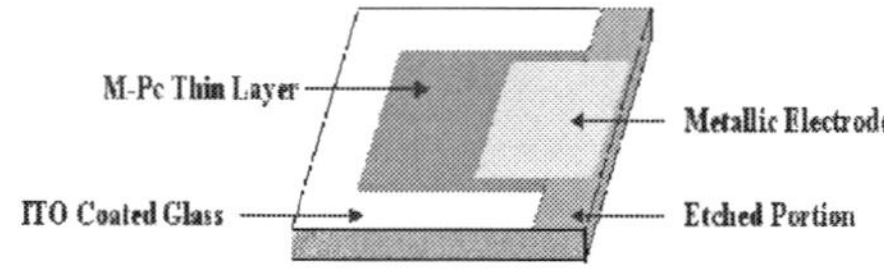

FIGURE 2: Shottky device having configuration ITO / PbPc: TiO2 / Ag

Effects of γ-ray on the Optical Transitions

Optical properties of the thin-film structure were recorded in visible region of the spectrum before & after exposure to radiation (Fig. 3). The absorbance of both B and Q bands increased linearly when exposed to γ-ray doses i.e. from 10 mR to 800 mR. It has been observed that exposure to radiation imparts an accelerated decrease in the optical band gap (i.e. 1.7 eV before exposure and 1.47 eV after exposure) for the device ITO/PbPc: TiO2/Ag, due to subsequent increase in the structural disorder.

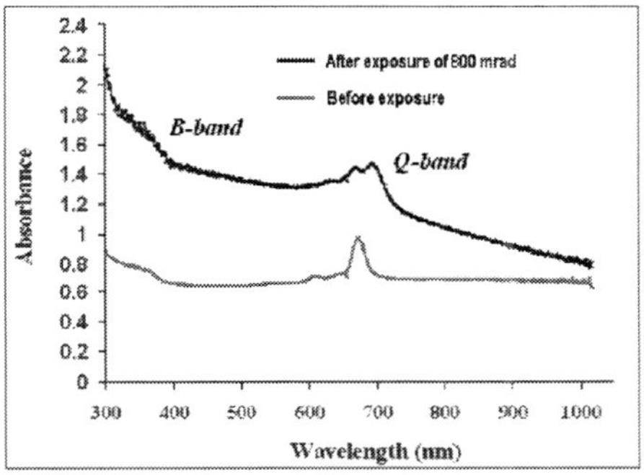

FIGURE 3: Absorbance characteristics of the device ITO/PbPc: TiO_2 / Ag

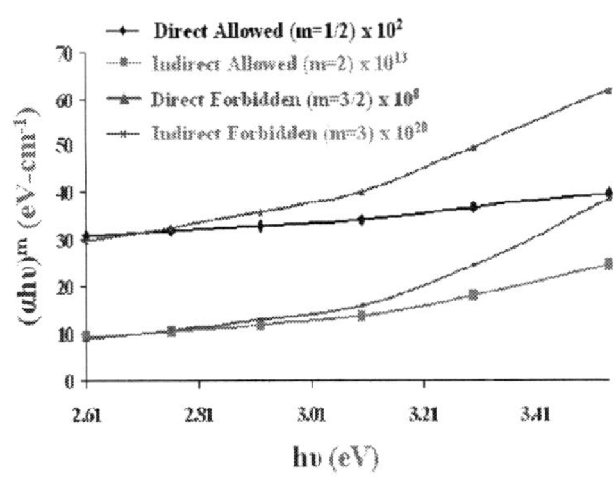

FIGURE 4:. The electronic transition of PbPc: TiO2 thin film before exposure

Fig. 4 represents plots of the four possible optically electronic transitions for the PbPc: TiO2 thin film; a) direct allowed transition, b) indirect allowed transition, c) direct forbidden transition and d) indirect forbidden transition. These transitions were obtained by analyzing the absorption edge of the B-band in view of the models proposed by the Mott & Davis [7]

$$\alpha(\nu)\, h\nu = B\,(h\nu - E_{opt})^m$$

Where $\alpha(\nu) = 2.303\ A/d$ (A is the absorbance and d is the film thickness), h: Planck's constant, ν: the frequency of the incident photons, B: constant obtained from the slope of the $(\alpha h\nu)^m$ versus hν (Fig. 4 & 5), Eopt: optical band gap and m = ½, 2, 3/2 or 3 for direct allowed, indirect allowed, direct forbidden and indirect forbidden respectively.

In Fig. 4, the only satisfactory straight line fit for $(\alpha h\nu)^{1/2}$ as a function of the hν is displayed. This indicates that the direct allowed transition is predominant transition for the PbPc: TiO2 thin film. The same direct allowed electronic transition has been established after exposure with γ-ray doses (Fig. 5).

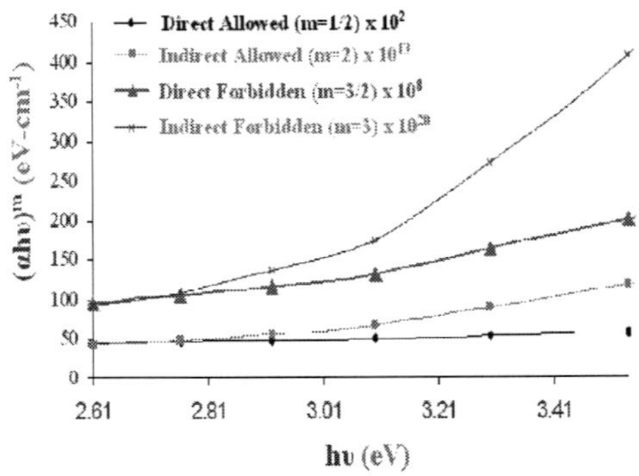

FIGURE 5: the effects of γ-rays on electronic transition for PbPc: TiO2

Effects of radiation over Current-Voltage characteristics

The current-voltage characteristics of ITO/PbPc: TiO_2/Ag was recorded before & after exposure to gamma radiation of variable doses ranging from 10 mR & 800 mR. (Fig.6). The biasing voltage was varied from 0 to 5 volts. The calibrated gamma radiation source Co-60 was used for exposing the device and the dose rate was maintained at 53 mrad/hour. The devices show rectification effects initially but successive decreases in rectification ratio were observed with increase in radiation dose. The rectification observed is due to the p-type conductivity of M-Pc, which forms Schottky-barrier with low work function Ag (4.2 eV) electrode and Ohmic contact with high work function Indium-tin-oxide (ITO) (4.8 eV) electrode.

It was found that forward bias current also decreases with increase in radiation doses and this can be attributed to an increase in distance of inter-chain hoping leading to widening of depletion layer width.

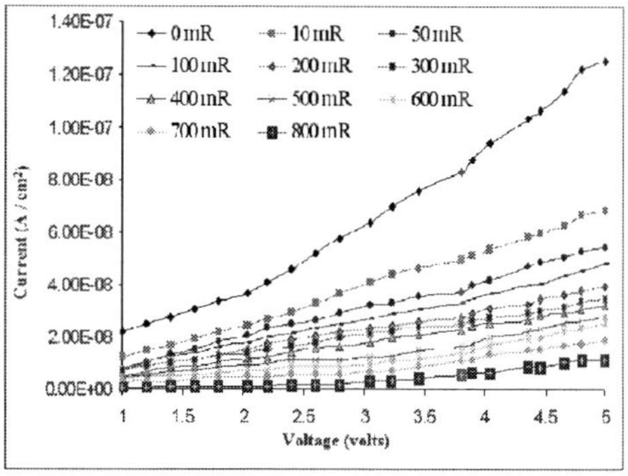

FIGURE 6:. J-V Characteristics of the device before & after exposure to radiation

It is also evidenced by subsequent increase in ideality factor and relative permittivity derived by re-plotting of ln (J) vs. V characteristics (Fig7) and ln (J) vs. $V^{1/2}$ characteristics (Fig8) respectively with increasing radiation doses (Table 1) and it reveals simultaneous increase in structural disorder resulted from recombination current.

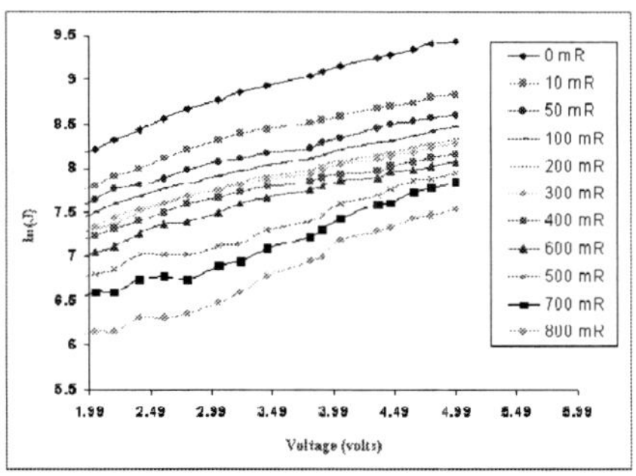

FIGURE 7: Variation of ln (J) versus V at different Doses

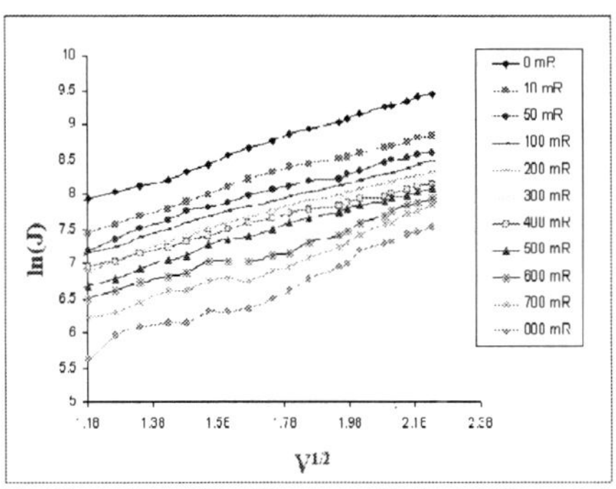

FIGURE 8:.Variation of ln (J) versus $V^{1/2}$ at different doses

To ascertain the potential of the device as radiation sensor or dosimeter, variation of current with absorbed dose was plotted at different biasing voltage as shown in Fig. 7 & Fig. 9. The observation of linear response between current vs. dose supports its possible application as radiation sensor or dosimeter.

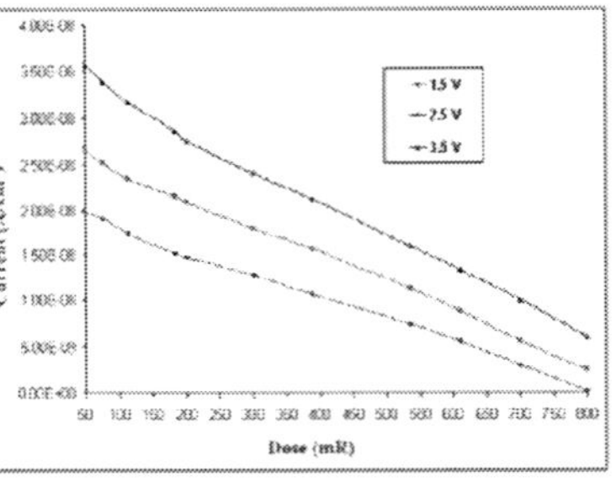

FIGURE 9:. Dose vs. Current characteristics with absorbed Dose

The variation of capacitance with increasing radiation doses (Fig. 10) reveals that capacitance decreases with increase in dose due to accelerated ionization taking place within the materials.

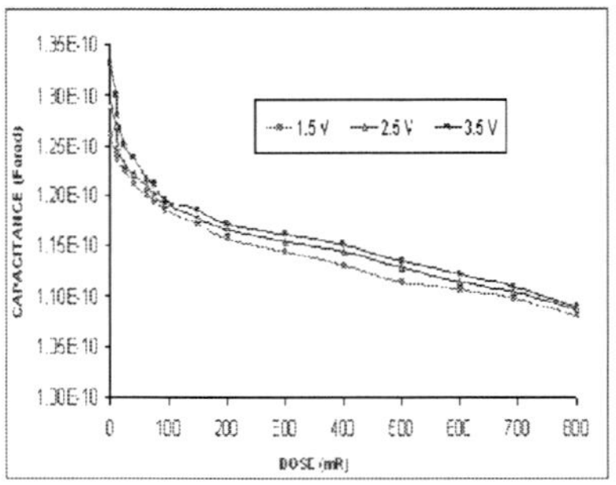

FIGURE 10:. Variation of capacitance with Absorbed Dose

TABLE: 1 Variation of Diode Ideality Factor and Relative Permittivity of the Device ITO/PbPc: TiO2/Ag

S.No.	*Dose (mR)*	*Ideality Factor*	*Relative Permittivity*
1.	Unexposed	1.83	1.99
2.	75	1.91	2.55
3.	100	1.92	2.76
4.	300	1.95	3.30
5.	500	2.01	3.96
6.	600	2.11	4.53
7.	800	2.20	4.79

CONCLUSION

The overall observations made with ITO/PbPc: TiO2/Ag Device reveals that exposure to radiation imparts successive removal of carriers and that further accelerated with increasing dose. This is supported by increase in ideality factor as well as decrease in optical band gap of the materials. In view of this, such devices have adequate potential in radiation monitoring.

REFERENCES

1. H. Spanggaard and F.C. Krebs., *Sol. Energy Mater. Sol.Cells* **83**, 125(2004)
2. F.T. Reis, D. Mencaraglia, S. Oould Saad, I. Seguy, M.Oukachmih, P. Jolinal, P. Destruel., *Synth. Met.* **138,** 33(2003)
3. M.S. Roy, Manish Kumar, Pratibha Jaiswal and G.D.Sharma *Radiation Measurement,* **38**, 205-209 (2004)
4. M. S. Roy, Anil Gautam, Manish Kumar, *Proceeding of International Workshop on Physics of Semiconductor Device (IWPSD-2005) Vol II* Page No. 935, Dec. 13-17 (2005)
5. J. Assaf, A.Al-Mohamadb, *Radiation Measurement,* 32 153-158 (2000)
6. H. S. Nalwa, A. Kakuta, *Thin Solid Films*, **254**,218-223 (1995)
7. Mott, N.F; Davis, *EA Electronic process in Non Crystalline Materials, 2nd Clarendon Press*, Oxford, U.K. (1997)

Investigation Of Ionic Character Of Alkali Oxide Glasses

Harikuttan Unnithan.C[a], Dhanya.I[a] and P.Predeep[b]
[a]Solid State Physics Laboratory
D.B.College, Sasthamcotta, Kollam- Kerala- 690 521
[b]Department of Science & Humanities, National Institute of Technology, Calicut

Abstract. The relations between the electro negativity difference of alkali oxide glasses X-O (X= Li,Na,K,Rb,Cs) and the ionic character of the bonds are discussed in detail. The ionic character studies on alkali oxide glasses using Pauling's equation, Hannay & Smyth relation and that of Gordy show that the ionic nature of alkali metals are controlled by the oxygen linkage and they posses Fractional Ionic Character (FIC). A comparative study on alkali oxide glasses is made using above three relations. The percentage of ionic character gives a clear idea about the colour of glass system. Further the packing fraction (PF) of Li-O and Na-O glasses show a structural change due to covalent packing while the other oxide glasses have almost similar range of PF of molecules having ionic bonds which are suitable for wide application in semiconductor technology.

PACS: 78.55.Qr

INTRODUCTION

Alkali Oxide Glasses are more important in the field of optoelectronics such that their electrical studies are more significant. This is an ongoing research work explains the effect of oxygen on alkali metals to form alkali oxide glasses. On comparing with highly doped glasses, binary glasses having more importance due to their wide glassy region in geometrical structures [1] .The series of glasses under this investigation includes X-O (X= Li,Na,K,Rb,Cs). In the above series of metal oxide glasses, the Na-O, Li-O glasses have large technological application that they can be used for manufacturing of optical modulators and optical switches [2]. In order to use them effectively in optical conductivity, it is essential to know the percentage of ionic character in terms of its electro negativity and Fractional Ionic Character (FIC). By knowing these two factors, one can determine the degree upto which the Non directional Bonding of Oxygen (NBO) [3]. Further the Packing Fraction (PF) of these glass systems is also a general measure of structure of oxygen modified glass systems. The PF shows that alkali oxide glasses having two types of packing, one having electro negativity smaller than oxygen corresponding to covalent packing (CP) (Li-O & Na-O) and the other having electro negativity larger than oxygen corresponds to Ionic Packing (IP) (K-O,Rb-O,Cs-O) [4].

CP1004, *Thermophysical Properties of Materials and Devices-NCTP '07*,
edited by P. Predeep, S. Prasanth, and A. S. Prasad

THEORY

Nuclear quadrupole coupling constants in molecules are closely correlated with molecular electronic structure and in favorable cases to give fairly clear evidences about the nature of chemical bonds [5]. The Alkali Oxide glass systems, with Non directional Bonding of Oxygen atom (NBO) can be explained by means of Fractional Ionic Character (FIC) and can be represented by Pauling's equation [6].

$$i(pauling) = 1 - \exp^{\frac{1}{4}(\chi_A - \chi_B)^2} \quad (1)$$

Hannay and Smyth relation [3] is given by,

$$i(Hannay\,\&\,Smyth) = 0.16(\chi_b - \chi_a) + 0.035(\chi_b - \chi_a)^2 \quad (2)$$

Gordy relation [7] between electro negativity and ionic character is given as,

$$i(Gordy) = 1 - \tfrac{1}{2}|\chi_A - \chi_B| \quad (3)$$

Here the electro negativity χ_B (cation) χ_A (anion) undergo the condition $\chi_A < \chi_B$.. The electro negativity of an atom is different from the electrode potential of an element which depends on the difference in free energy of the element in its standard state and in ionic solution. The difference in electro negativity of atoms for bonds in alkali oxide is χ_X- χ_o then the extra ionic energy is given by the relation,

$$\Delta'(\chi - O) = 30(\chi_x - \chi_o)^2 \quad (4)$$

The bond energy in terms of geometric mean (in Kcal/mol) is given by,

$$D(\chi - O) = \{D(\chi - \chi) * D(O - O)\}^{\frac{1}{2}} + 30(\chi_x - \chi_o)^2 \quad (5)$$

In the absence of the knowledge of bond energy of the elements, the arithmetic mean method is employed such that the above relation can broadly stated as,

$$D(A - B) = \tfrac{1}{2}\{D_{(A-A)} + D_{(B-B)}\} + 23(\chi_A - \chi_B)^2 \quad (6)$$

For Oxide glasses the above relation takes the form

$$D(x - o) = \tfrac{1}{2}\{D_{(x-x)} + D_{(o-o)}\} + 23(\chi_x - \chi_o)^2 - 26n_O \quad (7)$$

n_o represents the number of oxygen atom in the glass material.

Further the Packing in Alkali Oxide glasses [8] are also specified by means of molecular volume V_m(x) and density $\rho_{(x)}$ in terms of molar fraction (x).

$$PF = \sum\left(\tfrac{4}{3}\pi x_i^{\,3} n_i / V_m\right) \quad (8)$$

x(i) denotes the radius of ith atom and n_i represents the number of corresponding atom[9,10,11,12].

RESULTS AND DISCUSSION

The optoelectronic application of Alkali oxide glass requires significant studies on ionic character of the materials. The alkali metals such as Li, Na, K, Rb, Cs are completely ionic. The oxygen linkage with them decreases the ionic character and the ionic package is controlled by covalent bonding of oxygen. It is difficult to formulate a reliable relation between the partial ionic character of the bond and the difference in electro negativity of the two atoms between which the bond is formed. This difficulty arises from the fact that the description of a bond as a hybrid between a normal covalent bond and an extreme ionic bond is only a rough approximation. The relationship between the electro negativity and the FIC of alkali oxide glasses calculated from the three famous relations are compared [Fig.1 (a&b)].

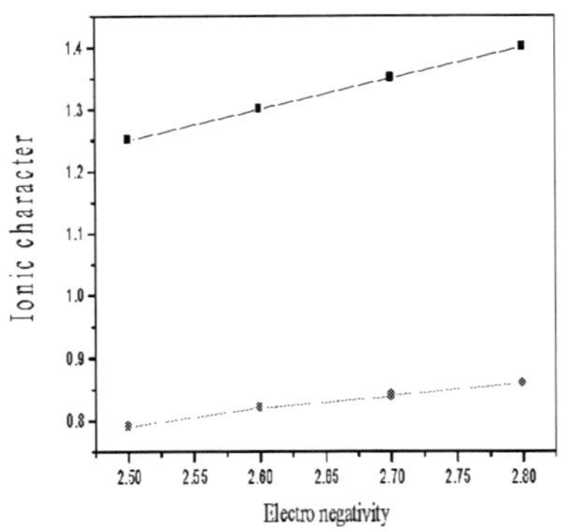

Fig 1.Variation of ionic character of Pauling and gordy with respect to electro negativity differenc

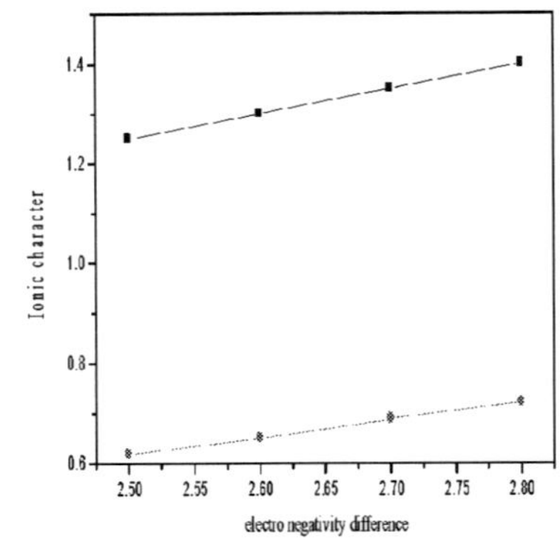

Fig.2. Variation of ionic character of gordy & hannay and smyth with respect to electro negativit

FIGURE 1. (a) Variation of ionic character of (a) Pauling and gordy with respect to electronegativity difference. (b) gordy & hannayand smith with respect to electronegativity difference.

The comparative study verifying fig.1a & b give the idea about their linear relation ship and the enthalpy of heat formation between the bonds. On applying the above result the bond energy of alkali oxide glasses can be determined and it gives a clear idea for the stability of oxygen in alkali oxide glasses. The Fractional Ionic Character (FIC) studies on the series of alkali oxide glasses also provide some interesting aspects as follows,

- On adding oxygen, the Fractional Ionic Character (FIC) increases in alkali metals from top to bottom in periodic table.
- Li-O bond is slightly less ionic and is more covalent in its character than Na-O linkage and is undisputedly due to a network forming bond in glass.

The percentage of ionic character also gives a correlation between colour of alkali oxide glass and bond character in them. The percentage of ionic character in alkali oxide glass is above 20, posses white colour similar to normal glasses. The Packing Fraction (PF) of Li and Na glasses is different in the series shows a structural change. Since the packing displays a structural trends across many systems. Li and Na with oxygen results in covalent packing while the others (K,Rb,Cs) with oxygen give rise to ionic packing. So it is clear that alkali oxides such as K-O, Rb-O, Cs-O are efficient materials in the field of semiconductor technology.

TABLE 1. Ionic character obtained by Gordy's relation, Pauling's equation, Hannay & smyth with respect to electro negativity of alkali oxide glasses.

Glass compositions	Electro negativity (X_a-X_b)	i(Gordy)	i(Pauling)	i(Hannay&Smyth)
Li-O	2.5	1.25	0.79	0.619
Na-O	2.6	1.3	0.82	0.65
K-O	2.7	1.35	0.838	0.687
Rb-O	2.7	1.35	0.84	0.69
Cs-O	2.8	1.4	0.859	0.722

TABLE 2. The bond energy and percentage of ionic character alkali oxide glasses

Glass compositions	Electro negativity (X_a-X_b)	Bond energy (Arithmetic mean)	% of ionic character (%)
Li-O	2.5	147.6	79
Na-O	2.6	155.08	82
K-O	2.7	164.87	84
Rb-O	2.7	164.47	84
Cs-O	2.8	176.27	86

TABLE 3. The Packing Fraction (PF) of alkali oxide glasses

Glass compositions	Density, ρ (gm/cc)	Radius, $r_i \pm 0.05 (A^0)$	Volume $V_m (A^{03}) \pm 1.0$	Packing Fraction(PF)
Li-O	2.013	0.73	1.63	0.43
Na-O	2.27	1.16	6.54	0.45
K-O	2.35	1.65	18.8	0.68
Rb-O	4.0	1.77	23.2	0.69
Cs-O	4.65	1.95	31.1	0.69

CONCLUSIONS

The present investigation on alkali oxide glasses derives the following conclusions:

1. The FIC and electro negativity give the clear idea about the activation enthalpy for glass transition of alkali oxide glasses.
2. The increasing percentage of ionic character in the series reveals that these materials are useful in optoelectronic application.
3. The Packing Fraction (PF) of alkali oxide glasses give clear understanding about the ionic packing in the glass network. Further it is clear that heavy alkali oxide glasses are efficient materials for the preparation of various optoelectronic devices. So by doping various elements on these glasses, the series become novel in the field of communication and technology.

ACKNOWLEDGMENTS

Sincere thanks to Kerala State Council for Science, Technology & Environment , Government of Kerala for financial support to this project.

REFERENCES

1. Z.G.Borizova, Leningrad state univ., Leningrad, USSR.
2. Stanford.R.Ovshinsky, Energy Conv. Devices, Inc., 1675, West maple road, Trog, U.S.A.
3. A.A Higazy, B.Bridge, Journal of non – crystalline solids, 72 (1985) 81-108.
4. sandeep Giri, Carl Garbler, Jonathan Helmus, Mario Affatigato, Steve Feller, Masao Kodama, Journal of non – crystalline solids, 347 (2004) 87-92.
5. B.P Dailey, C.H Townes, Journal of Chem.Physics, Vol. 23(1) (1995).
6. Linus Pauling, The nature of chemical bond and the structure of molecules and crystals, Third Ed.
7. J. Sheridan and W. Gordy, J. Chem. Phys. 19,965 (1951).
8. R.D Shannon, Acta Crystallogr. A. 32 (1976) 751.
9. R.L. Mozzi, B.E. Warren, J. Appl.Crystallogr.3 (1970) 251.
10. A.C. Wright, Private Communication.
11. W.H. Zachariasen, Acta Crystallogr. 17(6) (1963)749.
12. J.L Shaw, Ph.D. thesis, Univ. of Reading, 2003.

Dynamic Mechanical Properties of Natural Rubber/ Polyaniline Composites

Najidha.S[a], P.Predeep[a*] and N S. Saxena[b]

[a]*Condensed Matter Physics Lab., Sree Narayana College, Kollam, Kerala, India 691 001.(*ppredeep@gmail.com)*

[b]*Semiconductor and Polymer Science Lab, Department of Physics, University of Rajasthan, Jaipur*

Abstract. The Dynamic Mechanical properties of polymer composite containing Natural Rubber (NR) as the matrix and polyaniline as filler has been studied. The composites were prepared by mechanical mixing in a roll mill and vulcanized in a hot press. The dynamic modulus such as tanδ, storage modulus and loss modulus of the composite were evaluated. The glass transition (T_g) temperature of the Natural Rubber phase in the composite was shifted to lower temperature indicating that the polyaniline content strongly affects the behavior of the composite. Addition of polyaniline lowered the crosslinking degree, but produced a reinforcing effect in the elastomer.

Keywords: Polyaniline, composite, Natural Rubber, crosslinking degree.

PACS: 83.85Vb, 81.05.Lg

INTRODUCTION

Generally the uniform mixing of the conjugated polymers like polyaniline, polypyrrole etc. with an insulating polymer such as elastomer or plastics with a suitable loading of the conductive polymers to reach the percolation threshold is crucial in obtaining a highly conductive polymer composite. Composites with reasonable conductivities and mechanical properties can be prepared by adjusting the ratio of the conductive component to the insulating component. However, too high loading of the conductive polymer has a negative effect on the mechanical properties particularly on the elongation or flexibility of the composite material. Besides, in elastomeric composites, the effects of fillers are quite unpredictable, since several factors such as distribution, dispersion and interaction of the filler with the components of the composite may vary widely. However, the processing problems and the cost effectiveness for obtaining high-quality homogeneous polymers are still challenging issues for many potential applications[1-3]. Because of this, there is a growing demand for dry mill mixing method compared to the solution mixing. The roll mill mixing provides more uniformity and hence convenient method compared to the solution-processable composite in which the use of organic solvents arise problem.

The Measurement of dynamical mechanical properties helps in understanding the behaviour of elastomers under cyclic loading and temperature. The modulus

CP1004, *Thermophysical Properties of Materials and Devices-NCTP '07*, edited by P. Predeep, S. Prasanth, and A. S. Prasad

information obtained from dynamic mechanical testing is of importance to manufacturers and users of polymers in structural applications. The dynamic mechanical properties of elastomers are strongly dependent on temperature, type and concentration of filler and the extent of deformation [4-6]. In view of these aspects in this paper an effort has been made to determine the dynamic mechanical behavior of Natural Rubber/Polyaniline composites prepared by dry mill mixing technique. Polyaniline has been used as the filler for natural rubber matrix which ensures good compatibility compared to the commonly used fillers such as metal powders, carbon black etc.

EXPERIMENTAL

Natural Rubber used for this study is Indian Standard NR; grade 5(ISNR-5) was supplied by the Rubber Research Institute of India. Aniline (Merck) was thrice distilled under reduced pressure and stored under nitrogen at 4^0 C. Ammonium per sulfate (APS) (Nice Chemical) as oxidant and hydrochloric acid (36.46%) (Merck) as dopant were used as received. Methanol and acetone (Merck) were used without further purification. Polyaniline was synthesized in powdered form by the conventional chemical method of oxidative polymerization of aniline with ammonium peroxy disulphate (APS) in an aqueous acidic medium [7]. 20 g of aniline hydrochloride (16.373 ml or 0.15 mol) was dissolved in about 200 ml of 1.5 M HCl. 8.166g of ammonium per sulfate (0.038 mol) was dissolved in excess (50 ml) of 1.5 M HCl and added dropwise to the stirred solution. After reacting for 24 h, the solution was filtered. The synthesized product was repeatedly washed with 1 M HCl, distilled water and then using methanol and acetone. The powder was dried overnight at 60^0 C under dynamic vacuum to constant mass. Protonated pernigraniline was responsible for the blue colour of the reaction mixture observed during polymerization. The color of the reaction mixture changes to green, corresponding to protonated emeraldiene, at the end of the polymerization.

The blends were prepared by mechanical mixing in a roll-mill at room temperature according to the formulations indicated in table 1. NR was masticated and then blended with polyaniline. The samples were then vulcanized at 150^0C and under a pressure of 10 Tonnes for 4 min using a Hot press. Dynamic mechanical measurements were carried out using a Tritec 2000 Dynamic Mechanical Analyser. For these measurements, samples with dimensions 11.1 x 0.5 x 0.6mm^3 were used. The experiment was conducted in uniaxial tension mode from –70 to +30 °C at a frequency of 1Hz at with a programmed heating rate of 4$Kmin^{-1}$. Liquid nitrogen was used to achieve subambient temperatures. The mechanical loss factor tanδ and the dynamic moduli such as storage moduli and loss moduli were calculated with a microcomputer.

TABLE 1. Formulation of mixes used for crosslinking

Formulation	**Composition (g)**
Natural Rubber	100
ZnO	5
Stearic acid	2.5
Sulfur	2.5
HS	1
CBS	1
TMT	0.12
Polyaniline + Carbon black	40

RESULT AND DISCUSSION

The dynamic mechanical behaviors of Natural Rubber / Polyaniline composite system are shown in figures 1 to 3 .The glass transition temperatures was selected at the peak position of tanδ when plotted as a function of temperature. The T_g values of all the system are given in table 2. Figure 1 represents the tanδ values of the composite as a function of temperature. It is seen that the sulfur crosslinked Natural Rubber exhibits a transition at -40.6^{0}C. On the addition of polyaniline the Tg showed a tendency of shifting to the lower temperature region. At about 60% of polyaniline the Tg was found to be -52.1^{0}C.

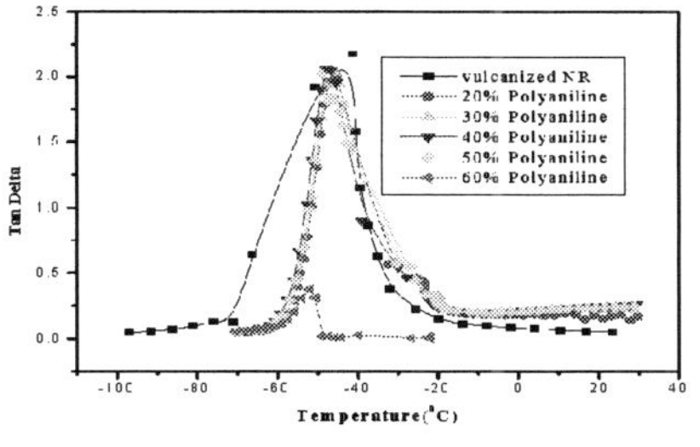

FIGURE 1. Tan δ as a function of temperature for Natural Rubber / Polyaniline composite system.

TABLE 2. Variation of Tg with volume percentage of Polyaniline

Volume percentage of Polyaniline	Glass transition temperature (T_g)
0	-40.6
20	-46.0
30	-46.6
40	-47.7
50	-48.8
60	-52.1

It has already been reported [8,9] that the hydrochloric acid doped polyaniline has an intrinsic acid content and an increasing the volume percentage of polyaniline in the elastomeric blend induces degradation process of the crosslinking agent by heterolytic decomposition and thus inhibiting the formation of the free radicals necessary for the vulcanization reaction. Thus we can assess that the vulcanization is precluded by the presence of acid because these react with the sulfur compounds, competing with their radicalar degradation during the vulcanization process. Thus the higher loading of polyaniline, the lower should be the crosslinking density due to the presence of the acid and hence the T_g shows a decreasing trend.

Figure 2 and 3 show the variation of the storage and loss moduli as a function of temperature for all the composites and for the Natural Rubber. Observing the storage modulus curves (modulus versus T), we note that a three orders of magnitude decrease of this variable with increase of temperature. This decrease represents a transition to the elastic state and is associated with the glass transition temperature. Comparing the composites with the pure Natural Rubber, an increase in the modulus in the elastic region with the addition of polyaniline is observed, confirming the reinforcement effect. However, the elastic region of 60% polyaniline showed some discontinuities which may be due to the fact that at high polyaniline loading, the mixture is susceptible to imperfections due to the large polyaniline agglomerates and since the adhesion is not perfect, it may induce sample break originate in the interfacial region. The absence of this adhesion in this region leads to an abnormal behaviour in storage and loss modulii.

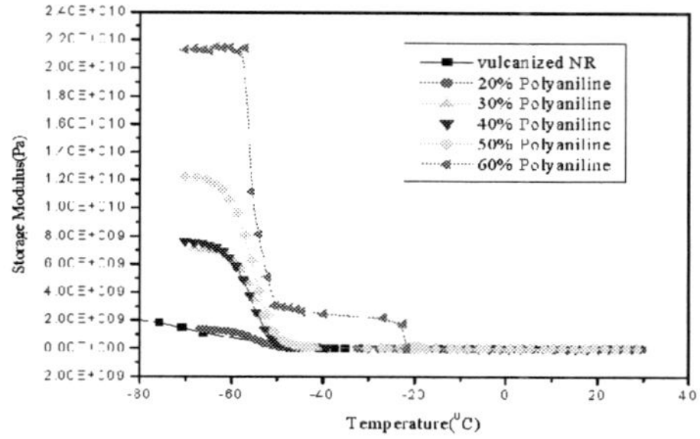

FIGURE 2. Storage modulus as a function of temperature for Natural Rubber / Polyaniline composite systems.

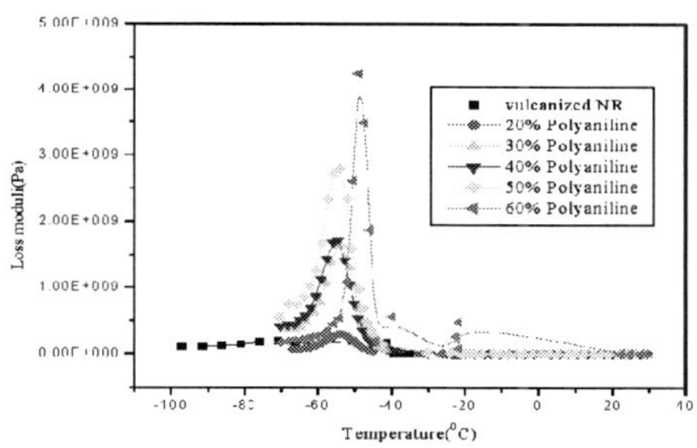

FIGURE 3. Loss moduli as a function of temperature for Natural Rubber / Polyaniline composite systems.

REFERENCES

1. A.M.Gessler,W.M.Hess and K.I.Medalia,Plast.Rubber Process.Appl.,**3**(1978)109.
2. L.E.Nelson, Trans.Soc.Rheol, **13**(1969) 141.
3. R.D.Bohme, J.Appl.Polym.Sci. **22**(1968) 1097.
4. A.K.Manna,P.P.De,D.K.Tripathy and S.K.De.Journal of Applied Polymer Science, **70** (1998) 723-730.
5. R.L.Blaine,P.S.Gill, R.L.Hassel and L.Wool.Journal of Applied Polymer Science, **34**(1998) 157-171.
6. U.K.Mandal, D.K.Tripathy and S.K.De. Journal of Applied Polymer Science, **55** (1995) 1185-1191.
7. A.G.Mac Diramid, J.C.Chiang, A. F. Richter, N.L.D.Somasiri and A.J.Epstein, Conducting Polymers.ed.L.Alcacer and D.Reidel Publishing Company,(1987) 105.
8. J.P.Van Drumpt.Rubber World, **197**(1988) 33.
9. R.C.Keller.Rubber Chem Technol, **61**(1988) 238.

Synthesis and Characterization of Polypyrrole Coated SBR Composites

Anisha Mary Mathew[1,3], Neena.P[1,3], Prasanth.S[2,3], N.S.Panwar[2]and P.Pradeep[1,3*]

1. NIT, Calicut, Kerala.

2. USIC, HNB University, Srinagar, Uttaranchal.

3. Centre for Advanced Scientific research,

Abstract: Conducting polymer of styrene butadiene rubber has been synthesized by chemical doping method. Subsequently the samples were coated with an intrinsically conducting polymer shell (ICP) of polypyrrole. The polymer shell was synthesized by chemical route method after the preparation of conducting polymer powder of SBR. The chemical structure of the composite was analyzed by taking the FTIR spectra. Electrical conductivity of the new hybrid material is found to be same as the conductivity of polypyrrole.

Key words: SBR, Polypyrrole, conducting composite, electrical conductivity.

PACS. 71.20Rv, 72.80Le, 72.80Tm, 73.61Ph

Corresponding author: ppredeep@gmail.com

INTRODUCTION

Intrinsically conducting polymers are an extensively studied class of materials due to their interesting properties including electrical conductivity, electrochromism and electro activity. Research in the field of such polymers aims mainly at some suitable modifications of existing polymers so that their applicability can be improved. Some of these modifications involve preparing hybrid materials in organic and inorganic materials combine in some special fashion with the conducting polymers to give rise to composites [1,7].

SBR is the world's most important rubber with a high market value because of its low cost and comparable mechanical properties. It belongs to the class of novel diene rubbers with many device applications due to its high flexibility, and good mechanical properties. Polypyrrole (PPy) is an important conducting polymer with high electrical conductivity and environmental stability. One of the properties which have restricted the development of PPy is sometimes its poor mechanical properties. This has been overcome to some extent by the use of composite structure. The present study deals with the synthesis and characterization of polypyrrole coated SBR composite and evaluation of electrical conductivity.

CP1004, *Thermophysical Properties of Materials and Devices-NCTP '07*,
edited by P. Predeep, S. Prasanth, and A. S. Prasad

The potential advantage of the composite polymer membrane approach is that the PPy can be dispersed through out the pore structure and yet the mechanical properties of the host polymer can be used to advantage [6].

EXPERIMENTAL

Materials

SBR with a molecular weight of 1.5 x 10^5 and styrene content of approximately 25 % was supplied from RRI Kottayam. The dopant $SbCl_5$ (Aldrich) a colourless liquid at room temperature and CCl_4 (Merck) as solvent were used. Pyrrole monomer was purified by distillation under reduced pressure and stored at 4^0C in the absence of light.

Synthesis of Conducting SBR/ PPy composite.

The preparation of SBR/ PPy composite can be carried out in two phases. In the first phase intrinsically conducting powder from SBR can be obtained by doping it with $SbCl_5$. The SbCl5 induced conjugation reaction was carried out by dissolving SBR (20 mmol / monomer unit) in anhydrous CCl_4 and then a solution of $SbCl_5$ (39 mmol) in definite quantity of CCl_4 was added dropwise. A deep black precipitate was formed immediately and after stirring 24 hrs the mixture was filtered off and rinsed thoroughly with CCl_4 to remove the residual unreacted dopants. Then the product was dried under vacuum to obtain a black insoluble powder.

In the second phase, for the preparation of SBR conducting powder- PPy composite, a fine dispersion of SBR conducting powder was made by adding a quantitative amount of SBR powder in 10 ml of de-ionized water under constant ultrasonic action for about 10 minutes. Afterwards a definite quantity of pyrrole was slowly syringed into the dispersion keeping pyrrole: SBR conducting powder mass ratio 8: 1 and continued sonication for about 10 minutes so that the monomer tends to react with the nanoparticles. Then the oxidant $FeCl_3$ dissolved in 5 ml de-ionized water was added dropwise under constant ultrasonic action keeping the pyrrole: $FeCl_3$ mole ratio 1:1.25. As soon as the pyrrole mixed with the $FeCl_3$ solution it turned to characteristic black colour, indicating the formation of polypyrrole. The solution was left for sonication about 1 hr for complete polymerization followed by centrifugation. The resulting nanocomposite was washed thoroughly with methanol first and then with de-ionized water several times.

MEASUREMENTS

The composites thus obtained were characterized by FTIR technique. The IR spectra of the composites were recorded in Nicolet IR spectrometer in KBr medium at room temperature in the region 4000-450 cm^{-1}. The prepared composite in the form of powder were pressed to form pellets by applying pressure of 100 Kg/cm^2 in hydraulic press. The dc electrical conductivity of the sample was measured by KEITHLEY 6514 electrometer and 2000 DMM by 2 and 4 probe methods [2].

RESULTS AND DISCUSSION

After doping at room temperature, an increase in electrical conductivity is observed which can be attributed to the formation of charge transfer complexes [3,4]. A black insoluble ultra fine powder is obtained as a result of doping indicating the formation of conjugated sequences in the back bone of the polymer [5].

FTIR Spectra

Introduction of conjugation along the backbone of polymer upon doping was confirmed by analyzing FTIR spectra of samples. Fig 1. (a) and (b) shows the FTIR spectra of pristine SBR and conducting SBR powder prepared by doping it with $SbCl_5$ respectively.

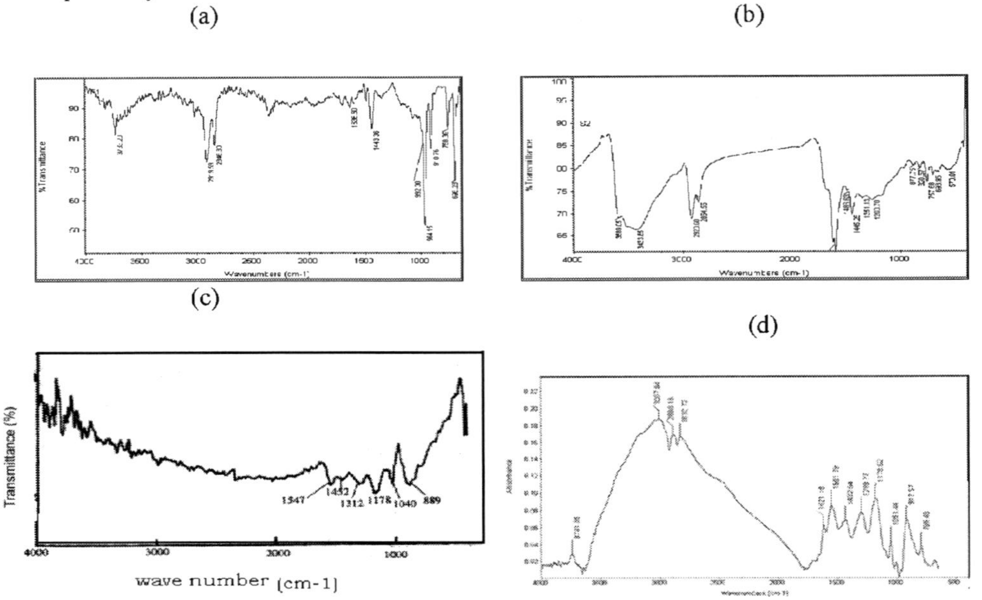

FIGURE 1: FTIR spectra of a. Pristine SBR, b. $SbCl_5$ doped SBR, c. Pure PPy, d. PPy coated Conducting SBR Composite.

The bands corresponding to the stretching isolated double bonds –C=C- seen in the spectrum of undoped sample at 1636.50 cm^{-1} is disappeared in the spectrum of conducting powder and new bands appeared at lower frequencies in between 1618-1595 cm^{-1}. This shift in the position of band is an indication of conjugated sequences of double bonds on the backbone of polymer due to doping. Intensity of bands at 964.15 and 698.28 cm^{-1} corresponding to =C-H out of plane bending vibration in pristine SBR gets decreased very much upon doping. The new band formed at 638.95 cm^{-1} can be attributed to stretching vibrations of newly formed C-Cl bond. The bands seen at2919.5 cm^{-1} , 2846.80 cm^{-1} and at 1443.38 cm^{-1} corresponding to symmetric and asymmetric C-H stretching vibration and bending vibration remains unshifted upon doping. The band at 757.68 cm^{-1} due to stretching vibration of –C=C- in benzene ring remains unshifted.

Fig 1.c shows the FTIR spectra of pure polypyrrole. The bands at 1312,1040 cm^{-1} may correspond to =C-H band in plane vibration .The bands at 1547 and 1452 cm^{-1} may be assigned to typical polypyrrole ring vibrations. The broad band at 1178 cm^{-1} may be assigned to N-C stretching band. The IR band peak observed at 889 cm^{-1} may be assigned to the =C-H out of plane vibration indicating polymerization of pyrrole [8,6].

Fig 1.d shows the IR spectrum of SBR conducting powder / PPy composite. The absorption peaks at 795.48 cm^{-1} may be assigned to stretching vibration of –C=C- in benzene ring in SBR. The peaks at 680 cm^{-1} can be attributed to stretching vibration of C-Cl bond. The bands at 2888.16 cm^{-1} and 2832.72 cm^{-1} are due to the symmetric and asymmetric stretching vibrations of –C-H bond. The bands at 1621.18 cm^{-1} are due to the stretching vibrations of –C=C- bond. The bands at 1551.79 cm^{-1} and 1432.64 cm^{-1} are due to the typical polypyrrole ring vibrations. The peak at 1178.62 cm^{-1} is assigned to N-C stretching band. The IR peak observed at 917.5 cm^{-1} may be assigned to the =C-H- out of plane vibration of polymerized pyrrole.

DC conductivity

The dc conductivity of the sample was measured using KEITHLEY 6514 electro meter and 2000 DMM. The conductivity of the composite prepared lies in the range of semi conductor.

Conductivity of pure SBR is of the order of 10^{-13} S/cm.
Conductivity of SBR Conducting powder = 8.3×10^{-6} S/cm
Conductivity of SBR conducting powder coated with poly pyrrole = 2×10^{-5} S/cm.

CONCLUSION

SBR conducting powder / poly pyrrole composites were prepared by chemical doping. The FTIR spectral peaks of the composites prepared shows the presence of polypyrrole in the polymer though PPy in the composites does not effectively involved chemically with the SBR. But weak interaction with PPy and SBR may contribute to the conductivity. The DC conductivity of the composites prepared lies in the semi conductor range. This composite powder has a potentially promising application as conducting fillers in rubber and plastic matrix for forming compatible polymer blends. The advantage is that it combines high electrical conductivity with mechanical properties of SBR.

REFERENCES

1. Hebert Naarmann, Polymer to the year 200 and beyond, Hohn Wiley and sons, chapter 4(1993)
2. Fifth edition Low Level measurements by Keiyhly.
3. N.C.Billingham, P.d.Calvert; electrically conducting polymers-A polymer science view point.1
4. David Bloor; electricalconductivity, Queen Mary College, London, U.K.
5. L.Dai, J.W. White, polymer, volume 32,Np.(1991)
6. T K Vishnuvardhan, V R Kularni, C Basavaraja and S C Raghavendra Bull.Mater. Sci., Vol.29,No. 1,Feb 2006, pp. 77-83.
7. Bhattacharya A, Ganguly K M, De A and Sarkar S 1996 Matter. Res Bull 31527
8. Chen W ,Xingwei L, Gi X ,Z haoquang W and Wenquing Z 2003 Appl. Surf.Sci.218216

Evaluation of a. c. conductivity Behavior of Conducting Rubber Based Composite

Neena.P[1,3], Prasanth.S[2,3], Anisha Mary Mathew[1,3], N.S.Panwar[2] and P.Pradeep[1,3*]

1. NIT, Calicut, Kerala.
2. USIC, HNB University, Srinagar, Uttaranchal.
3. Centre for Advanced Scientific research

Abstract. In this work an endeavor has been made to prepare conductive polymer composites using natural rubber as the matrix and conducting NR powder as the filler. The intrinsically conducting cis1,4 polyisoprene powder was prepared by treating the materials with Lewis acid treatment. The filler concentration was varied from 10phr to 50phr. The current voltage characteristics of the composite were studied using an impedence meter in the range of 5Hz to 13MHz.Due to the easy availability and cost effectiveness, it is considered as an attractive potential candidate in the optoelectronic fields with applications including polymer light emitting diodes.

Key words: Cis1,4 polyisoprene, ac conductivity.
PACS: 71.20Rv, 72.80Le, 72.80Tm, 73.61Ph

INTRODUCTION

Elastomers have been among the most recently studied region because of their high flexibility, good mechanical properties and above all easy processibility. Several compounds including polyisoprene, polybutadiene and synthetic elastomers such as styrene butadiene etc have been studied in the past few years. In particular a significant effort has been put in to understanding the properties of polyisoprene, which is considered to be a prototype of the classes of materials.

The excellent physical properties of natural rubber include resilience, strength and fatigue resistance and these together with the fact that it is renewable resource; mean that it is a very important electrometric material. The current and potential applications of polyisoprene span a wide range of areas going from simple flexible rubbery conductors to photovoltaic devices. The possibilities seem to be unlimited due to ease with which polymers can be tailored and confirmed. However, natural rubber having isolated double bonds with no conjugation does not satisfy the criterion and hence for a long time have been used as a component in conducting polymer composites or copolymers together with conjugated polymers, such as polyacetylene. Such composites have improved physical properties compared with most conjugated conducting polymers.

CP1004, *Thermophysical Properties of Materials and Devices-NCTP '07*,
edited by P. Predeep, S. Prasanth, and A. S. Prasad

cis1, 4 polyisoprene belongs to the class of novel diene rubbers with which many device applications due to its high flexibility, good mechanical properties and above all easy processibility. Natural rubber as fabricated in rubber products combines high strength with outstanding fatigue.

From the chemical point of view, doping should be considered as a redox reaction between the polymer's π electron system and the dopant. The high degree of electronic density delocalization in these polymers lowers the energy gap between valence and conduction bands so that upon doping with donor or acceptor species the material exhibits a drastic change in the electrical conductivity. It was discovered by Thakur that the electrical conductivity could also be induced into the nonconjugated polymer backbone by doping with halogens such as I_2, Br_2 etc

EXPERIMENTAL

Materials used

The Natural Rubber used for this study was Indian Standard NR, grade 3 (ISNR-3) with a number average molecular weight of 3 x 10^5.The dopant $SbCl_5$ (Aldrich), a colorless liquid at room temperature and Carbon tetra chloride (Merck) were used without further purification.

Sample preparation

The method used to conjugate polyisoprene by $SbCl_5$ has been described elsewhere [1,2]. The $SbCl_5$ –induced conjugation reaction was carried out by dissolving Natural Rubber (6mmol per monomer unit) in anhydrous carbon tetra chloride and then a solution of $SbCl_5$ (12 mmol) in carbon tetra chloride was added drop wise . In this work the conducting powder from polyisoprene obtained through solution phase doping with SbCl5.

The first phase was the preparation of intrinsically conducting filler material. Ultra fine conducting powder prepared from polyisoprene using Lewis acid like SbCl5 by liquid phase doping. For that, NR solution was made by dissolving NR in anhydrous carbon tetrachloride through continuous stirring. Then a solution of sbCl5 (dopant)in CCl4 was added drop wise into it. A deep black precipitate was obtained immediately and then stirring for 6 hrs the mixture was filtered out. In order to remove the unreacted dopant particle the precipitate was thoroughly washed out with CCl4.

In the second phase the intrinsically conducting powder has to be incorporated into the NR matrix. The NR matrix obtained by adding cross linking agents such as accelerators, activators, antioxidants and vulcanizing agent etc in proper fraction.. By compounding various percentages of the conducting polymer filler into the NR matrix using to roll mill mixing machine the composite was made.

In the third phase the composite molded into self standing films using die and hydraulic press specially designed for this purpose. All the time the hydraulic press is kept at a temperature of 150Kg/cm2. After 7 minutes the sample is taken by releasing the hydraulic press and allows it to cool.

Measurements

The smooth flexible films were subjected to AC conductivity studies using impedance meter in the range of 5Hzto 13Hz. For electrical conductivity measurements, the polymer samples were pressed in to pellets (13mm diameter) and was measured at room temperature using a Keithly 2000 DMM and 6514 Electrometer by four probe and two-probe technique

RESULT AND DISCUSSION

A deep black precipitate was formed immediately and after stirring for 24h, the mixture was filtered off and rinsed thoroughly with carbon tetrachloride to remove residual unreacted dopant. Then the product was dried under vacuum to obtain a black, insoluble powder. The graph which plotted by the data obtained from impendence analyzer gives the following information that as the percentage of filler increases the conductivity also increases. Since the material is predominantly amorphous and the backbones conjugated, the charge transport involves and intersite doping mechanism. The site of charge transfer was identified to be carbon-carbon double bond positions of the isolated double bond structure. The doping results in the formation of radical cations, subsequent to formation, these radical cations can be stabilized by the coulomb interaction with the dopant counter ions in the solid state.

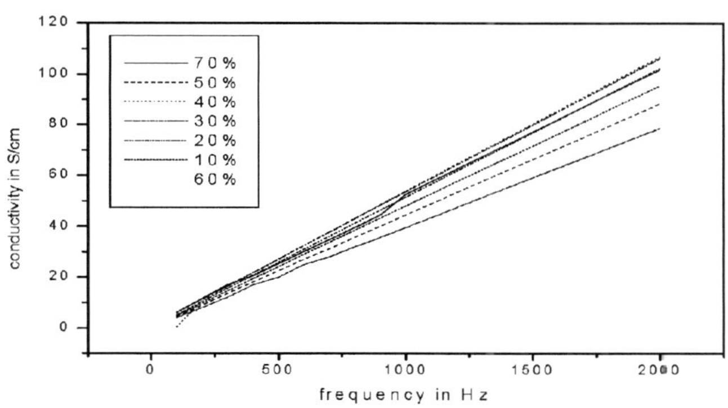

FIGURE 1: Variation of ac conductivity with frequency

Due to the absence of the conducting media pure natural rubber matrix is poor conductor of electricity. By blending fillers in different proportions the conducting properties of NR matrix were shown to be varied. The conductivity increases with increase in filler concentration and vice versa.

The spreading pattern of filler is quite dispersive in NR matrix. Due to vibrant behavior of alternating current its passage is more ratified than the passage of direct current in the NR matrix. The graphical analysis of ac conductivity pattern reveals that the conductivity proportionally increases with the amount of the filler.

CONCLUSION

Natural rubber continues to be important attractive engineering elastomers because of its several advantages including easy processibility, extreme flexibility, good mechanical properties and ozone resistance. This work is an attempt towards the systematic development of electrical conductivity in cis 1, 4 polyisoprene. This composite powder has a potentially promising application as conducting fillers in rubber and plastic matrix for forming compatible polymer blends. The advantage is that it combines high electrical conductivity with mechanical properties of NR.

REFERENCE

1. L.Dai, J.W. White, polymer, volume 32,Np.(1991)
2. L.Dai .J.W.White, soluble conducting polymers from polyisoprene; Research school of chemistry, Australian National University (1990)
3. Hebert Naarmann, Polymer to the year 200 and beyond, Hohn Wiley and sons, chapter 4(1993)
4. Fifth edition Low Level measurements by Keiyhly.
5. M.R. vallin.M.I.Felisberti, M.A.De. Paoli, blends of polyanilene with nitrile rubber, Journal applied polymer size.
6. N.C.Billingham, P.d.Calvert; electrically conducting polymers-A polymer science view point.1
7. David Bloor; electricalconductivity, Queen Mary College, London, U.K.

Band structure and Optical properties CdTe and $CdSn_3Te_4$ thin films

T. Venkatachalam[a] S. Velumani[b] S. Ganesan[c] K.Sakthivel[d]

[a,d] Department of Physics, Coimbatore Institute of Technology, Coimbatore-641 014, India.
[b] Department of Physics, Aulas 2-cub 111, ITESM-Campus Monterry, Mexico.
[c] Department of Physics, Government College of Technology Coimbatore-641 013, India.

[a] Corresponding Author Email: atvenkatachalam@yahoo.com

Abstract. CdTe and $CdSn_3Te_4$ compounds were prepared by direct reaction of their high purity elemental constituents using rotating furnace. Optimal deposition conditions for the deposition of CdTe and $CdSn_3Te_4$ thin films in hot wall evaporation setup were simulated using Monte Carlo technique. Thin films of CdTe and $CdSn_3Te_4$ were deposited on glass substrates by hot wall evaporation method. From the XRD measurements it was found that the films of CdTe and $CdSn_3Te_4$ were of cubic zinc-blende and rock salt structures respectively. The lattice parameters were determined as a = 6.476 Å (CdTe) and a = 6.238 Å ($CdSn_3Te_4$) from the XRD data. The UV-Vis-NIR optical transmittance spectra of thin films of different films were obtained and it was found that the direct optical band gaps were 1.4 eV (CdTe) and 0.8 eV ($CdSn_3Te_4$). Electronic structure, band parameters and optical spectra of CdTe and $CdSn_3Te_4$ were calculated from ab initio studies within the GGA approximation. The experimental results were in good agreement with the theoretical values.

Keywords: Thin films, CdTe, $CdSn_3Te_4$, Band gap, hot wall system, GGA
PACS: 61.10.NZ; 71.15.Ap; 71.15.Mb; 71.20Nr

INTRODUCTION

The II-VI and II-IV-VI group semiconductors attract extensive experimental and theoretical interest because of their technological applications in microelectronic, micro mechanic, optoelectronic, photonic, smart sensor, actuator and many more devices. CdTe has a direct band gap of 1.5 eV at room temperature and hence it is a suitable material for photovoltaic applications [1]. In order to cover the broad spectrum of wavelengths from 0.75 μm to 1000 μm, a variety of IR detectors have been developed over the past forty years since the first synthesis of HgCdTe in 1958. $Hg_{1-x}Cd_xTe$, $Zn_xCd_{1-x}Te$ and $Cd_xSn_{1-x}Te$ alloys have the unique feature of having direct band gap, tunable from the visible to the far infra red by adjusting the 'x' composition of the alloy. The advantage of this II-IV-VI alloy system is the narrow range of lattice constants. Theoretical works of this alloy $Cd_xSn_{1-x}Te$ with various values of 'x' (0 to 1) study the parameters such as structure, bonding character of individual valence bands, lattice constants, electronic charge density and bulk moduli. In the mixed system $Cd_xSn_{1-x}Te$, the transition from a ten-electron (x = 0) to an

CP1004, *Thermophysical Properties of Materials and Devices-NCTP '07*,
edited by P. Predeep, S. Prasanth, and A. S. Prasad

eight-electron system (x = 1), a change in the crystal structure from rock salt (SnTe) to Zinc blende (CdTe) and the calculated results confirm Vegard's rule and give a crossover from rock salt to zinc blende. The energy band gap can be adjusted to any value between 1.5 eV and 0.08 eV by varying the concentration of tin. The hot wall vacuum evaporation system is concerned with the fabrication of thin films under conditions close to thermodynamic equilibrium and with minimum loss of material. The hot wall vacuum evaporation method has been employed for preparation of the CdTe and $CdSn_3Te_4$ thin films. This paper presents experimental and theoretical studies aimed at understanding the crystal structure, electronic properties, band parameters and optical properties of hot wall evaporated CdTe and $CdSn_3Te_4$ thin films. The hot wall system is concerned with the fabrication of thin films under conditions close to thermodynamic equilibrium with minimum loss of material [2].

Cadmium Telluride And Cadmium Tin Telluride

CdTe thin films possess zinc blende structure and have a direct band gap which is optimal for absorbing the terrestrial solar spectrum. The maximum efficiency corresponding to such a band gap is about 27%. CdTe of few micron thickness absorbs more than 90% of the light having photon energy above the band gap. X-ray diffraction patterns of CdTe films deposited at 473 K show that the average grain size of the films corresponding to the [111] direction is of the order of 260 Å. The lattice constant of CdTe films estimated from Nelson–Riley plots [3] is found to range from 6.4999 Å to 6.5291 Å. CdTe presents tetrahedral coordination of both Cd and Te species in Zinc blende structure and their degrees of oxidation are +2 and −2, respectively. It is evident that the charge distribution around Te shows symmetry related to the tetrahedral bonds. The Cd electron density includes a full semi-core 4d level and a depopulated 5s level, which produce an overall spherical symmetry around Cd. CdTe has an iono-covalent chemical bond with ionicity coefficient 0.55. Polycrystalline CdTe films show the most prominent reflection along the [111] direction together with [220] and [311] reflections. The hexagonal phase becomes more and more prominent with increasing substrate temperature. The mixed crystal system $Cd_xSn_{1-x}Te$ is unique in that it realizes the transition from a ten (x = 0) to an eight-electron system (x = 1) or between a IV-VI and a II-VI compound. At the same time the crystal structure changes from rocksalt (SnTe) to cubic zincblende (CdTe) [4, 5]. The rocksalt structure is the stable structure upto a constituent $Cd_{0.8}Sn_{0.2}Te$. For larger x the samples grown under high pressure decompose into a two component system consisting of CdTe and $Cd_{0.8}Sn_{0.2}Te$ when the pressure is released [6, 7]. As far as the authors' knowledge go there is not much work done in $CdSn_3Te_4$ alloys.

EXPERIMENTAL DETAILS

Deposition and Optical Characterization

With the help of molecular flow dynamic study of hot wall system, the parameters such as dimensions of the tube, source temperature and wall temperature are optimized and the quartz tubes of diameter 1 cm and length of 10 cm (for CdTe) and 8 cm (for $CdSn_3Te_4$) are shaped in such a way that the bottom end rests inside a tungsten conical basket (source heater). The source and wall temperatures are maintained at 780 K and 760 K (CdTe), 720 K and 700 K ($CdSn_3Te_4$) respectively. The deposition is carried out at a pressure of 4×10^{-5} Torr. CdTe and $CdSn_3Te_4$ films with thickness approximately 760 Å and 2260 Å are deposited with the substrate temperature around 350 K. The structural analysis of the evaporated CdTe thin films is carried out using a computer controlled X-ray diffractometer system (model JDX 8030, Japan) with Ni filter and Cu Kα radiation. The scanning is carried out using the θ–2θ scan coupling mode, the ratings being 40 kV, 20 mA. The particle size, lattice parameter and lattice spacing values for the various films are calculated. The optical studies are made on the films deposited on glass substrates in the wavelength range from 190 nm to 2500 nm at room temperature by using Spectrophotometer (JASCO-370V). The transmission vs wavelength is recorded for all the thin films. The optical band gap values for the films deposited are determined by plotting the graph ($(\alpha h\gamma)^2$ Vs $h\gamma$) where h is the Planck's constant, γ is the frequency of the light and α is the absorption coefficient.

Ab Initio Calculations

CdTe structure (zinc blende) is constructed as shown in figure 1 with the lattice parameter 6.476 Å. $CdSn_3Te_4$ structure (rock salt) is constructed as shown in figure 2 with the lattice parameter 6.238 Å. The calculations of the electronic, optical and structural properties of CdTe and $CdSn_3Te_4$ are performed using a first principle plane wave pseudo potential technique based on the density functional theory with the Cambridge Serial Total energy Package code [8]. The density functional is treated by the generalized gradient approximation with exchange-correlation potential parameterization of PW-91. The ultra soft pseudo potentials description is used for the elements that are present in the crystal structures. The ultra soft pseudo potentials description is used for the elements that are present in the crystal structures. It describes the electron ion interaction and represents electronic wave functions using a plane wave basis set. For the crystal lattice, integration over the symmetrised Brillouin zone is performed using k-points generated via the Monkhorst-Pack scheme [9]. The crystal optimization is done by BFGS method and density mixing is done by Pulay scheme. The cutoff energy of 240 eV is assumed in the plane wave basis set. Optimization of the atomic positions and crystal cell parameters are performed before the main calculation of electronic characteristics: total electronic energy, band energy dispersion and density of electronic states.

RESULTS AND DISCUSSIONS

All the CdTe thin films deposited are uniform, smooth and have good adhesion with the glass substrate. The x-ray diffraction patterns of the CdTe films deposited at different substrate temperatures exhibit crystalline structure with reflections from (111), (220) and (311) planes. The lattice parameter and particle size of thin films deposited are 6.476 Å and 440 Å respectively. Similarly the $CdSn_3Te_4$ films deposited at different substrate temperatures exhibit crystalline structure with reflections from

(100), (110) and (111) planes. The lattice parameter and particle size of thin films deposited are 6.238 Å and 220 Å respectively. The optical band gap values of CdTe and $CdSn_3Te_4$ thin films are estimated from the extrapolation to zero, the optical band gap values for the films are in the order of 1.4 eV (CdTe) and 0.8 eV ($CdSn_3Te_4$) and they are found to be direct band semiconductors. The band structure of the CdTe and $CdSn_3Te_4$ are simulated using material studio software and the energy levels of the electrons in the molecules are analyzed as shown in figures 3 and 4. The energy gap between the valence band and the conduction band at and near the gamma valley are measured and it is found to be nearly 1.4 eV (CdTe) and 0.8 eV ($CdSn_3Te_4$). Similar band gaps are measured from the simulated density of states plots of CdTe and $CdSn_3Te_4$, and the results are same. The effective mass of the electrons and holes of CdTe and $CdSn_3Te_4$ are determined by taking the energy levels nearer to the gamma valley. The effective mass of electrons is -0.0147 m_e and that of the holes is 0.00132 m_e for CdTe and that of electrons and holes in $CdSn_3Te_4$ are -0.00116 m_e and 0.00172 m_e respectively. For CdTe material, 72 electrons (36 up spins and 36 down spins) and 40 bands are chosen as electronic parameters. The total energy per atom convergence, eigen-energy convergence tolerance, smearing width and Fermi energy convergence tolerance are taken as 0.2×10^{-5}eV, 0.4×10^{-6}eV, 0.1 eV and 0.4×10^{-7}eV respectively which are taken as electronic minimization parameters. Band structure of the crystal was calculated of 48 band of the Brillouin zone (BZ) with band convergence tolerance of 0.1×10^{-4}eV and 4 k-points are considered for Brillouin zone sampling. The calculations of the electron energy characteristics and optical spectra of CdTe crystal have been performed using the CASTEP code. The pseudo atomic calculations performed for Cd (core (Rb) + $4d^{10}5s^2$), Sn (core (Pd) + $5s^25p^2$) and Te (core (Pd) + $5s^25p^4$) are converged to a total energy of -1175.588 eV, -90.401 eV and -216.758 eV respectively. The electronic minimization parameters obtained in the simulation are i) total energy / atom convergence tolerance of 0.2000×10^{-5} eV, ii) eigen energy convergence tolerance of 0.5714×10^{-6} eV iii) Fermi energy convergence tolerance of 0.5714×10^{-7}eV.

FIGURE 1 Structure of CdTe

FIGURE 2 Structure of $CdSn_3Te_4$

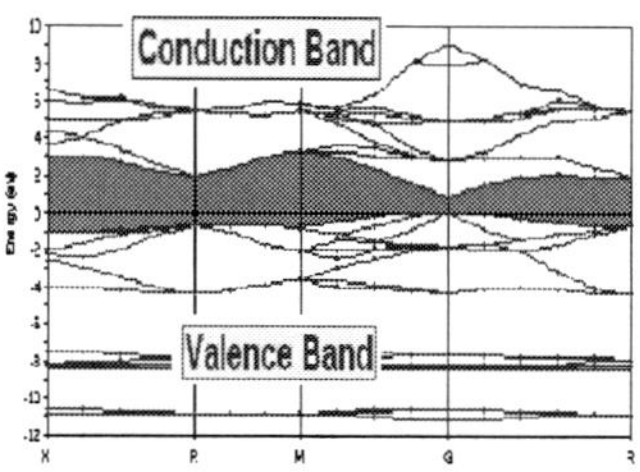

FIGURE 3 Electronic band structure of CdTe

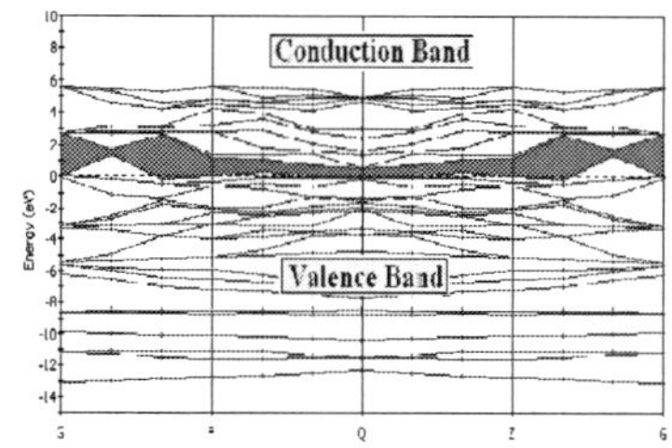

FIGURE 4 Electronic band structure of $CdSn_3Te_4$

CONCLUSIONS

Monte Carlo simulation technique is used in optimizing the hot wall evaporation setup for the deposition of CdTe and $CdSn_3Te_4$ thin films and they are deposited on a glass plate under a pressure of 4×10^{-5} Torr. By analysis of X-ray diffractogram it is found that the CdTe film exhibits a cubic zinc blende structure and $CdSn_3Te_4$ thin film exhibits rock salt structure. The measured lattice parameter and particle size of CdTe are 6.476 Å and 440 Å respectively and that of $CdSn_3Te_4$ are 6.238 Å and 220 Å respectively. Using the measured lattice parameters, band structure parameters for CdTe and $CdSn_3Te_4$ have been obtained using first-principles calculations with the generalized gradient approximation. The band structure of CdTe and $CdSn_3Te_4$ are simulated using the module Cambridge Serial Total energy Package code. The optical transmission spectra of the CdTe and $CdSn_3Te_4$ are used to find their optical band gaps. It is found that the results are in good agreement with the simulated results. The effective mass of the electrons and holes of CdTe and $CdSn_3Te_4$ are determined at and near the gamma point in the Brillouin zone. The characterization studies of CdTe thin films confirm the formation of films crystalline in nature and it is suitable for photovoltaic applications. Similarly $CdSn_3Te_4$ shows rock salt structure and it is suitable for infra red detector applications.

REFERENCES

1. S. Lalitha, S. Zh. Karazhanov, P. Ravindran, S. Senthilarasu, R. Sathyamoorthy and J. Janabergenov *Physica B* **1387**, 227-239(2007).
2. T. Venkatachalam, S. Ganesan and K. Sakthivel *J. Appl Phys D* **39**, 1650-57 (2006).
3. Nelson J B and Riley D B *Proc. Phys. Soc. (Lond.)* **57**, 160-70 (1945)
4. Okamoto T, Harada Y, Yamada A and Konagai M *Sol.Energy Mater. Sol. Cells* **67**, 187–97 (2001).
5. B. Freytag, U. Rossler, and K. Karch, H. Grosch and K. J. Range J. Chem... Phys **99**, 6751-55 (1993).
6. G. H. Grosch, B. Freytag, K. J. Range, U. Rossler J. Chem. Phys **101,** 6782 – 89 (1994).
7. Jose Menendex and Hartmut Hochst Thin solid films **111**, 375 – 379 (1984)
8. M.D. Segall, P. J. Lindan, M. J. Probert, C. J. Pickard, P. J. Hasnip, S. J. Clark and M. C. Payne J. Phys. Cond Matt. **14**, 2717- 2743 (2002).
9. H. J. Monkhorst, J. D. Pack, Phys. Rev B **13**, 5188- 95 (1976).

Pressure-Induced Phase Transition in Zinc-Blend CuI

Deoshree Baghmar[a], Purvee Bhardwaj[b], N. K. Gaur[a], Sadhana Singh[a] and D.C. Gupta[c]

[a] *Department of Physics Barkatullah University, Bhopal-462026, India.*
[b] *Centre for Basic Sciences M. P. Bhoj (open) University, Bhopal-462016, India.*
[c] *School of Studies in Physics, Jiwaji University Gwalior-474 011, India.*

Abstract. In the present investigation the Zinc-Blende (ZnS) — Tetragonal (Intermediate Phase) — Rock-Salt (NaCl) phase transition pressure of Copper Iodide (CuI) has been predicted using an interionic potential theory. The calculated compression curves and values of different high-pressure properties for CuI are reported and compared with the available experimental values. The calculated values for the phase-transition pressures (P_t) and the associated volume collapses [$V(P_t)/V(0)$] are generally in good agreement with their measured data.

Keywords: High Pressure, Structural Phase transition, Gibbs free energy, Volume Collapse.
PACS: 64.70Kb, 62.20Dc.

INTRODUCTION

Investigation on structural phase transformations and behaviour of materials under high pressure [1-3] has been a subject of interest since past few decades. The copper halides are the borderline case between the ionic and covalent compounds [4]. These copper halides CuX (X= Br, Cl, I) crystallize in the tetrahedrally co-ordinated Zinc-Blende (ZnS) structure under normal conditions. They exhibit an ionic conductive phase at high temperature [5] and non-conductive phase at extremely low temperature [6]. Besides, they undergo number of structural phase transitions under pressure. Copper iodide (CuI) is expected to transform from four fold co-ordinated Zinc-Blende (ZB) space group $F\bar{4}3m$ to six fold co-ordinated Rock-Salt (RS) space group $Fm\bar{3}m$ [7] structure passing through an intermediate Tetragonal structure space group $P4/nmm$ [8] under static compression. Various experimental [8, 9] attempts have been made to obtain phase transition pressure from Zinc Blende to Tetragonal and from Tetragonal to Rock-Salt structure. An attempt has been made by one of us [10] to predict the structural properties by means of Three Body Potential (TBP) model from ZnS to NaCl structure but ignored the intermediate tetragonal phase. We thought it pertinent to apply the interionic potential to study the structural phase transition of copper iodide from parent (ZB)- intermediate (Tetragonal)- adopted (RS) structure.

CP1004, *Thermophysical Properties of Materials and Devices-NCTP '07*,
edited by P. Predeep, S. Prasanth, and A. S. Prasad

In the present work, we have investigated the effect of high pressure on the structural properties of CuI for all the three phases, under elevated pressure. An interionic potential comprises of both the long-range (L-R) Coulomb interactions and van der Waals interaction; short-range (S-R) overlap repulsive interactions have been used for the determination of structural properties of CuI. The interionic potential is discussed in the next section and results and discussed in the followed section.

METHOD OF COMPUTATION

The effective interionic potential used for the investigation of the structural properties is expressed [10, 11] as:

$$U_B(r) = -\sum_{ij} \frac{\alpha_M Z^2 e^2}{r_{ij}} + b\sum_{ij} \beta_{ij} \exp\left[\frac{r_i + r_j - r_{ij}}{\rho}\right] - \sum_{ij} c_{ij} r_{ij}^{-6} \tag{1}$$

Here, first term represents long-range Coulomb attraction, second term corresponds to Hafemeister and Flygare [12] type short-range overlap repulsion and the last term represents to vdW interactions. Ze is the ionic charge; b and ρ is the model parameters calculated by using equilibrium condition and bulk modulus [13, 14] The stability of a particular lattice structure is decided by the minimum value of the Gibbs's free energy (G) given as:

$$G = U + PV - TS \tag{2}$$

where, U is the internal energy, which at 0 K is equivalent to the cohesive energy; S is the vibrational entropy [15] at absolute temperature (T), pressure (P) and volume (V). In our calculation, the cohesive energies for ZnS, Tetragonal and NaCl phases are uniquely defined as:

$$U_{B_3}(r) = \frac{-\alpha_M e^2 Z^2}{r} + n_1\phi_{ij}(r) + \frac{n_2}{2}\left[\phi_{ii}(r) + \phi_{jj}(r)\right] \tag{3}$$

$$U_{B_T}(r') = \frac{-\alpha'_M e^2 Z^2}{r'} + n_3\phi_{ij}(r') + \frac{n_4}{2}\left[\phi_{ii}(r') + \phi_{jj}(r')\right] \tag{4}$$

$$U_{B_1}(r'') = \frac{-\alpha''_M e^2 Z^2}{r''} + n_5\phi_{ij}(r'') + \frac{n_6}{2}\left[\phi_{ii}(r'') + \phi_{jj}(r'')\right] \tag{5}$$

Here r, r' *and* r'' are the nearest neighbors (nn) separation; α_M, α'_M *and* α''_M are the Madelung constants and n_1-n_6 are the nearest and next nearest neighbour for all the phases respectively.

The short-range potentials is given as:

$$\phi_{ij}(r) = b\sum_{ij} \beta_{ij} \exp\left(\frac{r_i + r_j - r_{ij}}{\rho}\right) - \sum_{ij} c_{ij} r_{ij}^{-6} \qquad (i, j = 1,2) \tag{6}$$

Here, c [16] is van der Waals coefficients and other symbols have their usual meaning as defined earlier.

RESULTS AND DISCUSSION

Using the values of input directly taken from our earlier [11] and using the values we have calculated model parameters hardness (b), range (ρ) and reported them in Table1.

TABLE 1. Input data and Model parameters of the Crystal CuI

Input data		Model Parameters	
r_+ (Å)	0.740	ρ (Å)	0.313
r_- (Å)	2.048	$b(10^{-12}$ ergs)	0.366
r_0 (Å)	2.620[a]		
$B_T\,(10^{-12}$ dynes/cm^2)	0.355[a]		

a- ref [11].

Using the values of model parameters and isothermal bulk modulus, we have minimized the energies for all the phases to obtain the equilibrium lattice separations. The corresponding values of the cohesive energies for all the phases at zero pressure are reported in Table 2 and compared with the available experimental data [9, 17, 18].

TABLE 2. Cohesive Energy of CuI in various phases in (Kcal/mol).

U_{B_3}		U_{B_T}		U_{B_1}	
Cal	Exp	Cal	Exp	Cal	Exp
-205.96	-225.8[a]	-227.87	-290[b]	-186.04	-

a- ref [9], b- ref.[17].

In an attempt to reveal the structural phase transition of the compound, we minimize the Gibbs's free energies of the different phases for the equilibrium interatomic spacing. As the pressure is increased, ΔG decreases to zero at phase transition pressure and becomes negative when pressure is further increased. The Gibbs's free energy difference (ΔG) and relative volume collapse Δ[V(P)/V(O)] have been plotted as function of Pressure (P) and displayed in fig 1(a,b) and fig 2. It is obvious from the fig that the energy differences ΔG have attained their values to be zero at the phase transition pressure (P_t = 7.5 GPa & P_t = 10.0 GPa.) for the respective phases. The values of phase transition pressures are in satisfactorily agreement with the available experimental results [8,9] (P_t = 4-10 GPa & P_t > 10.0 GPa.). We have calculated the % of relative volume changes using the compression curves and presented them in Table 3. Our comments on the reliability of percentage of volume collapses are restricted until the report of experimental data on them.

TABLE 3. Cohesive Energy of CuI in various phases in (Kcal/mol).

CuI	ZnS-Tetragonal		Tetragonal-NaCl	
	Cal	Exp	Cal	Exp
P_t (GPa.)	7.50	4-10[a]	10.00	10.00[a]
$\Delta[V(P_t)/V(O)](\%)$	2.88	-	2.37	-

a- ref [8,9].

In view of the overall achievements, it may be concluded that the TBP model is capable of describing the high-pressure behaviour of copper iodides with the same success at it done in the curves of compound semiconductors [13-15].

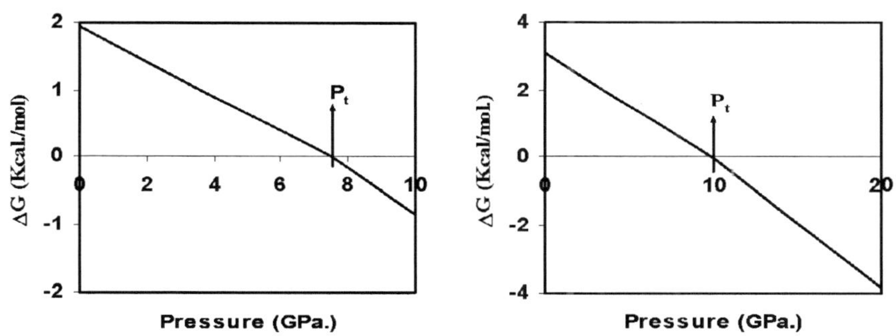

(a) (b)

FIGURE 1. Variations of Gibbs free energy as a function of pressure (P) form : (a) ZnS-Tetragonal, (b) Tetragonal to NaCl structure.

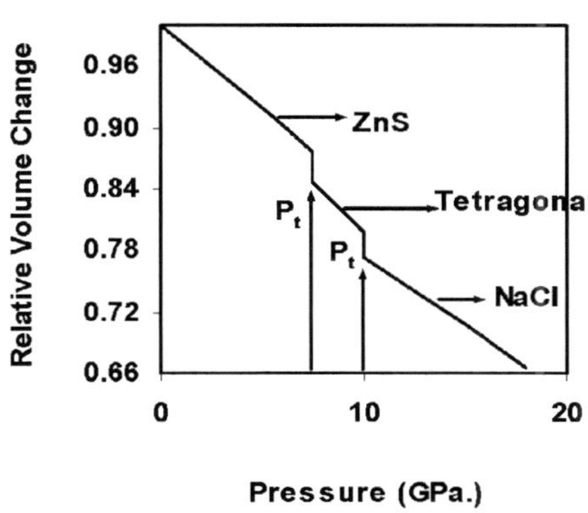

FIGURE 2. Relative Volume Collapse as a function of Pressure (P) (ZnS-Tetragonal-NaCl structure).

REFERENCES

1. N.V. Chandra Shekhar and P.C. Sahu, *J. Mater. Sci.* **41**, 3207-3583 (2006).
2. S. Shang, Y. Wang. and Zi-Kui Liu., *Phys. Rev. B.* **75**, 24302-11 (2007).
3. H. F. El Haj, Z. Mouhamed and H. Abbas, *Physica A* **14**, 245-255 (2005).
4. J. C. Philips, *Rev. Mod. Phys.* **42**, 317-356 (1970).
5. J. B. Boyce, T. M. Hayes and J. C. Mikkelsen, *Phys. Rev. B.* **23**, 2876-2896 (1981).
6. Z. Verdney and O. Brafman, *Phys. Rev. B.* **19**, 3276-3289 (1979).
7. M. Yanming, T. S. John and K. D. Dennis, *Phys. Rev. B.* **69**, 64102-1-6(2004).
8. M. Hofmann, S. Hull and D. A. Keen, *Phys. Rev. B.* **51**, 12022-12025 (1995); *Phys. Rev. B.* **50**, 5868-5885 (1994).
9. A. Blacha, N. E. Christensen and M. Cardona, *Phys. Rev. B.* **33**, 2413-2421 (1986).
10. R. K. Singh, D. C. Gupta, *Phys. Rev. B.* **40**, 11 278-11 283 (1989).
11. R. K. Singh and D. C. Gupta, *I L. Nuovo Cimento D.* **2**, 1253-1264 (1987).
12. D. W. Hafemeister and Flygare, *J. Chem. Phys.* **43**, 795- (1965).
13. R. K. Singh, D. C. Gupta and S. P. Sanayal, *Phys. Stat. Sol. (b)* **149**, 121-125 (1988).
14. V. Srivastava and S. P. Sanayal, *J. of Alloys and Compd.* **36**, 15-20 (2004).
15. S. Singh, R. K. Singh, B. P. Singh, S. K. Singhal and R. Chopra, *Phys. Stat. Sol. (a)* **180**, 459-466 (2000).
16. B. Mouffok, H. Feraoun and H. Aourag, *Phys. Stat. Sol. (b)* **243**, 1182-1192 (2006).
17. S. Ves, D. Glotzel, M. Cardona and H. Overhof, *Phys. Rev. B.* **24**, 3073-3085 (1981).
18. A. Blacha, S. Ves and M. Cardona, *Phys. Rev. B.* **27**, 6346-6362 (1983).

Corresponding author: deoshreebaghmar@gmail.com

Semiconducting Properties of $HfSe_2$ Single Crystals Grown by CVT Technique

K.Radhakrishnan[a] and K and Mohanan pilla[b]

[a]*Department of Physics , N.S.S. Hindu College, Changanasserry – 686102, Kerala , India*

[b] Department of Physics , D.B.College Parumala , Kerala ,India

Abstract : Single crystals of Hafnium di selenide ($HfSe_2$) which are a new class of solarcell material have been grown by chemical vapour transport (CVT) technique , with iodine as the transporting agent using a two-zone horizontal furnace. Electrical resistivity of the grown crystals were measured using two probe and four probe methods. Activation energy of charge carriers were determined.Hall mobility,Hall coefficient and carrier concentrations were also determined. All crystals were found to be of n-type conductivity. The electrical conductivity,Hall effect and their temperature dependence concludes the characteristic features of the semiconducting properties of these compounds.

Keywords : Transition metal dichalcogenide ; hafnium di selenide ; chemical vapour transport .
PACS : 81.10.Bk ,71.55.Gs

INTRODUCTION

Electrical properties of Group IVA transition metal dichalcogenides have been reviewed by Wilson and Yoffee (1) . $HfSe_2$ are semi conductors with energy gap in the range 1.3 eV. (2). The band gap of these compounds are ideal for photovoltaic applications (3). Anomaly in the resistivity was identified (4). It was reported that $HfSe_2$ crystallizes with lattice parameters a = 3.748 A° and c = 6.159 A° (5). The main aim of the present work is to investigate the electrical resistivity, Hall effect and temperature dependence in order to conclude the characteristic features of the semi conducting properties of this compound.

EXPERIMENTAL

Single crystals of $HfSe_2$ were grown by chemical vapour transport (CVT) method using iodine as the transporting agent. The charges were taken in stoichiometric proportions and were placed in shaped quartz ampules (length 200 mm and diameter 28 mm) along with transporting agent iodine (5 mg/cm^3). The ampoules were evacuated to a pressure of 10^{-5} torr using oil diffusion pump. It is then placed in a muffle furnace for 72 hours for poly crystallization and then transferred to a specially designed two zone horizontal furnace with appropriate stationary temperature profile. Temperature of the furnace is increased in steps of 70° C/ hour. After 432 hours of uninterrupted heating crystals were found. The growth parameters were shown in Table I.

CP1004, *Thermophysical Properties of Materials and Devices-NCTP '07*,
edited by P. Predeep, S. Prasanth, and A. S. Prasad

TABLE 1: Growth parameters of $HfSe_2$ Single crystals grown using chemical vapour transport technique

Crystal	Hot Zone °C	Cold Zone °C	Temperature gradient °C	Growth time (h)	Size mm^3
$HfSe_2$	820	770	50	432	12×10×.20

Lattice parameters were determined using powderXRD. Selected crystals were examined by electron probe microanalysis.Energy dispersive analysis (EDAX) confirms the stoichiometry of the grown crystals.Electrical resistivity of single crystals was measured using van der pauw method and two probe methods at room temperature. Low temperature resistivity measurements were performed for a constant current in the temperature range 93 K to 293 K using a pyrex cryostat. High temperature resistivity measurements was also performed in the temperature range 303 - 413 K using a four probe setup. Anisotropy of different samples are determined in the temperature range 303 – 423 K. The sign, mobility and concentration of charge carriers were determined by Hall effect measurements in the magnetic field range 0 to 10 KG.

RESULTS

Typical data are presented in figure1. which show the resistivity as a function of square of temperature (93–293K). From the plots it is seen that resistivity increases with increase in temperature and propotional to T^2. Figure 2. depicts the variation of resistivity with temperature in the range 323 - 423 K. The present plot has two distinct straight line regions.

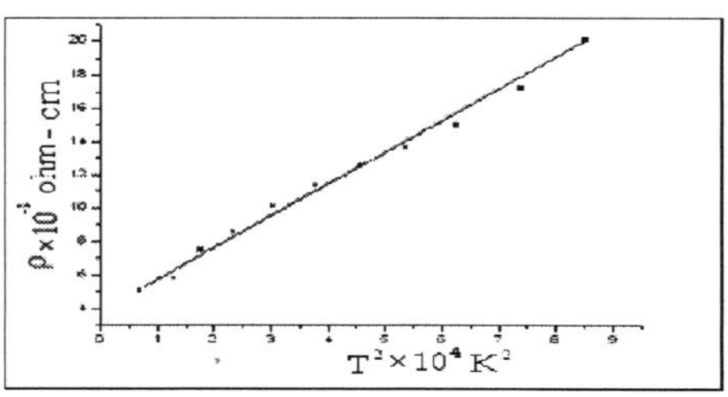

FIGURE 1. Variation of resistivity with square of temperature

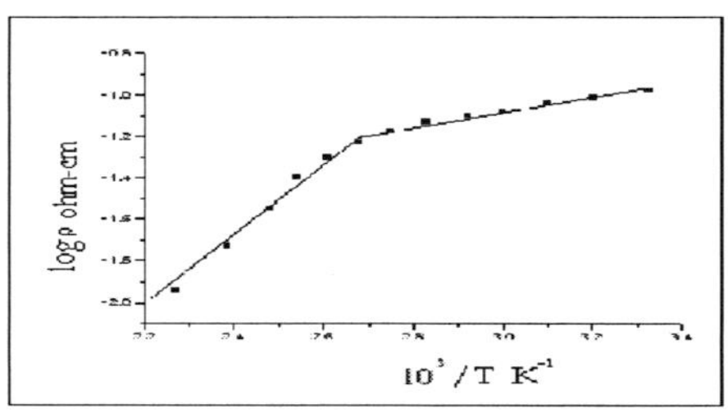

FIGURE2: Variation of resistivity with temperature

The activation energy of charge carriers werecalculated using the formula :
$Ea = 2.303 \times K \times 10^3 \times$ slope, were $K = 8.62 \times 10^{-5}$ $eVk^{-1}$and are seen to be 0.0416the temperature range 303-373 K and 0.126eV in the temperature range 373-423 K . Fig. 3 shows thevariation of anisotropy with temperature in the temperature range 303 – 423K. It is seen that anisotropy increases with temperature.

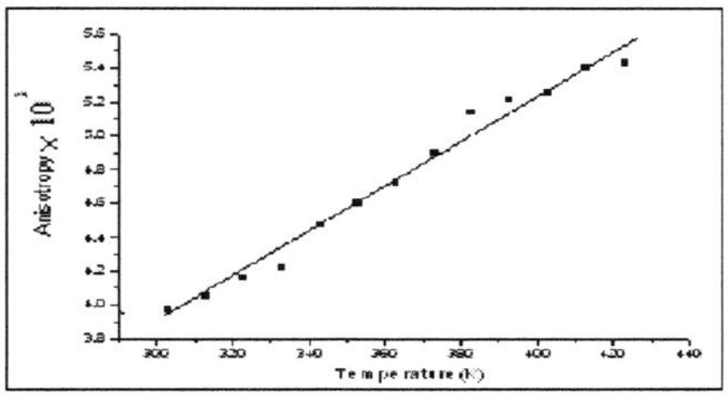

FIGURE 3: Variation of anisotropy with temperature

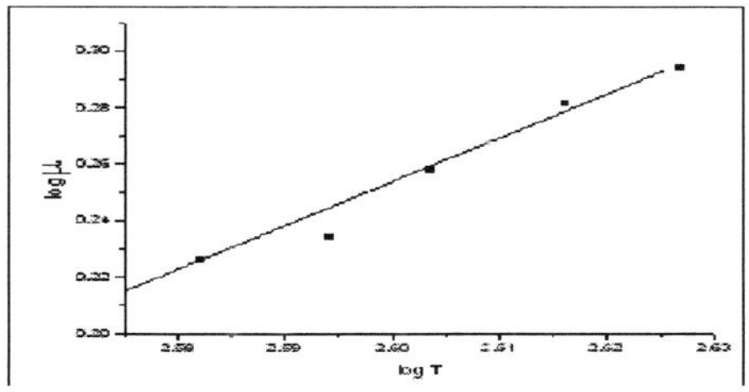

FIGURE 4: Variation of Hall mobility with temperature.

The Hall mobility was found to be 1.589 $cm^2V^{-1}s^{-}$.
The values of Hall coefficient (R_H) and carrier concentration (n) at room temperatures were determined. The variation of Hall mobility with temperature is shown in figure 4. The graph shows a positive temperature dependence of mobility due to scattering mechanism in the crystals..The behavior of carrier concentration as a function of 10^3 / T is shown in figure 5.

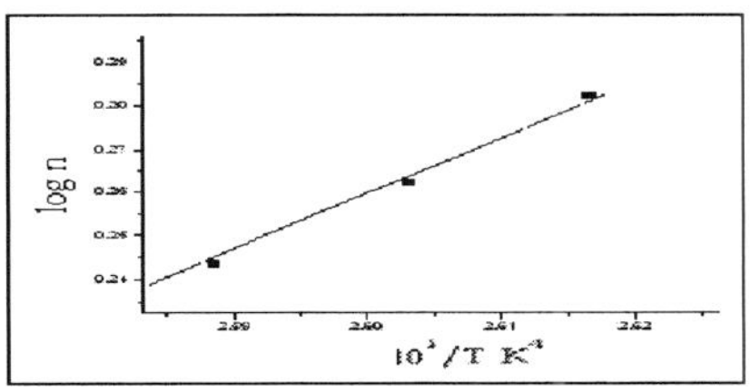

FIGURE 5: The temperature dependence of carrier concentration

DISCUSSION

The low value of resistivity along the basalplane and high value resistivity perpendicular to it confirms the anisotropic bahaviour of the crystals. The low value is due to covalent bonding and high value due to Van der waal forces. The T^2 dependence of electrical resistivity may be due to carrier scattering.The positive temperature dependence of mobility implies that scattering mechanisms may be due to the ionized impurity scattering in the observed temperature range. From the behaviour of carrier concentration with temperature, it is observed that as temperature increases, the carrier concentration also increases. The slope of the curve is relatively low due to the presence of deep donor like levels in the material.

CONCLUSIONS

Single crystals of Hafnium di selinide grown using iodine as the transporting agent exhibits metallic behaviour in the range of 93 - 293 K and semi conducting behaviour in the range of 303 – 423 K. The crystals were of n type semi conductors

REFERENCES

1. JA Wilson and AD Yoffe, Advances in Phy., **18**, 193 (1969)
2. C Gaiser, T Zandt, A Krapf, R Serverin, C Janowitz and R Manzke, Phys. Rev. B **69**, 075205 (2004)
3. H Tributsch, Faraday Discuss. Chem. Soc., **70**, 189 – 205 (1980)
4. I Taguchi, J. Phys. C: Solid State Phys. **14**, 3221 – 3229 (1981)
5. BL Evans, RA Hazelwood, Brit. J. Appl. Phys. (J. Phys. D.), **2** (1969)
6. HP Vaterlaus, F Levy, J. Phys. C: Solid State Phys. **18**, 2351-2359 (1985)

Nonlinear Optical Properties Of Lanthanum Gallogermanate Glasses

Harikuttan Unnithan.C[a], Dhanya.I[a] P.Predeep[b] and S.Jayakumar[c]

[a]*Solid State Physics Laboratory*
D .B. College, Sasthamcotta, Kollam- Kerala- 690 521
[b]*Department of Science & Humanities, National Institute of Technology, Calicut*
[c]Department of Physics, MSM College, Kayamkulam

Abstract. The oxide glasses are of important class of materials, especially because of their high transparency in mid IR region.. Lanthanum Gallogermanate glass is having greater interest for IR transmission applications due to their long IR cut off wave length in comparison with phosphate, borate or silicate glasses. Various optoelectronic parameters such as Nonlinear Refractive Index (Nω), Polarisibity (P), Bandgap Energy(Eg), Susceptibility (χ_0)are determined for series I glass systems consist of x $La_2O_3 - GeO_2 - 0.25\ Ga_2O_3$ with x =(0.2, 0.25, 0.33 & 0.5) and for series II glass systems with 0.25 La_2O_3 - GeO_2- x Ga_2O_3 with x = (0.2. 0.25 & 0.5) using the available data. The values obtained are compared with pure SiO_2 glass system. It is observed that with the addition of La_2O_3 in series I, enhance the non linearity in the glass system while with the addition of glass modifier Ga_2O_3 in Series II, reduce the non linearity factor. With proper addition of dopants such as La_2O_3 or the conditional glass former like Ga_2O_3, in Lanthanum Gallogermate glassy network, the non linearity factor can be maximized, which will be useful in producing second harmonic generation leading to different optoelectronic applications such as optical modulation, optical rectification and frequency doubling.

PACS: 78.55.Qr

INTRODUCTION

Recently multi component glasses have received much attention due to their versatile applications in optics and optoelectronics. Investigations focused on chalcogenide glasses especially on oxide glasses have very low phonon energies and higher refractive indices are expected to be efficient as host materials for fiber optic amplifiers and infrared lasers when doped with rare earth ions such as (La_3^{2+}) The oxide glasses are of important class of materials, especially because of their high transparency in mid IR region. [1]. Lanthanum Gallogermanate glass is having greater interest for infra red (IR) transmission applications due to their long IR cut off wave length in comparison with phosphate, borate or silicate glasses[2]. Dumbaugh found that the addition of Ga_2O_3 into PbO - Bi_2O_3 and Cd O-Bi_2O_3 improved glass stability and these glasses are transparent to about 7μm[3] .Kokubo etal, alkali and alkaline earth Ta_2O - Ga_2O_3 or Nb_2O - Ga_2O_3 glasses also have optical transmission to about

CP1004, *Thermophysical Properties of Materials and Devices-NCTP '07*, edited by P. Predeep, S. Prasanth, and A. S. Prasad

7μm [4.5]. Glasses based on Ge O_2 offer favorable fiber core material because of their longer wavelength cut off at which the Ge O_2 stretching vibration occurs. A recent study on Barium Gallo germanate glasses proved their potential for both IR optical wave guides and bulk optic components in the 3-5 μm region [6]. Lindquist and Shelby have investigated the properties of rare earth Gallogermanate glasses [7]. The IR cut off occurs at 6.05μm. These glasses containing rare earth elements also have excellent chemical durability in water and a very high Verdet constants due to the rare earth ion concentration. The second property makes these glasses suitable for use in Faraday rotation devices [7].This paper reveals on a segment of on going research programme consists of the non-linear optical properties of Lanthanum Gallo Germanate glasses in terms of IR transmission and IR reflectivity from the known refractive indices [2].

THEORETICAL CONSIDERATION

The optical devices such as wave guides and ultra fast switching possibilities require efficient non-linear materials. The best inorganic glasses such as oxide glasses have a figure of merit substantially higher than that of any other non-linear optical materials [8]. The non-linear coefficient of refractive index n_2 can be 50 times for La_2 O_3, than that of Si O_2 .On the other hand, an attempt has been made to explain the origin of the non-linearity in such materials and the validity of the molecular orbitals bonding (MO) theory extended to glasses containing Titanium oxide polyhedra [9,10].

The non-linear optical properties of Lanthanum Gallo germanate glasses are due to their high value of refractive indices. When light propagates through the dielectric medium like glass, it will induce the motion of charged particles in glass system. Subsequently the bound charges in glass will start to oscillate in the applied electric field of light and these oscillating electric dipoles, causes the formation of quadrupoles consisting of two positive charges and two negative charges. This is meant by electric dipole approximation. The induced electric dipole in Lanthanum based glass system is due to the displacement of electron cloud with respect to the center of positive nucleus of an atom, which gives rise polarization

$$P = \varepsilon_0 \chi_o E \qquad (1)$$

with χ_o as electric susceptibility.

The reflectivity coefficient (r_ω) of a multicomponent glass system is a complex function defined in the case of glass surface is the ratio of reflected electric field $E|ref|$ to incident electric field $E\ |inc|$

$$ie, r_\omega = \frac{E|ref|}{E|inc|} = \rho_\omega e^{i\theta(\omega)} \qquad (2)$$

with $\rho(\omega)$ and $e^{i\theta(\omega)}$ as the amplitude of the optical wave and the representation of phase factor $\theta(\omega)$ of reflective coefficient respectively.

The reflectance due to normal incidence

$$r(\omega)=\frac{n+ik-1}{n+ik+1} \quad (3)$$

Here 'k' is the extrinsic coefficient

The refractive index N(ω) related to dielectric function E(w) through relation

$$\varepsilon(\omega)=[N(\omega)]^2 \quad (4)$$

Also reflectivity coefficient or Reflectance $r(\omega)$ of lanthanum based glasses can be expressed using the relation,

$$r(\omega)=\frac{N(\omega)-1}{N(\omega)+1} \quad (5)$$

The magnitude of energy gap in first order diffraction of light waves in Lanthanum based glasses is given by :Eg = U - Fourier component of glass potential, ie U ^{2} $=(\lambda-\varepsilon)^2$

$$U=\pm(\lambda-\varepsilon), \quad \therefore Eg=\pm(\lambda-\varepsilon) \quad (6)$$

There is also an intrinsic carrier concentration given by $\frac{Eg}{k_B T}$ is the ratio of band gap energy to Absolute temperature.

By Lyddance – Sachs – Teller (LST) relation,

$$\frac{\omega_L^2}{\omega_T^2}=\frac{\varepsilon_0}{\varepsilon}$$

$\therefore$ The static dielectric constant,

$$\varepsilon_0=\frac{\varepsilon\omega_L^2}{\omega_T^2}=\frac{\varepsilon 2\pi V_L^2}{2\pi V_T^2}=\frac{\varepsilon V_L^2}{V_T^2} \quad (7)$$

The dielectric constant €of the glasses relative to the vacuum, is defined in terms of the macroscopic field E is given by relation :

$$\varepsilon=1+\frac{P}{\varepsilon_0 E} \quad (8)$$

$$ie, (\varepsilon-1)=\frac{P}{\varepsilon_0 E}$$

Also Susceptibility, $\chi_0=(\varepsilon-1)$

$$\therefore \chi_0=\frac{P}{\varepsilon_0 E} \quad (9)$$

Based on the above theoretical considerations, various parameters are determined for series I glass systems consist of x La_2O_3 – GeO_2 – 0.25 Ga_2O_3 with x =(0.2, 0.25, 0.33 & 0.5) and for series II glass systems with 0.25 La_2O_3 - GeO_2- x Ga_2O_3 with x = (0.2. 0.25 & 0.5) using the available data[2, 13]. The values obtained are compared with pure SiO_2 glass system [13].

RESULTS AND DISCUSSION

Lanthanum Gallogermanate glass possess a high refractive index than that of air [2] shows its structural complexity and non linearity of glass structure. The variation of complex refractive index (Nω) is plotted against the atomic percentage(x) of La_2O_3 content in figure 1 (with x = 0.2, .25, 0.33 & 0.5). The exponential increase of Nω implies that the complexity of glass structure increase with La_2O_3 content.

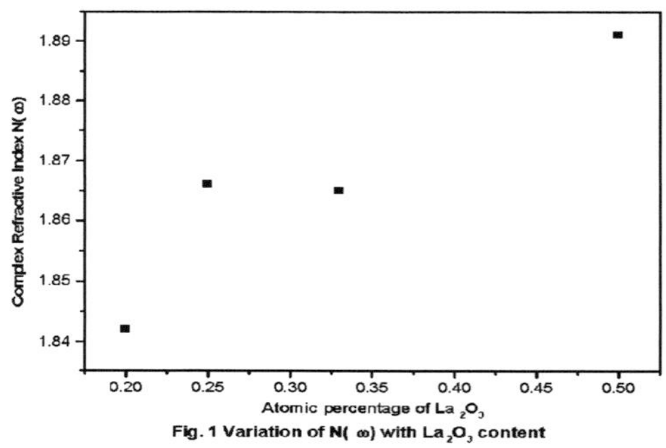

FIGURE 1. Variation of N (ω) with La_2O_3 content

Further, from table-1, it is evident that both the dielectric constant (E) and band gap energy (Eg) have values approximately equal to 3.5 eV as mentioned in case of Germanium based glass system [13]. I. R diffraction technique is applied for the investigation of this glass systems [13] .Raman scattering technique can also be applicable in La- based glass systems for the electrical as well as optical studies due to its varying electrical polarisabilities [14]. The optically active modes are not present in them since λ is of nm range (514.5nm) and not in several A^0 range. The wave vector k having a considerable value rather than zero and there is no lattice vibrational modes present in these glass systems. So the crystal dimensions will not be allowed to go to infinity. [14].

The static dielectric constant (ε_0) is finite in the present glass systems since the ion-ion interaction in it is negligibly smaller, which give rise to semiconducting property to the Lanthanum based glass systems. The static dielectric constant describes the screening of ion-ion forces at distances larger compared to a nearest neighbor distance. In semi conductors the screening factor is just ($1/\varepsilon_0$), the value is so small and thus the dielectric constant assumes finite value. Since the wave vector- k corresponding to the valence band and conduction band are being different, these series of Lanthanum Gallo Germanate glasses possess a definite indirect band gap. In pure SiO_2 glass systems, the direct band gap can be seen, while in x La_2O_3 - Ge_2O_3 -

Ga_2O_3, an indirect band gap is present due to the presence of conditional glass former, Gallium oxide. ie, it does not form glass by it self but does so when combined with other suitable network forming oxides [13].

The covalent bonds in semi conductors are characterized by the accumulation of change between nearest neighbours, which produces a well- localized bond charge. With increasing ionicity, the bond charge. With increasing ionicity, the bond charges are displaced towards the anion. When bond charge is centered on anion, the covalent structure in Lanthanum Gallogermanate glass structure becomes unstable.

In pure Ge O_2, Si O_2 etc., the energy in conduction band (Ec) is greater than that of Valence band (Ev) resulted positive value for Eg. On adding glass conditioning modifiers (such as Ga_2 O_3) etc., they can be either more positive or negative Eg according to the nature of modifiers. In the present, the addition of Ga2 O3 modifies, to Lantham based glass systems, produce more negative value (see table –1) for Eg. [15].

In pure GeO_2 glass system there is a distorted tetra hedral units [16]. In present multi component glass structure there is an inter connection of non symmetrical Non Bridging Oxygen (NBO) atom in between different elements [17]. So the position of the highest phonon band is important because the multiphonon decay of rare earth ions in this glass depends on maximum phonon energy of the host glass such as Ge O_2. The rate of multi phonon decay is sensitive to maximum vibrational energy of the glasses [17].

Further, the (P/E) ratio of present glass system possess higher value corresponding to the maximum La_2 O_3 content in series I (see figure-2)

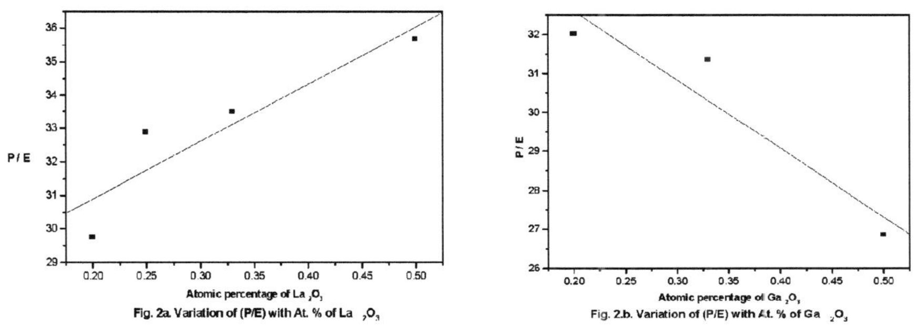

FIGURE 2. Variation of (P/E) with At % of (a) La_2O_3 (b) Ga_2O_3

In Series I glass, the values of (P/E) ratio increases with La_2 O_3 content. The increase of La_2O_3 content results in the formation of more Non-Bridging Oxygen atoms due to the ion-ion interaction and leads to the breaking of Oxygen bonds. This ultimately causes the formation more ionicity in the network. On the contrary in Series II, the value of (P/E) ratio decreases and resulted in the poor ionicity with the increase of Ga_2O_3 content.

The above analysis reveals that the polarizability as well as non-linearity in Series I glasses have linear relationship with La_2O_3 content. On the other hand, the Series II glass systems show has inverse relationship of polarizability and non linearity with the increase of atomic percentage of conditional glass former such as Ga_2O_3.

TABLE 1. Related parameters

Glass sample	N(w)	E = $(Nw)^2$	χ_0	r(w)	Band gap Energ y (Eg)	VL	VT	Eo	P/E
Scrics – I									
0.2 La_2O_3-GeO_2-0.25Ga_2O_3	1.866	3.393	2.393	0.545	3.393	5203.2	2717.8	12.436	29.759
0.25La_2O_3-GeO_2-0.25Ga_2O_3	1.842	3.482	2.482	0.554	3.481	5224.4	2678.4	13.248	32.882
0.33La_2O_3-GeO_2-0.25Ga_2O_3 0.5	1.865	3.478	2.478	0.553	3.478	5242.3	2659.0	13.519	33.500
La_2O_3-GeO_2-0.25Ga_2O_3	1.897	3.599	2.599	0.565	3.598	5272.5	2699.5	13.729	35.682
Series –II									
0.25La_2O_3-GeO_2-0.2Ga_2O_3	1.852	3.429	3.429	0.549	3.429	5280.2	2693.2	13.1805	32.015
0.25La_2O_3-GeO_2-0.33Ga_2O_3	1.856	3.445	2.445	0.5501	3.445	5280.2	2735.9	12.832	31.374
0.25La_2O_3-GeO_2-0.5Ga_2O_3	1.790	2.137	2.204	0.5243	3.204	5390.8	2764.3	12.1851	26.856
For pure silica glass									
SiO_2	1.462	2.137	1.137	0.3625	2.137	5944.2	3748.7	5.373	6.1091

CONCLUSION

The theoretical investigation on two series of glasses are made in terms of polarizability and non linearity with the addition of La_2O_3 in series I and the glass modifier Ga_2O_3 in Series II . In series I, the addition of more La_2O_3 content up to x = 0.50 enhances the non linearity in the glass system while the addition of Ga_2O_3 in series II glasses reduces the non linearity factor. With proper addition of dopants such as La_2O_3 or the conditional glass former like Ga_2O_3, in Lanthanum Gallogermate glassy network, the non linearity factor can be maximized, which will be useful in producing second harmonic generation leading to different optoelectronic applications such as optical modulation, optical rectification, frequency doubling etc.

ACKNOWLEDGMENTS

Sincere thanks to Kerala State Council for Science, Technology & Environment Government of Kerala for financial support to this projec

REFERENCES

1. V.S Vassilev etal, J.of Phy and Chem. of Glasses b3 (2002) 815-819
2. Luu – Gen Hwa, Yi – Rui – Chang, Sung – Ping Szu, J.of. Non – Cryst. Solids, 231 (1998) 222-226.
3. W.H. Dumbaugh, Phys.Chem.Glasses, 27(1986) 119.
4. T. Kokubo, Y. Inaka, S. Sakka, J.Non – Cryst. Solids, 80 (1986) 518.
5. T. Kokubo, Y . Inaka, S. Sakka, J. Non-Cryst. Solids 81 (1987) 337.
6. P.L Higby, I.D. Agarwal, J. Non – Cryst . Solids, 163 (1993) 303
7. B.H Lind quist, J.E Shelby, Phy.Chem.Glasses, 35 (1994) 1.
8. E.M. Vogel, M.J. Weber and D.M Krol; Phy. and Chem. of Glasses, 32(6) – 231 (1991)
9. M.E Lines, Phys. Rev, B 43, 1411978 (1991)
10. L. Canioni, L. Sarger, P . Seconds, A. Ducasse etal, Solid State Comm., 84 – 11 1065 (1992)
11. A. Berthereau, Y. Le Luger etal , Pergamon , 0025 – 5408 (94) 0048 – 4
12. Sungping Szu, Chanping Shu, Luu – Gen Hwa, J. of Non – Cryst . Solids, 240 (1998) 22-28
13. Young Wang, Mistsutaka Nakamura ,Osamu Matsuda etal, J. Non – Cryst . Solids , 226 – 269 (2000) 872 – 875.
14. Sindu, Karthan etal, J. of Mat. Sci. 22 (2003) 9 – 11.
15. H.Hagiwara etal. J. of Non – Cryst. Solids. 22 (2003) 9 – 11
16. Zhengela Pan – J. of Non – Cryst. Solids 210 (1998) 130 – 135

Optical Computing and Solitons

K.S.sreelatha[a], Lakshmi Parameswar[b], and K.Babu Joseph[c]

[a]*Department of Physics, Amrita Vishwa Vidyapeetham, Amrita puri Campus, Clappana P.O., Kollam-690525, Kerala*
[b]*Department of Physics, D.B.Pampa College, Parumala, Pathanamthitta-689626*
[c]*SCMS School of Technology and Management, North Kalamasseri P.O., Kochi*
E-mail:sureshlakshmi@sancharnet.in; kssreelatha@yahoo.com; babu_65@hotmail.com,

Abstract. An optical computer is a computer that uses light instead of electricity (i.e. photons rather than electrons) to manipulate, store and transmit data. Most research projects focus on replacing current computer components with optical equivalents, resulting in an optical digital computer system processing binary data. Astudy of using a physical medium for computation-solitons is presented here. As particle like nonlinear waves, solitons carry and exchange information, thus acting as computing agents. Oprical solitons have pulse widths measured in pico seconds or femtoseconds. The general idea of implementing logic usng using solitons is investigated.

Keywords:optical computer, solitons nonlinear schrodinger equation.
PACS: 42.65.Tg, 42.81.Dp, 42.79.Ta

INTRODUCTION

Photons have fundamentally different physical properties than electrons, and researchers have attempted to make use of these properties to produce computers with performance and capabilities greater than those of electronic computers[1]. Optical computer technology is still in the early stages: functional optical computers have been built in the laboratory, but none have progressed past the prototype stage. Optical computers would transport information ten times faster than traditional electronic devices, smashing the intrinsic speed limit of silicon technology.

Timing jitter caused by nonlinear interactions can be described mathematically by a simple summation of all frequency shifts which occur during pairwise soliton collisions, and the resulting variance in pulse arrival times depends not only on perturbations to the fiber but also on the encoding of information[2]. A new physical medium for computation: colliding solitons is proposed and is studied. As particle like nonlinear waves, solitons can carry and exchange information, thus acting as computing agents[3]. Optical solitons have pulse widths measured in picoseconds or femtoseconds, and this approach could ultimately offer an alternative to electronics. In a fiber with loss and amplification, it was shown that extensive wavelength multiplexing with solitons is feasible, and possibly preferable to other alternatives for achieving high aggregate throughput in computer communication systems[4].

CP1004, *Thermophysical Properties of Materials and Devices-NCTP '07*,
edited by P. Predeep, S. Prasanth, and A. S. Prasad

A generalized interaction geometry between orthogonally polarized optical spatial solitons in which a weak signal soliton induces a small angular deflection of a stronger power supply, or pump, soliton, resulting in a spatially resolved shift of the pump at the gate output. This geometry allows for the all-optical realization of true three-terminal, inverting and restoring logic devices with gain, which can serve as building blocks for more complex logic operations [5]. This new research could be a major breakthrough in the quest to create super-fast computers that use light instead of electrons to process information.

The propagation of soliton envelopes in nonlinear optical media has been predicted and demonstrated experimentally [1]. This prediction arises from a reduction the Maxwell equations which govern the electro magnetic fields in the medium to a single, completely integrable, partial differential equation. The well-studied example is the nonlinear Schrödinger (NLS) equation [6]. Two independent NLS equations must then be replaced by a coupled pair of equations which have the form

$$iq_{1x} + c_1 q_{1tt} + 2\,(\alpha|q_1|^2 + \beta|q_2|^2)\,q_1 = 0 \qquad (1.a)$$

$$iq_{2x} + c_2 q_{2tt} + 2\,(\beta|q_1|^2 + \gamma|q_2|^2)\,q_2 = 0 \qquad (1.b)$$

Here c_1, c_2, α, β and γ are real parameters, variables x and t are normalized space and time, c = +1 for anomalous GVD and c= -1 for normal GVD, $q_1 = q_1(x,t)$ and $q_2 = q_2(x,t)$ are the complex amplitudes of two interacting components.

This equation known as the Coupled-Nonlinear Schrödinger (CNLS) system and has been obtained from Maxwell equation including nonlinear polarization effects.

Although a large number of investigations exists in the literature on the study of soliton propagation in CNLS equations, exact results are scarce, except for a special version of the CNLS, namely the Manakov model. Recently, Radhakrishnan etal have obtained the explicit two-soliton solution of the integrable Manakov system and identified a novel shape changing property in it which can be used to construct logic gates in an optical computer [7].The ultimate aim should be the realization of all-optical digital computers interfaced, without the need for opto-electronic converters, of soliton communication system in which the data rates are so high that conventional electronic logic is too slow.

In this paper we consider the nonlinear cubic-quitic Schrödinger equation (CNLCQS) equation. The CNLCQS equation model the propagation of solitons through photorefractive crystals and multimode fibers. We study the basic integrability properties of this equation. This equation is integrable and soliton solution is obtained by the Hirota method.[8] . The solution may be used for construction of logic gates.

Bright Soliton Solutions of Coupled Nonlinear Quintic Schrödinger Equation (CNLQSE)

Here we consider the CNLCQSE which is obtained by taking into account the quintic nonlinearity also. Along with cubic nonlinearity, which given by

$$iq_{1x} + q_{1tt} + 2\mu\ (|q_1|^2 + |q_2|^2)\ q_1 + 2\mu\ (|q_1|^4 + |q_2|^4)\ q_1 = 0 \quad (2.a)$$

$$iq_{2x} + q_{2tt} + 2\ \mu\ (|q_1|^2 + \beta|q_2|^2)\ q_1 + 2\mu\ (|q_1|^4 + |q_2{}^4)\ q_2 = 0 \quad (2.b)$$

Based on the Painleve property, the above CNLS equations are integrable in these cases. A partial differential equation (pde) is said to posses Painleve property if the solution is single valued in the neighbourhood of a non-characteristic movable singularity manifold. We can represent the solution of the equation (2) in the form of Laurent series.

$$q_i\ (x, t) = \phi^{-1}\sum_{j=0}^{\infty} q_{ij}(x,t)\phi^j(x,t) \qquad i = 1, 2 \quad 3.16 \quad (3)$$

U sing Hirota's bilinearization technique, we write the equation (2) in the form;

$$f^3(iD_x + D_t^2)g.f - gf^3D_t^2(f.f) + 2\mu gf^2(gg^* + hh^*) + 2\mu g((gg^*)^2 + (hh^*)^2) = 0 \ (4.a)$$

$$f^3(iD_x + D_t^2)h.f - hf^3D_t^2(f.f) + 2\mu hf^2(gg^* + hh^*) + 2\mu h((gg^*)^2 + (hh^*)^2) = 0 \ (4.b)$$

where, the Hirota operator

$$D_x{}^m\ D_t{}^n\ g(x, t) = \left(\frac{\partial}{\partial x} - \frac{\partial}{\partial x'}\right)^m \left(\frac{\partial}{\partial t} - \frac{\partial}{\partial t'}\right)^n g(x, t)\ f(x', t') \quad \text{at } x = x', t = t'$$

Decoupling the above equation and collecting coefficients of equal powers, the bright soliton solution for this equation is obtained as

$$q_1 = \frac{e^{k_1(t+ik_1x)+\eta_1^0}}{1 + \dfrac{e^{\varepsilon+\varepsilon^*}}{1 - e^{\varepsilon+\varepsilon^*}}} \quad (5)$$

This is the one soliton solution of the CNLCQSE.
The propagation of one soliton solution is shown in figure 1.

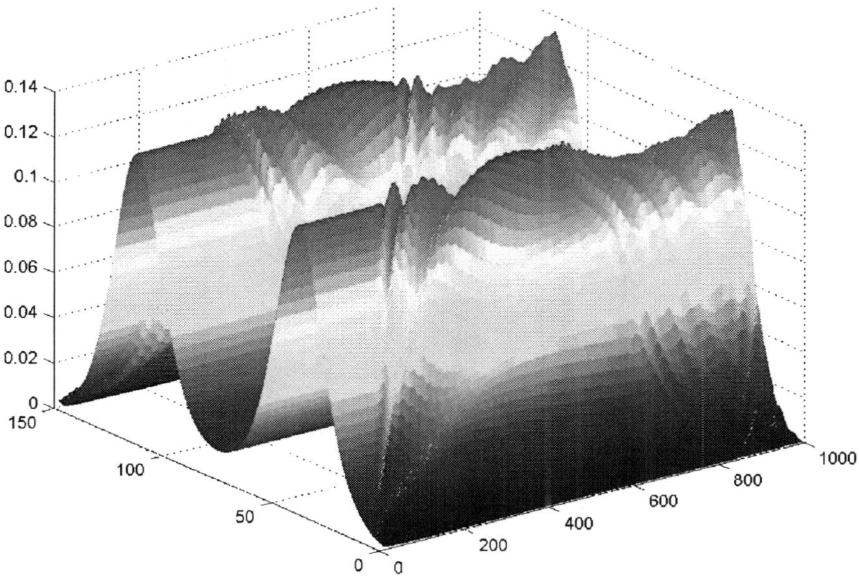

FIGURE 1: The propagation of solitons of the coupled system

CONCLUSION

We studied the the integrability of coupled nonlinear cubic quintic Schrödinger equation and found the bright soliton solution of this equation. We used Hirota method to find the soliton solutions. The soliton solutions may be used as the inputs for computing and their collision properties may be used for constructing logics gates.

ACKNOWLEDGMENTS

We thank the authorities of the Amrita VishwaVidyapeetham, Amritapuri Campus and D.B Pampa.College, Thiruvalla for extending to us the facilities of these Institutions. One of us (KSS) is grateful to the Department of Science & Technology, Government of India, for the award of a fellowship under the Women Scientist Scheme(WOS-A).

5. REFERENCES

1. Mohammed A Karim and Abdul A.S.Awwal, *Optical Computing: An Introduction*, John Wiley and sons(1992).
2. G.P Agrawal, *Nonlinear Fibre Optics*,Academic, Newyork,(1989).
3. Ken Stegigllitz, *Phys.Rev.E* **63**(2000)016608
4. K.Porzerian and V.C.Kuriakose(Editos), *Optical Solitons* ,Indian Acsdcmy of Sciences,(2001).
5. Steve Blair and Kelvin Wagner, Applied Optics, Vol. 39, Issue 32, pp. 6006-6018
6. K.Lakshmanan and T.Kanna, Pramana-JPhys.,**57** no.5&6(2001)
7. M.H.Jakubowki,K.Steiglitz and R.Squier, *Phys, Rev* **E58**(1998)6752
8. R.Hirota,*J.Math.Phys.*(NY)**14**(1973)805

Structural Properties of CsI

Deoshree Baghmar[a], Varsha. R. Mehto[a], N. K. Gaur[a] and D.C. Gupta[b]

[a] *Department of Physics Barkatullah University, Bhopal-462026, India.*
[b] *School of Studies in Physics, Jiwaji University Gwalior-474 011, India.*

Abstract. We have investigated the pressure induced phase transition of Cesium Iodide using Effective Interionic Potential. The cubic phase (B_2) of the compound Cesium Iodide (CsI) was found to undergo a phase transformation to an Orthorhombic (B_0) through intermediate tetragonal phase under high pressure. The calculated values of the structural properties, the phase transition pressure (P_t) and relative volume collapse [$V(P_t)/V(O)$] obtained by us are in good agreement with the available experimental data.

Keywords: High Pressure, Structural Phase transition, Gibbs free energy, Volume Collapse.
PACS: 64.70Kb, 62.20Dc.

INTRODUCTION

Alkali halides being most typical solids have attracted a lot of interest by many investigators [1-3]. Among the alkali halide crystals, the Cesium halides CsX (X=I, Cl, Br) form a special group not only because of their crystallographic structure (CsCl, B_2) but also because of the complexity of their electronic band structure compared with the lighter alkali halides [4]. CsI a prototypical insulator and ionic crystal, has been the object of intensive high-pressure investigations in the past, since it eventually metallizes under compression and enters a superconducting phase when compressed further [4-6]. Cesium iodide is found to undergo a phase transition from Cesium chloride (B_2) to Tetragonal (B_T) and from Tetragonal (B_T) to Orthorhombic (B_O) structure. In a number of previous investigations [7-11] of two successive phases transition of CsI has been reported experimentally but no theoretical attempts have been to explain the phase transition.

In the present paper we have reported the structural properties of CsI using an interionic potential, which has been quite successful in predicting the high-pressure behaviour of several alkali halides. The essentials of the present theory and method are given in section 2. The results are presented and discussed in section 3.

ESSENTIAL THEORY AND METHOD

The interionic potential for CsI in the framework is expressed as:

$$U_B(r) = -\sum_{kk'} \frac{\alpha_M Z^2 e^2}{r_{kk'}} + b\sum_{kk'} \beta_{kk'} \exp\left[\frac{r_k + r_{k'} - r_{kk'}}{\rho}\right] - \sum_{kk'} c_{kk'} r_{kk'}^{-6} - \sum_{kk'} d_{kk'} r_{kk''}^{-8} \qquad (1)$$

Here, first term represents the long-range coulomb interaction and second term represents the short-range interaction. The third and fourth terms represents the van

CP1004, *Thermophysical Properties of Materials and Devices-NCTP '07*,
edited by P. Predeep, S. Prasanth, and A. S. Prasad

der Waals coefficients due to dipole-dipole (d-d) and dipole-quadrupole (d-q) interactions [12,13] respectively. ρ and b are the range and hardness model parameters, which are calculated from the equilibrium condition and bulk modulus [14,15].

$$\left|\frac{dU}{dr}\right|_{r=r_o} = 0 \quad and \quad \left[\frac{d^2U}{dr^2}\right]_{r=r_o} = 9Kr_oB \qquad (2)$$

where, r_o is the ionic radii and K is the crystal constant.

The stability of the structure is decided by the minimum value of the Gibbs energy, which is given as:

$$G = U + PV - TS \qquad (3)$$

At absolute temperature (T), pressure (P) and volume (V), U the internal energy is equivalent to the cohesive energy and S is the vibrational entropy. The Gibbs's free energy for CsCl, tetragonal and orthorhombic structure are expressed as:

$$\begin{aligned} G_{B_2} &= U_{B_2}(r) + PV_{B_2} \\ G_{B_T} &= U_{B_T}(r') + PV_{B_T} \\ G_{B_o} &= U_{B_o}(r'') + PV_{B_o} \end{aligned} \qquad (4)$$

Here r, r' and r'' are the nearest separation for CsCl, Tetragonal and orthorhombic phases respectively and the cohesive energy for the respective phases are given as:

$$U_{B2}(r) = \frac{-\alpha_M e^2 Z^2}{r} + b\sum_{kk'} \beta_{kk'} \exp\left[\frac{r_k + r_{k'} - r_{kk'}}{\rho}\right] - \sum_{kk'} c_{kk'} r_{kk'}^{-6} - \sum_{kk'} d_{kk'} r_{kk''}^{-8} \qquad (5)$$

$$U_{B_T}(r) = \frac{-\alpha_M e^2 Z^2}{r} + b\sum_{kk'} \beta_{kk'} \exp\left[\frac{r_k + r_{k'} - r_{kk'}}{\rho}\right] - \sum_{kk'} c'_{kk'} r_{kk'}^{-6} - \sum_{kk'} d'_{kk'} r_{kk''}^{-8} \qquad (6)$$

$$U_{B_O}(r) = \frac{-\alpha_M e^2 Z^2}{r} + b\sum_{kk'} \beta_{kk'} \exp\left[\frac{r_k + r_{k'} - r_{kk'}}{\rho}\right] - \sum_{kk'} c''_{kk'} r_{kk'}^{6} - \sum_{kk'} d''_{kk'} r_{kk''}^{8} \qquad (7)$$

with, $\alpha_M(\alpha'_M/\alpha''_M)$ as the Madelung constants; $c(c'/c'')$, $d(d'/d'')$ are vdW coefficients for the respective phases.

RESULTS AND DISSCUSION

We have calculated the model parameters and cohesive energies for different phases of CsI by using eq. (2, 3) and (5-7) and listed them in Table 1 and Table 2 respectively. The cohesive energies for different phases of CsI Compound confirm that at ambient conditions compound is stable in CsCl (B_2) structure and transforms from CsCl (B_2) to Tetragonal (B_T) and from Tetragonal (B_T) to Orthorhombic (B_O) at high pressure.

TABLE 1. Input data and Model Parameters of CsI

Input data		Model Parameters	
r_0 (Å)	3.956	ρ (Å)	0.315
$B(10^{12}$ dynes/cm)	0.135[a]	b (10^{-12} ergs)	0.183

a- ref [5].

TABLE 2. Cohesive energies of different structures of CsI (in Kcal./mole)

Crystal	U_{B_2}	U_{B_T}	U_{B_O}
CsI	-153.83	-183.23	-197.95

For the computation of transition pressure of CsI we have computed the different values of inter-ionic separation for different pressure using the technique of minimization of Gibbs free energy for the respective phase. The Gibbs free energy difference ΔG and the relative volume collapse have been calculated as a function of pressure (P) and displayed them in fig 1 (a, b) and in fig 2. It is clear from fig. that ΔG tends to zero at the phase trasition pressure 40 GPa and 58 GPa for respective phases. We have also calculated the percentage of relative volume change using compression curve (fig. 2). The calculated values of structural properties (Phase transition pressure and relative volume collapse) for both the phases (Table 3) are in good agreement with the available experimental data [10, 11, 16].

TABLE 3. Cohesive Energy of CuI in various phases in (Kcal/mol).

Crystal CsI	CsCl-Tetra.		Tetra-Ortho	
	Cal	Exp	Cal	Exp
P_t (GPa.)	40.8	40.0[a]	58	56[a]
$\Delta[V(P_t)/V(O)](\%)$	2.54	-	2.77	-

a - ref. [16].

Finally, we conclude that during the crystallographic transition from Cesium chloride (CsCl) to Tetragonal and from Tetragonal to orthorhombic structure, the discontinuity in pressure volume (P-V) phase diagram follows the same trend as reported experimentally.

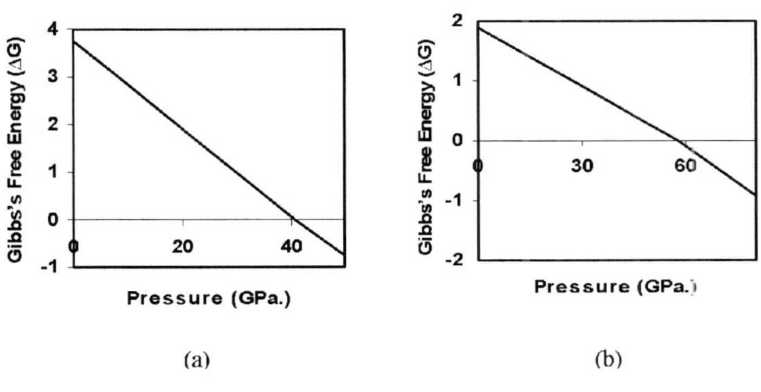

(a) (b)

FIGURE 1: Gibbs's Free Energy as a function of Pressure (P): (a) CsCl-Tetragonal and (b) Tetragonal-Orthorhombic structure.

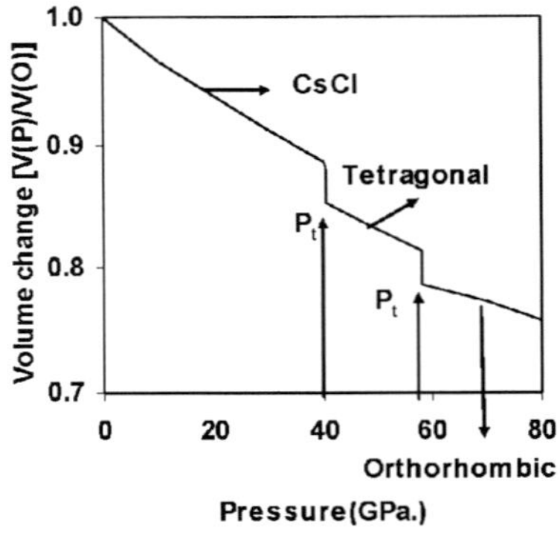

FIGURE 2: Volume collapse as a function of of Pressure (CsCl-Tetragonal-Orthorhombic)

REFERENCES

1. S. Dixit and R. Niwas, Phys. Stat. Sol (b) **161**, 583-596(1990).
2. H. Schilosser, J. Phys. Chem. Solids **53**, 855-856 (1992).
3. C. E. Sims, G. D. Barrera and N. L. Allan, Phys. Rev. B. **57**, 11164-11172 (1998).
4. C. H. Yoo, M. J. Lipp, D. Strachan and W. B. Daniels, Phys. Rev. Lett. **84**, 3875-3878 (2000).
5. R. Reichlin, M. Ross, S. Martin and K. Goettel, Phys. Rev. Lett. **56**, 2858-2860 (1986).
6. M. I. Eremets, K. Shimizee, T. C. Kobayashi and K. Amaya, Science **281**, 1333-1335 (1998).
7. K. Asaumi, Phys. Rev. B. **28**, 1118-1120 (1984).
8. E. Knittle and R. Jeanloz, Science **223**, 53-56 (1984); E. Knittle and R. Jeanloz, J. Phys. Chem. Solids **46**, 1179-1184 (1985).
9. T. L. Huang and A. L. Ruoff, Phys. Rev. B. **29**, 1112-1114 (1984).
10. Y. K. Vohra, S. J. Duclos and A.C. Rouff, Phys. Rev. Lett, **54**, 570-573 (1985).
11. Y. K. Vohra, S. T. Weir, K. E. Bristles and A. L. Ruoff, Phys. Lett. **55**, 977-979 (1985).
12. N. V. K. Prabhakar, N.K. Gaur, R. K. Singh and N. N. Sharma J. Phys. Condens. Matter **2**, 3445-3449 (1990)
13. R. K Singh and Sadhna Singh, Phys Rev B. **39**, 671-676 (1989); ibid **45**, 1019-1022 (1992); Phase Transit, **15**,127-134 (1989); J Phys. Soc. Jpn. **66**, 1714-1717 (1997).
14. R. K. Singh and D. C. Gupta, Phys. Rev. B. **40**, 11278-11283 (1989); IL Nuovo Cimento D **9**, 1253-1264 (1987).
15. V. Shrivastava and S. P. Sanayal, J. of Alloys & Compd., **366**, 15-20 (2004).
16. M. B. Nardelli, S. Baroni and P. Giannozzi, Phys. Rev. Lett. **69**, 1069-1072 (1992).

Corresponding author: deoshreebaghmar@gmail.com

POSTER PRESENTATIONS

Synthesis And Structural Studies Of CdTe Thin Films Formed By Vacuum Evaporation Technique

Ajit Mahadkar[a]; Alka Chauhan[b]; Madhavi Thakurdesai[c] and Deepak Gaikwad[b]

[a]*Tata Institute of Fundamental Research, Homi Bhabha Road, Colaba, Mumbai – 400 005, India*
[b]*Department of Physics, University of Mumbai, Santacruz (E), Mumbai - 400 098, India.*
[c]*Department of Physics, Birla College, Kalyan- 421304, India*

Abstract. Te/Cd/Te trilayer has been successfully synthesized by using vacuum evaporation technique onto cleaned glass substrate at room temperature under the vacuum of the order of 10^{-5} mbar. Furnace annealing has been carried out at different temperatures (573 K, 623 K, 673 K and 723 K) in Argon ambient. CdTe compound was evaporated under same conditions for comparison studies. X-ray diffraction (XRD) studies were done for structural analysis of these as-deposited and annealed samples. These studies reveal that the phase formations are highly dependent on the annealing temperature. Optical band gap measurements show sharpening of the band edge at 723 K annealing, band gap value were found to lie in the range of 1.45- 1.5 eV. X-Ray Fluorescence (XRF) studies were carried out for elemental analysis of these trilayer.

Keywords: Cadmium Telluride; vacuum evaporation; X-ray diffraction; Annealing
PACS: 81.15.Ef, 78.66.w.

INTRODUCTION

CdTe is one of the efficient materials for fabrication of semiconductor devices. The high optical absorption coefficient for visible spectrum and near optimum band gap (1.45 eV) makes this material most suitable for low cost application of solar cell [1]. The literature survey shows that various techniques such as spray pyrolysis [2], electro deposition [3], closed space sublimation [4], vacuum evaporation method [5] have been investigated for obtaining device grade CdTe thin films. Understanding of microstructure and morphology evolution of CdTe films is important as this affects performance of active devices such as solar cells made from these layers. The grain size and surface morphology are often defined during the film growth period. These characteristics may also be modulated through high temperature annealing carried out after deposition [6]. A suitable experimental approach will help in improving CdTe based devices such as solar cells, γ and IR detectors and Field Effect transistors (FET's) [7-8]. The efficiency conversion of these thin film CdTe device are strongly dependent on the structural, optical and electrical properties [9]. In this work, Te/Cd/Te trilayer thin films are deposited on a glass substrate and are annealed to elevated temperatures. Effects of the as-deposition and post deposition anneal

CP1004, *Thermophysical Properties of Materials and Devices-NCTP '07*,
edited by P. Predeep, S. Prasanth, and A. S. Prasad

temperature on the film properties such as grain size; phase formation and optical band gap and elemental content were studied using various techniques.

EXPERIMENTAL

Te/Cd/Te trilayer samples were prepared by evaporating Te (99.99%) and Cd (99.99%) powder onto cleaned glass substrate. The glass substrates were cleaned with isopropyl alcohol. The samples with the total thickness of 2382 Å was made. The deposition rate and thickness of the film was controlled and monitored using conventional quartz crystal monitor. The post deposition furnace annealing was carried out at 573 K, 623 K, 673 K, and 723 K in argon ambient for two hours at each temperature. The structure of these samples were investigated using X-ray diffractometer (XRD) using Cu-Kα radiation with λ= 1.5418 Å on a Phillips Panalytical Xpert Pro MPD spectrometer. The samples were scanned for the 2θ of 20-80 degree and were rotated at a speed of 0.0006°/Sec. The crystallite size (D) were calculated from full width at half maximum (FWHM) (w) using the Scherrer formula [10]

$$D = \left(0.94\lambda / w\cos\theta\right) \quad (1)$$

The strain (ε) and dislocation density (δ) [11] was calculated. The dislocation density (δ) is defined as the length of dislocation lines per unit volume of the crystal.The lattice parameter 'a' can be evaluated from the relation;

$$a = d\sqrt{h^2 + k^2 + l^2} \quad (2)$$

where h,k,l are Miller indices.

The optical measurements were carried out using Varian CARRY 500 UV-VIS-NIR Spectrometer. Baseline correction with respect to glass substrate was done. The elemental analysis was carried out on a Jeol JSX-322 EDXRF with rhodium as a target.

RESULTS AND DISCUSSION

Structural Studies

X-ray diffraction (XRD) studies were made on the Te/Cd/Te trilayer deposited onto glass substrate and furnace annealed at different temperature in argon ambient condition. The compound CdTe powder (99.99%) was also deposited on glass substrate at room temperature for the comparative studies. Fig 1. (a), (b), (c), (d) and (e) show the XRD spectra of Te/Cd/Te trilayer deposited on glass substrate and annealed at 573 K, 623 K, 673 K ,723 K and compound respectively. The structure of as-deposited film shows hexagonal Cd and Te phases. Fig.1. (b) shows XRD pattern of the tri -layer annealed at 573 K. The initiation of the formation of CdTe structure is observed with feature peaks at 2θ= 23.74, 39.38, 46.48, 62.3 and 76.02 corresponding to (111), (110), (102), (331) and (511) planes of cubic CdTe as confirmed from JCPDS X-ray powder file data. At this particular temperature presence of Cd and Te is also observed along with the compound. Further increase in the annealing temperature to 623 K and 673 K shows the increased intensity of the featured peak of CdTe compound with less intensity peak from Cd and Te indicating formation of CdTe films with some traces of Cd and Te. The diffraction intensity of CdTe peaks is found to be increasing with increasing annealing temperature and peak intensity of Cd and Te is extremely low as seen in Fig 1. (c) and (d). The diffraction intensity increases due to the growth of the incorporated material with annealing temperature [12]. At all annealing temperatures, the peak intensity of (111) was high compared to other peaks. This indicates preferential orientation of the micro-crystallites is in (111) direction which is perpendicular to the surface [13]. The annealing of these trilayer at 723 K (Fig 1.(e)) shows, pure CdTe film formation with cubic phase corresponding to zinc blend structure, with no diffraction peaks from Cd and Te. The CdTe films forms a cubic phase having a lattice constant of 6.52 Å. At this particular temperature all peaks corresponding to cubic phase are observed.

Fig 1.(e and f) shows comparison of CdTe thin film formed by trilayer deposition and annealed at 723 K and CdTe compound deposition onto a cleaned glass It is seen that the intensity of the CdTe peaks is more in case of Te/Cd/Te sample annealed at 723 K then the CdTe compound. Both the films form cubic phase with preferential orientation of micro crystallites in (111) direction i.e zinc blend structure. The full width at half maximum (FWHM) of the cubic (111) peaks were calculated and was found to decrease with increase in annealing temperature. The decrease in FWHM is due to decrease in the concentration of lattice imperfection as the micro-strain decreases within the film and the crystallite size increases. Table 1. shows the comparison of the crystallite size (D), strain (ε) and dislocation density (δ) of the samples annealed at different temperatures. It is seen that at an annealing temperature of 573 K, the grain size increases but dislocation density decreases. This indicates initialization of the formation of a high quality film. At a temperature of 623 K and 673 K the average grain size decrease with increased strain and dislocation may be attributed due to formation of smaller grains on larger grains [14]. The increase in crystallite size and the significant decrease in dislocation density indicate high purity film formation at annealing temperature of 723 K.

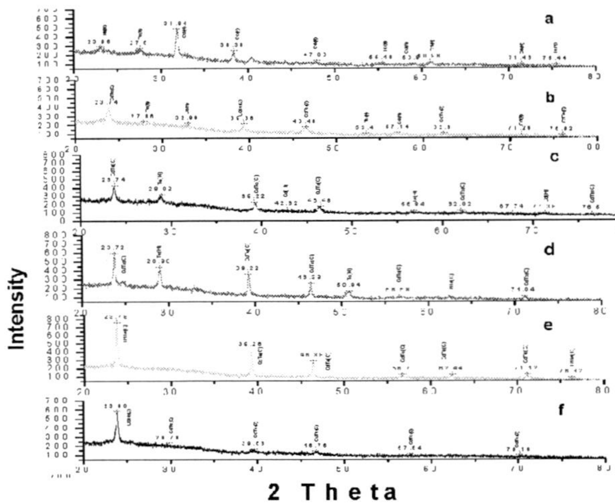

FIGURE 1. XRD patterns for Te/Cd/Te samples, (a) As-deposited, (b) annealed at 573 K, (c) annealed at 623 K, (d) annealed at 673 K , (e) annealed at 723 K and (f) CdTe compound.

TABLE.1. Comparative look of the crystallite size (D), strain (ε) and dislocation density (δ) of the Te/Cd/Te trilayer annealed at different temperature for 2 hours.

Annealing Temperature	2θ	D(Å)	ε x10^{-4} (lin^{-2} m^{-4})	δx10^{-13} (lin/m^2)
573 K	23.74	345	10.51	84.23
	39.38	531	6.82	35.47
623 K	23.74	463	7.82	46.65
	39.22	453	7.99	48.75
673 K	23.72	644	5.62	24.10
	39.22	616	5.88	26.36
723 K	23.78	706	5.13	20.09
	39.28	855	4.23	13.66

Optical Studies

The optical band gap values of the films were obtained from absorption measurements and by plotting $(\alpha h\nu)^2$ versus hν. Fig.2 shows the data at near band edge for CdTe films formed at different annealing temperature. The extrapolated values of the optical energy gap; E_g are 1.5 eV, 1.48 eV, 1.47 eV and 1.45 eV for the samples annealed at 523 K, 623 K, 673 K and 723 K respectively. It is seen from Fig.2. that the absorption edge moves towards lower energies with increased annealing temperature which is appropriate for single crystal CdTe. The change in the sharpness of the absorption edge took place on annealing at a critical temperature of 723 K. The decrease in energy band gap on annealing may be due to an increase in particle size and decrease in strain and dislocation density. The optical band gap value for the CdTe compound deposition onto glass substrate is found to 1.46eV.

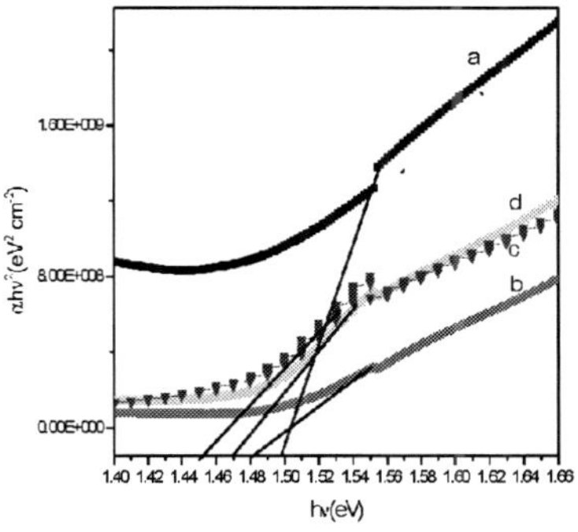

FIGURE 2. Absorption spectra of Te/Cd/Te films annealed at (a) 573 K, (b) 623 K, (c) 673 K and (d) 723 K.

Elemental Studies

The chemical/elemental composition of the CdTe films formed by trilayer deposition and annealed at 723 K and a CdTe compound sample deposited onto glass substrate were investigated using X-Ray Fluorescence (XRF) and is shown in Table. 2. The elemental composition for trilayer is seen to be quite close to the CdTe compound. The ratio of the atomic percentage of Cd and Te in trilayer as well as as-deposited CdTe compound is less than 1 which makes these films to be slightly p-type conducting due to formation of Cd vacancies acting as an acceptor center in CdTe lattice thus making the film Te rich. [4].

TABLE 2. Comparison of the elemental composition of CdTe thin films.

Sample Condition	Cd (at%)	Te (at%)
Te/Cd/Te annealed at 723 K	46.7	53.3
CdTe compound	46.48	53.52

CONCLUSION

We have studied the structural, optical and elemental properties of evaporated Te/Cd/Te thin films at post-deposition annealing temperatures. It is observed that a pure CdTe film is formed at an annealing temperature of 723 K and has zinc blend structure with orientation along the (111) direction. Optical band gap of these films became sharper on 723 K annealing with band gap value of 1.45 eV. The elemental analysis revealed that the deposited films were found to be Te rich.

ACKNOWLEDGEMENT

Authors would like to thank Mr. Nilesh Kulkarni, Mr. Rudhir Bapat, Mrs. Bhagyashri Chalke and Mr. Meghan Patankar from TIFR, Mumbai for their help in carrying out various experiments.

REFERENCES

1. T. Aramoto, S. Kumazawa, H.Higuchi, T. Arita, T. Shibutani, T. Nishio, J. Nagajima, M.Tsuji, A. Hanakusa, T. Hipino, K. Omura, H. Ohyama, M. Murazono, Jpn. *J. Appl. Phys.* 36 (1997) 6304.
2. K. Vamsi Krishna, V.Dutta, *Solar Energy Materials and Solar cells* 80 (2003) 247-256.
3. M. Takahashi, K. Uosaki, H. Kita, *J. Appl. Phys.* 55 (1984) 3879.
4. N. Abbas Shah, A. Ali, Z. Ali, A. Maqsood, A. K. S. Aqili, *Journal of Crystal Growth* 284 (2005) 477-485.
5. S. Gongoi, K. Barua, *Thin solid films* 92 (1982) 227.
6. B. Bansol, Int. *J. Sol. Energy* 12 (1992) 25.
7. G. Landwehr, A. Wagg, K. Hoffmann, R.N. Bicknell- Tassius, *Proc. SPIE* 1362 (1991) 382.
8. O. Zelaya-Angel, A. Picos-Vega, R. Ramirez-Bon, F.J. Espinoza-Beltran, *Vacuum* 52 (1999) 99.
9. D. Polcman, J. Vcnnik, *J. Phys. D* 21 (1988) 1004.
10. G. Gordillo, J. M. Florez, L.C. Hernandez, *Sol. Energy Mater. Sol. Cells* 37 (1995) 273.
11. G.B. Willamson, R.C. Smallman, Philos. Mag. 1 (1956) 34.
12. N. El. Kadry, A. Ashour, S.A. Mahmoud, *Thin Solid films* 269 (1995) 112.
13. A.H. Eid, S. Mahmoud, M.M. Sayed , Fizika 4 (1987) 397.
14. A.Lopez- Otero *Journal of Crystal Growth* 42 (1977) 157

Mechanical Characterization of Polymethyl Methacrylate and Polycarbonate Blends

Manasvi Dixit*, Vinodini Shaktawat, Kananbala Sharma, Narendra S. Saxena* and Thaneshwar P. Sharma

Semiconductor and Polymer Science Laboratory, Department of Physics, University of Rajasthan, Jaipur 302 004, INDIA

*Corresponding E-mail: n_s_saxena@rediffmail.com, manasvi1584@yahoo.co.in

Abstract. Blends of Poly methyl methacrylate (PMMA) and Polycarbonate (PC) were prepared using solution casting method in five different concentrations (0%, 25%, 50%, 75%, and 100% by weight) using dichloromethane (CH_2Cl_2) as a solvent. The mechanical properties were studied by Dynamic Mechanical Analyser (DMA). Stress-strain scan have been used to study the tensile strength and Young's modulus for these blends. It has been observed that complex modulus decreases with temperature due to the thermal expansion of the film of polymer blend. The result obtained for variation of Tan δ and modulus with concentration indicates about the degree of miscibility of this blends. Glass transition temperature of pure PMMA is found to be at 136^0C because it is highly isotactic in nature. It is observed that T_g of the blend shifts towards the lower temperature with the increase of concentration of PC. Blending of PMMA and PC improves the mechanical properties such as Young's modulus and tensile strength. The Young's modulus, ultimate strength and tensile strength increases with the increase in concentration of PC upto 50% and then decreases on further increase in concentration. The value of Young's modulus and tensile strength is found to be highest at 50% concentration.

Keywords: Polymer blends, Poly methyl methacrylate, Polycarbonate, Glass transition temperature, Tensile strength, Young's modulus.
PACS: 61.25.hk, 64.70.pj

INTRODUCTION

In recent years, polymer blends have become an increasingly important area of research activity because they lead to an end product that possesses promising properties that may be absent in individual polymer. These blends are looked upon due to their simple and low cost preparation. In addition, the resulting materials may present unique characteristics as a consequence of a synergic effect. In industry different techniques are used for blend production such as dry mixing (in room or higher temperature), precipitation process and sinter, however in laboratory practice, the solution blending is usually applied [1].

From an exclusively thermodynamic point of view, the blends may be classified as miscible, partially miscible and immiscible. Miscibility is linked to the thermodynamic concept of homogeneity, to the existence of a single phase and to property isotropy. Mechanical properties show that a miscible blend is more desirable than an immiscible one for higher modulus application.

CP1004, *Thermophysical Properties of Materials and Devices-NCTP '07*,
edited by P. Predeep, S. Prasanth, and A. S. Prasad

In order to investigate miscibility and phase behavior of polymer blends through the determination of the glass transition temperature (T_g), Dynamic Mechanical Analyzer (DMA) has been frequently used. Generally, the use of T_g in determination of polymer miscibility is based on the judgment that the observation of a single T_g between those of pure components is taken as the evidence of miscibility, although strictly speaking, a single Tg is only indicative of state of dispersion, whereas the appearance of two T_g's suggests the occurrence of phase separation.

Poly (methyl methacrylate)(PMMA) exhibits excellent mechanical properties and good performance at various processing conditions. It is an amorphous polymer having high optical properties, good chemical resistance and high tensile strength [2]. But the use of PMMA presents some practical disadvantages, e.g. brittleness, low elongation at break and high water absorption. To circumvent these drawbacks, many efforts have been made through copolymerization and polymer blending. Among a number of PMMA blends studied for miscibility, the mixture of PMMA and PC is one of the most investigated polymeric systems. This may be attributed to the excellent properties of PC, including outstanding ductility, low water absorption.

Gardlund [3] and Viville [4] showed that PC/PMMA blends are partially miscible, due to the n–p complex formation between the ester group of PMMA and the phenyl ring of PC. Agari et al [5] reported miscible PC/PMMA blends obtained by rapid solvent casting method. Since then, the miscible blend has been investigated by NMR, DSC and DMA measurements. The result of these studies showed that this blend is a single-phase system under low critical solution temperature (LCST).

A lot of work has been done on the PMMA/PC blend to study the effect of mixing conditions and solvents on miscibility and compatibility of blend but it is observed that not much information is available as regards the mechanical properties of PMMA/PC blend. In this present work, films of blend in different mass ratio of PMMA/PC were prepared and their mechanical properties and glass transition temperature were studied by DMA.

EXPERIMENTAL

Films of PMMA and PC blends were prepared by solution casting technique [6]. Dichloro methane (CH_2Cl_2) was used as solvent. The blend of PMMA/PC had compositions of 100/0, 75/25, 50/50, 25/75, and 0/100 by weight.

The Dynamic Mechanical Analysis (DMA) measurements were carried out by using Tritec2000 DMA. All measurements were performed from 30–150^{0}C temperature at a heating rate/ramp rate of 2°C min^{-1} for all the samples. The test samples used for DMA were about 50μm thick. Frequency of oscillation was fixed at 1 Hz and strain amplitude 0.01 mm within the linear viscoelastic region. All experiments are performed in tension mode. Dynamic mechanical test methods have been widely employed for investigating the structures and viscoelastic behaviour of polymeric materials for determining their relevant stiffness and damping characteristics for some kinds of applications. The details about Dynamical mechanical analyser have been discussed elsewhere [7, 8].

RESULTS AND DISCUSSION

The variation of the tan δ of the polymer films as a function of temperature is shown in Fig.1. This figure shows the T_g of PMMA, PC and PMMA/PC blend at different concentrations. It is observed from Tan δ peak that the blend of each composition exhibits only single T_g, which changed with PC content. Thus it is confirmed that the blend is miscible over all compositions. The T_g of pure PMMA was found to be at 136.4^{0}C because it is highly isotactic in nature [9] and T_g of pure PC is found to be at 125^{0}C.

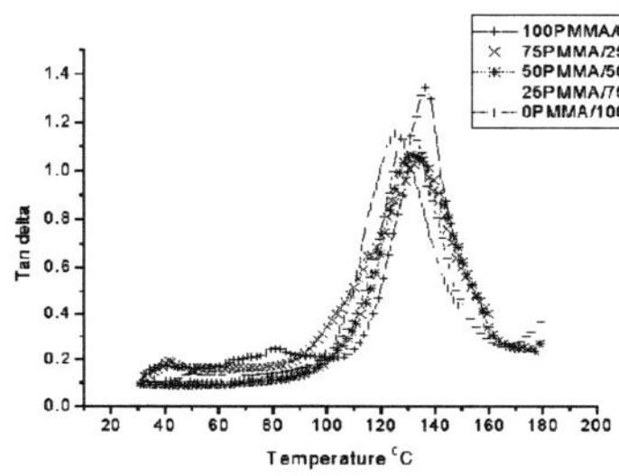

FIGURE 1. Variation of Tan δ with temperature

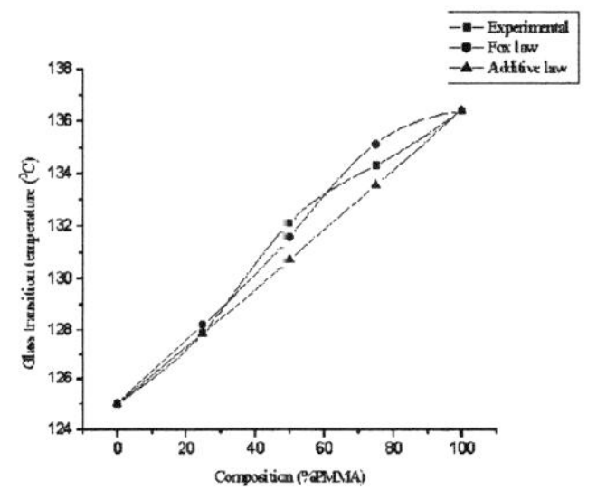

FIGURE 2. Variation of glass transition temperature with blend composition

As the polymers are mixed with each other at various composition the free energy of the blend will vary and hence the miscibility and mechanical properties will also vary. It is observed that T_g of the blend shifts towards the higher temperature with the increase of concentration of PMMA (Fig.2).

The glass transition of a miscible blend is usually described by either the Fox law:

$$1/T_g = w_1/T_{g1} + w_2/T_{g2} \tag{1}$$

or by the additive law which is linear relationship described by the equation:

$$T_g = w_1 . T_{g1} + w_2 . T_{g2} \tag{2}$$

where w_1 and w_2 are the weight fraction of the blend constituents and T_{g1} and T_{g2} are the corresponding T_g's [10]. The calculated Tg of a miscible PMMA/PC (50/50 w/w) blend is 130.7^{0}C from the Additive law and 131.57^{0}C from the Fox law. Experimental values show a slight positive deviation than the Fox law and Additive law. There is an overall good agreement between the theoretical and experimental results (Table 1).

TABLE 1. Glass transition temperature from different law

Blend Composition	Experimental value of T_g	Fox law (T_g)	Additive law (T_g)
75PMMA/25PC	134.3	135.13	133.55
50PMMA/50PC	132.1	131.57	130.7
25pmma/75PC	127.9	128.20	127.85

Tensile Strength and Young's Modulus of the Polymer Blends

The stress-strain curves of the polymer films for the different concentration of PMMA/PC blend are shown in fig.3.

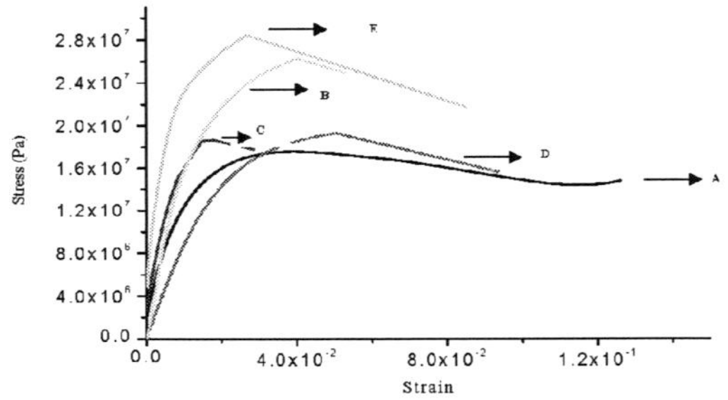

FIGURE 3. Stress-strain curves of the polymer films (A) Pure PC (B) Pure PMMA (C) PMMA/PC (75/25) blend (D) PMMA/PC (25/75) blend (E) PMMA/PC (50/50) blend

All polymer films show that the Young's modulus increases slightly with the increasing of the strain rate. It is observed from fig.3 that the values of tensile strength, Ultimate strength and Young's modulus for the blend of PMMA/PC (50/50%) content are highest. At this concentration more and more molecules of PMMA are linked with molecules of PC i.e. crosslinking of the PMMA/PC blend is maximum. Young's modulus is a fundamental measure of the stiffness of a material. Stiffness increases with increasing value of young's modulus i.e. higher its value (Young's modulus), the more resistant the material is to being stretched.

In all cases of stress-strain curve, at low strain, there exists in the materials the so-called elastic energy, which is stored in the form of strain energy of the chemical bonds prior to break. For greater strain the stored energy is termed plastic energy and is used to describe the amount of energy residually stored in the material as internal and free (orientation) energy after break [11].

The blend is completely miscible in the composition of 50/50%, which is evident from the highest value of Young's modulus, tensile strength at this 50/50% composition. The single phase, which occurs in this composition, is energetically favoured and is most stable range in this study [12].

The mechanical properties and the glass transition temperature of different polymer blends of different concentration of PMMA/PC are shown in table 2.

TABLE 2. Mechanical properties of different compositions of polymer films of PMMA/PC blend

Composition (wt%)	Young's modulus (GPa)	Glass transition temperature (^{0}C)	Ultimate tensile strength (MPa)	Tensile strength at break (MPa)	Elongation at break (%)
100PMMA/0PC	1.799	136.4	26.26	24.94	5.287
75PMMA/25PC	1.74	134.3	18.6	17.76	2.87
50PMMA/50PC	2.61	132.1	28.4	21.72	8.52
25PMMA/75PC	0.914	127.9	19.28	15.63	9.36
0PMMA/100PC	0.872	125	17.5	14.75	12.62

Properties of miscible polymer blends may be intermediate between those of the individual components (i.e. additive behaviour). In other case, blend properties may exhibit either positive or negative deviation from additivity [13]. The curve of the tensile strength and young's modulus versus composition is displayed in fig.4 (A) and 4(B). Here, both modulus and tensile strength of miscible polymer blends exhibit a small maximum at some intermediate blend composition. This shows a synergistic behaviour of the blend films.

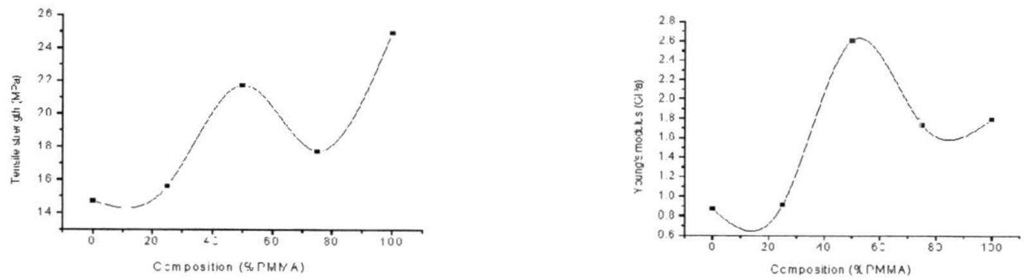

6(A) 6(B)

FIGURE 4. Variation of (A) tensile strength and (B) Young's modulus with blend composition

CONCLUSIONS

The miscible PMMA/PC blend was prepared, confirming its miscibility by DMA. The result shows that the T_g values of the blend films increases with the increase of concentration of PMMA. A single T_g of PMMA/PC blend suggests the existence of miscibility in the complete composition range, which increases the application as high performance materials. For PMMA/PC (50/50%) blend the value of Young's modulus, ultimate strength and tensile strength is highest due to good crosslinking.

ACKNOWLEDGMENTS

sAuthors are thankful to Mrs. Rashmi Saxena, Ms. Deepika and Mr. Vishal Mathur for their help in various ways.

REFERENCES

1. H. Kaczmarek, D.L. Zagorski and M. Siatek, Proceedings, *SPM-2002*, pp.214-216 (2002).
2. R. Bonzanini, E.M. Girotto, M.C. Gonc¸alves, E. Radovanovic, E.C. Muniz and A.F. Rubira, *Polymer* **46**, 253-259 (2005).
3. G.J. Garlund, *ACS Polymer Prep.* **23**, 258-263 (1982).
4. P. Viville, D. Thoelean, S. Beauvois, R. Lazzaroni, G. Lambin, J.L. Bredas, K. Kolev and L. Laude, *Appl. Surf. Sci.* **86**, 411-416 (1994).
5. Y. Agari, A. Ueda, Y. Omura and S. Nagari, *Polymer* **38**, 801-807 (1997).
6. W. P. Hsu, *Journal of Applied Polymer Science* **80**, 2842-2850 (2001)
7. V. Shaktawat, N. Jain, N. S. Saxena, K. B. Sharma and T.P. Sharma, *J. Poly. Sci. Series B* **49 (10)** 236-239 (2007).
8. Kevin P. Menard, "*Dynamic Mechanical Analysis: A practical introduction*", Boca raton: CRC press LLC, 1999).
9. http://www.cem.msu.edu/~reusch/VirtualText/polymers.html.
10. S.S. Ray and M. Bousmina, *Macromol. Rapid Commun.* **26**, 450-455 (2005).
11. A. Linares and J.L. Acosta, *Eur. Polym. J.* **33 (4)** 467-473 (1997).
12. http://www.pslc.ws/mactest/blend.html.
13. Joel R. Fried, "*Polymer Science and Technology*", New Delhi: Prentice-Hall of India, 1999.

Optical and Electrical Properties of In_2O_3: Mo Films Prepared by Activated Reactive Evaporation Technique

S. Kaleemulla[a], A.Sivasankar Reddy[b], S. Uthanna[a] and P.Sreedhara Reddy[a*]

[a]*Department of Physics, Sri Venkateswara University,Tirupati 517 502, India*
[b]*Department of Ceramic Engineering, Yonsei University, Seoul 120-749, Korea*

Abstract. Transparent conducting molybdenum doped indium oxide (In_2O_3: Mo) thin films with high performance have been deposited on glass substrates at 300 °C by activated reactive evaporation technique and the effect of molybdenum content on the physical properties of the deposited films had been studied using spectrophotometer and van der Pauw methods. The films exhibited a lower electrical resistivity of 7.2×10^{-4} Ωcm and an average spectral transmittance of 83 % in visible region with an optical band gap of 3.68 eV.

Keywords: Transparent conducting oxide, Thin films, Photoluminescence
PACS: 73.61.-r, 78.20.-e, 78.55.-m

INTRODUCTION

Now a days thin films and thin film technologies are playing an important role in our daily life as they found in number of applications i.e., from microelectronics to automobile parts. These applications highly depend on the optical and electrical properties of the material chosen. Many efforts have been put on the materials and techniques to produce transparent and electrically conducting films. Since then these TCO films have received great attention as they exhibited excellent properties of low electrical resisitivity (< 10^{-3} Ωcm), high optical transmittance (> 80%) and high infrared reflectance [1,2]. The electrical resistivity of these TCOs can be decreased by controlling stochiometry or by adding proper dopants [3]. Various techniques have been employed for the preparation of molybdenum doped indium oxide thin films (IMO) such as reactive thermal evaporation, sputtering, etc.

In the present work, highly transparent and conducting molybdenum doped indium oxide (IMO) films were prepared by a novel thermal activated reactive co-evaporation technique and studied the influence of Mo doping on the optical, electrical and photoluminescence properties of IMO films.

CP1004, *Thermophysical Properties of Materials and Devices-NCTP '07*,
edited by P. Predeep, S. Prasanth, and A. S. Prasad

EXPERIMENTAL

Thin films of molybdenum doped In_2O_3 (IMO) were prepared using activated reactive
co-evaporation technique in which high density plasma enhances the reaction rate between the evaporants. The metal In and MoO_3 were evaporated from two separate sources separately.

The structural, optical, electrical and room temperature photoluminescence properties were systematically studied using X-rat diffractometer, van der Pauw method, Hitachi U-3400 UV-VIS-NIR double beam spectrophotometer and fluorescence spectrophotometer (model: YVON Florolog-3), respectively.

RESULTS AND DISCUSSION

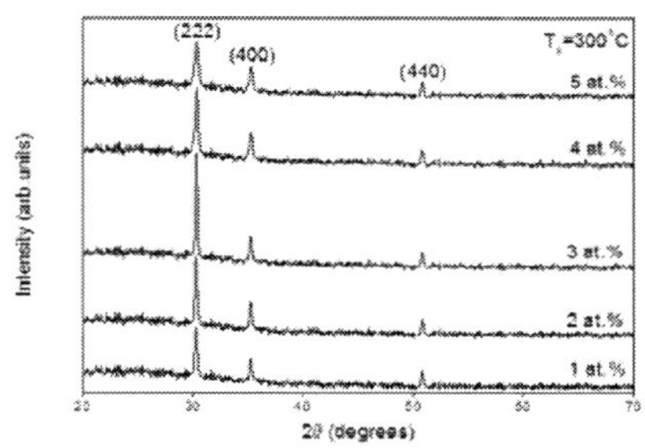

FIGURE 1. X-ray diffraction patterns of IMO films prepared at different Mo-doping levels.

Fig. 1 shows the X-ray diffraction pattern of the IMO films formed at Mo doping of 0 - 5 at. %. The films formed at lower substrate temperatures (≤ 200 °C) were amorphous in nature. The amorphous nature of the IMO film is confirmed by the analysis of surface morphology of the films as shown in Fig. 2.

FIGURE 2. SEM pictures of IMO (3 at. % of Mo) films formed at substrate temperatures of (a) 30 °C and (b) 300 °C

These films have very small and uniform grains over the whole surface of the grown layers. The diffraction peaks observed for the films, prepared at a temperature above (≥ 250 °C and have larger grains (Fig. 2 (b)), indicating that crystallinity of the films had started at 250 °C and the peaks shifted slightly to a higher diffraction angle as Mo content increased from 1 - 5 at. %.

All the diffracted peaks as labeled were coincided with In_2O_3 cubic structure and we did not observe any impurity phases along with In_2O_3 phase. The (222) plane grew more predominately than other peaks and intensity of the diffracted peaks increased upto 3 at. % of Mo doping level after that slightly decreased with further increasing of Mo doping level. The mean crystallite size L was calculated by using Debye Scherrer's relation,

$$L = k\lambda / \beta\cos\theta \qquad (1)$$

where, λ is the wavelength of the X-rays, β is the full width at half maximum (FWHM) and θ is the Bragg's diffraction angle. The crystallite size (L) of the films increased from 35 nm to 42 nm as the Mo-doping varies from 0-3 at. % and lattice constant of the films decreased from 10.120 Å to 10.110 Å with the increase of Mo content from 0 – 5 at. %.

Fig. 3 shows the variation of electrical resistivity (ρ), carrier concentration (n) and Hall mobility (μ) of Mo doped indium oxide thin films (IMO) films deposited at various doping levels of Mo and at a constant substrate temperature of 300 °C. In general, it is known that for TCO films, the conduction electrons are basically supplied by doping donars and / or oxygen vacancies [4]. In the process of doping, the donor ions substitute the ions in parent oxides and provide extra electrons in the band. However when Mo is doped in In_2O_3 matrix, three extra electrons, which are loosely bonded with the parent atom will act as free carriers. Hence the conductivity of TCO films increases due to increase in carrier density. The electrical resistivity of the films decreased with the increase of Mo doping from 0 - 3 at. % after which, it increased. The pure In_2O_3 films exhibited an electrical resistivity of 1.1×10^{-3} Ωcm and then it decreased to lowest value of 7.2×10^{-4} Ωcm when Mo doping level increased from 0 - 3 at. %. After that it increased to 8.3×10^{-4} Ωcm when the Mo doping level increased from 4 – 5 at. %. It may be due to the increase of lattice boundary scattering. Similar increase in electrical resistivity was observed in Zr doped In_2O_3 films (IZO) by Asikainen et al. [5].

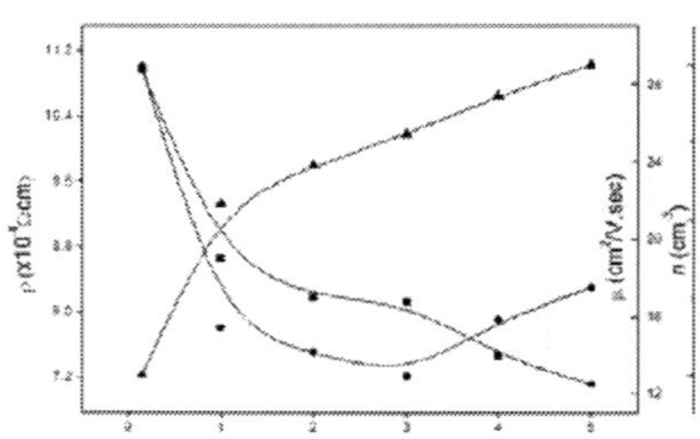

FIGURE 3. Variation of electrical resistivity (ρ), Hall mobility (μ) and carrier concentration (n) of IMO films with Mo-doping level.

The optical transmittance spectra of IMO films for different Mo doping levels is shown in Fig 4. The pure In_2O_3 films exhibited an average transmittance of only 75 %

after which it increased with the increase of Mo doping level from 1 - 3 at. %. It may be due to increase in crystallinity as evidenced from the X-ray diffraction pattern. The films exhibited an optical transmittance of 85 % at 3 at. % of Mo doping; thereafter it decreased to 75 % when the doping level increased from 4-5 at. %. It may be due to structural defects and / or electron boundary scattering. The increase in transmittance may be due to increase in carrier concentration which increases light scattering. The optical band gap Eg of the IMO films was determined from optical absorption coefficient (α) and incident photon energy (hυ) using the relation [6],

$$(\alpha h\upsilon) = (E_g - h\upsilon)^n \qquad (3)$$

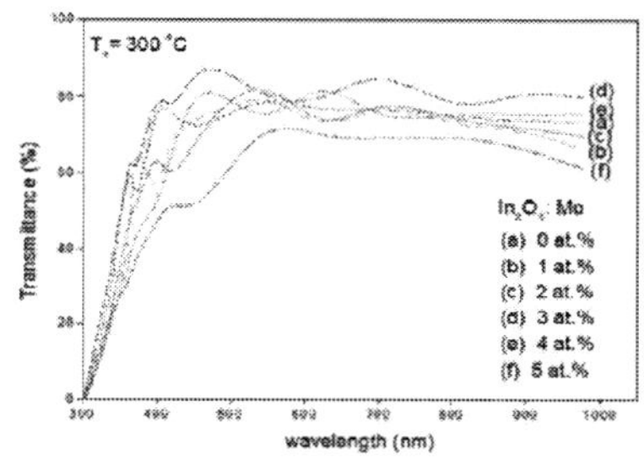

FIGURE 4. Optical transmittance spectra of IMO films with different Mo-doping levels.

where, n depend on the kind of optical transition that prevails. Here n= ½, as IMO films are directly allowed n-type degenerate semiconductors. The optical bang gap E_g is obtained by plotting $(\alpha h\upsilon)^2$ versus the photon energy (hυ) and by extrapolating of the linear region of the plots to zero absorption (α = 0). An optical band gap (Eg) of 3.68 eV was observed for the films prepared at 300 °C and Mo doping of 3 at. %. The undoped In_2O_3 films exhibited an optical band gap of 3.61 eV after which it increased to 3.68 eV as the Mo doping level increased from 1-3 at. %. It may be due increase in crystallinity of the films which is evidenced from X-ray diffraction pattern. Thereafter it decreased to 3.60 eV for higher Mo-doping level. It may be due to deviation in the structure.

The performance of transparent conducting oxides can be judged using Haacke's figure of merit defined as,

$$\Phi_{TC} = T^{10} / \rho_s \qquad (2)$$

where, T is the transmittance at 550 nm and ρ_s is the sheet resistance of the film. The figure of merit increased from $1.27\times10^{-3}\ \Omega^{-1}$ to $6.8\times10^{-3}\ \Omega^{-1}$ with the increase of Mo doping from 1 - 3 at. %. The increase in figure of merit with Mo doping level was due to the decrease in the electrical resistivity as well as increase in optical transmittance. Higher is the figure of merit, better is the performance of the transparent conducting film. The best performance of the samples was obtained at a Mo doping of 3 at. % with figure of merit $\Phi_{TC} = 6.8\times10^{-3}\ \Omega^{-1}$. The variation of figure of merit with Mo-doping level is shown in table 1.

Fig. 5 shows the photoluminescence emission spectra of Mo doped indium oxide thin films (IMO) formed. Two distinct peaks in the PL spectrum were observed in blue region at 415 nm and 440 nm wavelength under the excitation wavelength of 375 nm. It is well known that bulk In_2O_3 cannot emit light at room temperature [7].

TABLE 1. Summary of sheet resistance (ρ_s), transmittance (T%) and figure of merit (Φ) with Mo (%) doping level

Mo (%)	(Ω/□)	(T%)	(Φ) ($x10^{-3}$ Ω^{-1})
0	44	75	1.27
1	31	77	2.34
2	30	80	3.51
3	28	85	6.80
4	32	75	1.71
5	33	70	0.84

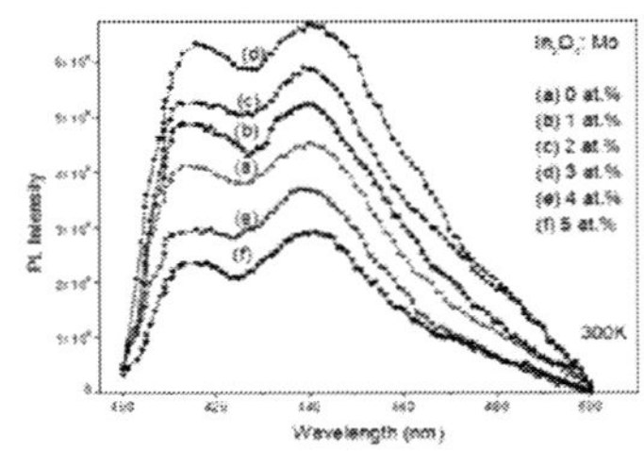

FIGURE 5. Photoluminescence spectra of the IMO films for different Mo-doping levels

CONCLUSIONS

Mo-doped In_2O_3 thin films were deposited by using activated reactive evaporation method and studied the influence of Mo-doping level on the structural, optical, electrical and photoluminescence properties of the films. The films were polycrystalline with cubic structure having an average optical transmittance of 83 %. The films exhibited a minimum electrical resistivity of $7.2x10^{-4}$ Ωcm with $6.8x10^{-3}$ Ω^{-1} of figure of merit. Photoluminescence emission peaks were observed at 415 and 440 nm in the blue region. These characteristics of the Mo doped indium oxide films (IMO) may have potential applications in electro-optical devices as a blue light source.

REFERENCES

1. H. Ohta, K. Nomura, K. Nomura, H. Hiramtsu, K. Veda, T. Kamiya, M. Hirano and H. Hosono, Solid State Electron., **47**, 2261-2267 (2003).
2. I. Hamberg and C. G. Granqvist, J. Appl. Phys., **60**, 123- 160 (1986).
3. K.L. Chopra, S. Major and D.K. Pandya, Thin Solid Films, **102**, 1-46 (1983).
4. H.Y. Yeom, N. Popovich, E. Chason and D.C. Paine, Thin Solid Films, **411**, 17-22 (2002).
5. T. Asikainen, M. Ritala and M. Leskela, Thin Solid Films, **440**, 152-154 (2003).
6. C.G. Granqvist, Sol. Energy Mater. Sol. Cells, **60**, 201-262 (2000).
7. Y. Ohhata and F. Shinoki, Thin Solid Films, **59**, 255-261 (1979).

* corresponding author: pamanji_reddy@rediffmail.com

Photoluminescence Properties of Tm^{3+} & Er^{3+}:$Ca_4GdO(BO_3)_3$ Powder Phosphors

B. Vengala Rao[a], G. Seeta Rama Raju[a], I. Omkaram[a] and S. Buddhudu[a*]

[a]Department of Physics, Sri Venkateswara University, Tirupati-517502, A.P.
*Email: drsb99@hotmail.com

Abstract. Luminescent phosphors of Tm^{3+} & Er^{3+}:$Ca_4GdO(BO_3)_3$ have been prepared by a solid state reaction technique. For these two powder phosphors, XRD, SEM and FTIR have been carried out to understand their structural properties. Emission spectra and lifetimes of the prominent visible emission bands have been measured. For Tm^{3+}: $Ca_4GdO(BO_3)_3$ phosphor, a blue emission band of ($^1D_2 \rightarrow {}^3F_4$) at 452nm with λ_{exci} = 356nm and for Er^{3+}: $Ca_4GdO(BO_3)_3$ phosphor, an intense green emission band of ($^4S_{3/2} \rightarrow {}^4I_{15/2}$) at 561nm with λ_{exci} = 377nm have been observed. Emission from these phosphors, have been explained in terms of relevant energy level schemes.

Keywords: Photoluminescence, Tm^{3+} & Er^{3+}:phosphors
PACS: 78.55-m, 78.55.Hx.

INTRODUCTION

Many lanthanide elements have been recognized to hold tremendous potential in the area of optoelectronic and photonic applications, ranging from solid-state lasers (Nd:YAG laser) and phosphors for color lamps and displays to optical fiber telecommunications[1,2]. These so called rare-earth elements have a partially filled inner ($4f^n$) shell shielded from its surroundings by completely filled outer ($5S^2$ and $5P^6$) orbitals. Due to the shielding, the intra $4f^n$ shell transitions result in very sharp optical emissions at wavelengths from ultraviolet to infrared[3,4]. The wavelengths of these emission lines are determined by the energy of the transition between 4f states of atomic rare-earth elements and are relatively independent of the host material. On the other hand, the host materials do have a very strong effect on the radiative transitions probability. With the development of science and technology, many nonlinear optical crystals have been found for UV and VUV generation. It is well known that the requirements for non-linear optical(NLO) crystals are as follows: wide transparency range for operating wavelength, high laser-induced damage threshold, low material cost and good chemical stability etc.[5-7] These requirements fulfill the preconditions for an ideal phosphors as well. Among NLO materials, borates have attracted much attention not only owing to their excellent chemical-physical properties and low cost, but also a number of compounds with different structure can be selected. Furthermore, the B-O covalent bond has no absorption in UV region because of its large covalent bond energy and the borate compounds have high damage threshold owing to their

CP1004, *Thermophysical Properties of Materials and Devices-NCTP '07*,
edited by P. Predeep, S. Prasanth, and A. S. Prasad

wide band-gap of crystal[8-10]. Recently, much work has been concentrated on the novel phosphors based on RE doped borate compounds. Lanthanide ortho borates($LnBO_3$) doped with rare-earth ions phosphors have been found to be optimal candidate in Field emission display (FED) and Plasma display panel (PDP) devices as they are sufficiently conductive to release electric charges stored on the phosphor particle surfaces[11-14]. With the guidance of above, borate based NLO materials ($Ca_4GdO(BO_3)_3$:RE^{3+}) were adopted as our research emphases. Upon realizing the relevance and importance of these RE doped borate phosphors, earlier we have carried out some preliminalry work on the emission properties of RE^{3+} (Eu^{3-}, Tb^{3+}, Sm^{3+}) doped in $Ca_4GdO(BO_3)_3$ powder phosphors[15-17]. Now in the present work, as a continuation we have selected with yet other rare-earth ions (Tm^{3+}, Er^{3+}) containing $Ca_4GdO(BO_3)_3$ powder phosphor for its systematic optical characterization.

EXPERIMENTAL STUDIES

The rare-earth oxides Gd_2O_3, Tm_2O_3 and Er_2O_3 (99.9 purity), $CaCO_3$ and H_3BO_3 (AR grade) were used as reactants to prepare the Tm^{3+} & Er^{3+} doped $Ca_4GdO(BO_3)_3$ powder phosphors by solid state reaction method. The general chemical equation can be expressed as follows

$$8CaCO_3 + 6H_3BO_3 + (1\text{-}x)Gd_2O_3 + xRE_2O_3\ (RE = Tm, Er) \rightarrow 2Ca_4Gd_{1\text{-}x}O(BO_3)_3{:}RE^{3+} + 9H_2O + 8\ CO_2$$

Thoroughly mixed reactants with a stoichiometric amounts were first and then powdered with acetone in an agate mortar for a homogeneous mixing and it was collected into an alumina crucible. The temperature was gradually raised from room temperature to 900^0c, where the mixture was kept for about 10h in an ambient atmosphere with an intermediate grinding.

To understand the phase purity of these samples, XRD pattern was measured at $2\theta=10^0$- 70^0 in steps of 0.02/s, on a powder X Ray Difractometer of Xpert PRO equipped with a Ni filter and CuK_α (1.54060 Å) radiation at an applied voltage of 40 kV and with 30 mA as an anode current, calibrated with Si. The morphology of the powder phosphor was observed on a JEOL JSM 840A Scanning Electron Microscope. The samples were gold coated using a sputter coater polaron SC7610 system. FTIR spectrum of the sample was recorded on a Nicolet IR-200 spectrometer with KBr pellet technique from $4000cm^{-1}$ to $400cm^{-1}$. The photoluminescence spectra of these phosphors were recorded on a YVON Fluorolog-3 Fluorimeter with a Xe-arc lamp (450W) as an excitation source and the lifetime of the prominent emission bands were measured with a phosphorimeter attachment to the main system which is a computer controlled with a Xe-flash lamp (25W).

RESULTS AND DISCUSSIONS

Fig.1 shows the X-ray diffraction patterns of a)$Ca_4GdO(BO_3)_3$ b)Tm^{3+} :$Ca_4GdO(BO_3)_3$ c) Er^{3+} :$Ca_4GdO(BO_3)_3$ powder phosphors. The standard XRD pattern for $Ca_4GdO(BO_3)_3$ is also shown as a reference. It can be seen clearly that all the diffraction peaks of the synthesized $Ca_4GdO(BO_3)_3$: RE^{3+} (=Tm^{3+}, Er^{3+}) phosphors are in good agreement with those of standard of $Ca_4GdO(BO_3)_3$. These powder phosphors belong to a monoclinic structure with a space group Cm. (a=8.078 ; b=15.98 ; c=3.55 ; β=101.26 A°). On comparing these XRD patterns, it is found that the Tm^{3+} & Er^{3+} ions could easily be substituted (in the place of Gd^{3+}) in the host matrix without any change in the host matrix structure[18-20].

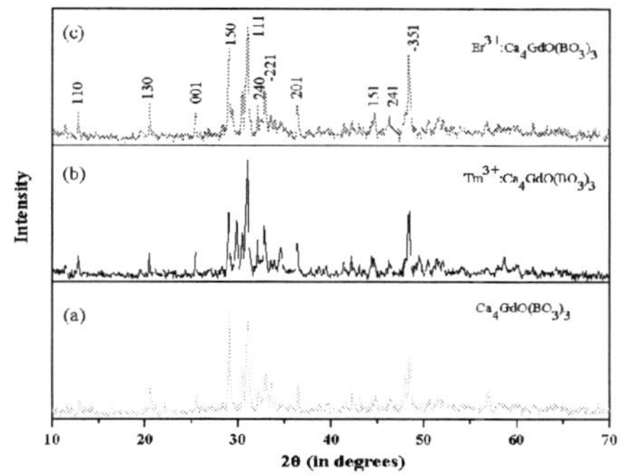

FIGURE 1.:XRD patterns of (a)$Ca_4GdO(BO_3)_3$ (b)$Tm^{3+}:Ca_4GdO(BO_3)_3$ (c) $Er^{3+}:Ca_4GdO(BO_3)_3$ powder phosphors

All the diffraction peaks of $Ca_4GdO(BO_3)_3:RE^{3+}$ (=Tm^{3+},Er^{3+}) could be indexed according to those of $Ca_4GdO(BO_3)_3$, as shown in Fig.1. and Fig.2 shows the SEM micrographs of a) Tm^{3+}: $Ca_4GdO(BO_3)_3$ and b) Er^{3+}: $Ca_4GdO(BO_3)_3$ powder phosphors to understand their morphology and from this figure, we have found that the particles are agglomerated. The particle size cannot be measured exactly from the SEM micrographs as shown in Fig.2(a&b). It is evident from Fig.2(a&b) that the particles are of 0.5μm. It may be mentioned that crystalline powder and micrometer dimension of the powder with a high strength could be useful for applications, as these micro crystalline phosphors can result in high luminescent intensities [21].

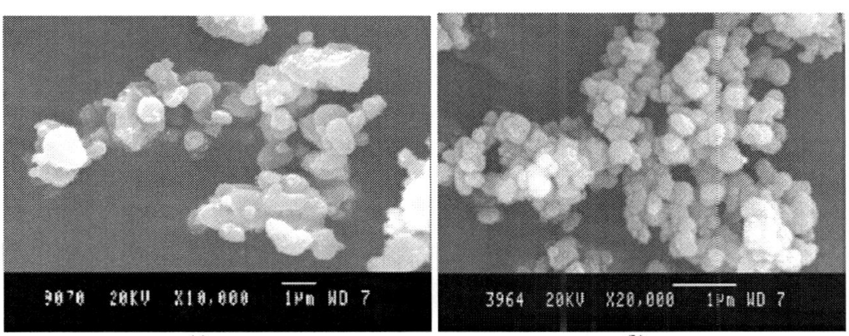

(a) (b)

FIGURE 2. SEM micrograph of (a)Tm^{3+}: $Ca_4GdO(BO_3)_3$ (b)$Er^{3+}:Ca_4GdO(BO_3)_3$ powder phosphors

Based on the infrared spectrum various B-O arrangements in RE^{3+}(=Tm^{3+},Er^{3+}) doped $Ca_4GdO(BO_3)_3$ powder phosphors have been understood. In the borate compounds containing $(BO_3)^{3-}$groups, the electronic delocalization in the planer borate anions has been predominant and which could induce NLO properties [22]. The FTIR spectral profiles of $Ca_4GdO(BO_3)_3$, Tm^{3+}: $Ca_4GdO(BO_3)_3$ and Er^{3-}: $Ca_4GdO(BO_3)_3$ powder phosphors have been shown in Fig.3.

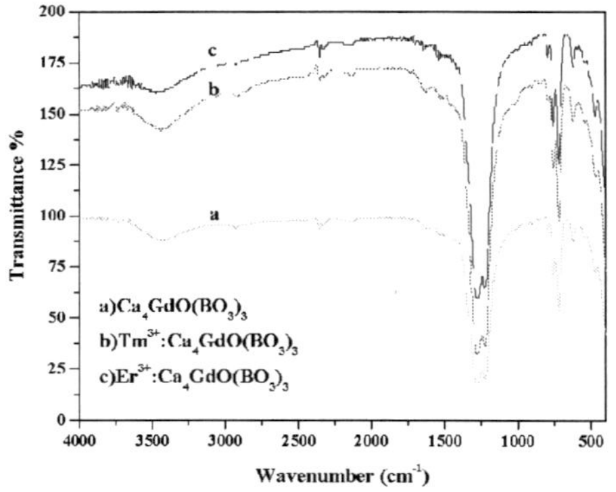

FIGURE 3. FTIR spectra of (a) $Ca_4GdO(BO_3)_3$
(b) Tm^{3+}: $Ca_4GdO(BO_3)_3$
(c) Er^{3+}: $Ca_4GdO(BO_3)_3$
powder phosphors

TABLE 1. FTIR spectrum & energy level assignments for $Ca_4GdO(BO_3)_3$ based powder phosphors from 4000 cm^{-1} to 400 cm^{-1}

FTIR peak energy (cm-1)	**Corresponding vibrational mode**
3430 2358 1638	Stretching vibrations of OH^- groups
1500	B-O Asymmetric stretching relaxations of $(BO_3)^{3-}$
1283 1221	$\nu_3(E^1)$ Stretching vibration of $(BO_3)^{3-}$
762 720	B-O-B bending vibrations
617	Ca (Translatory lattice modes)
480	Vibration mode of Gd/Tm/Er-O

The O-H stretching vibrations were found at 3430, 2358 and 1638cm^{-1}. The absorption bands recorded around 1500 cm^{-1} could be due to asymmetric stretching relaxation of the B-O bond of trigonal units. The B-O bond $\nu_3(E^1)$ stretching vibrations of $(BO_3)^{3-}$ units have been found at 1283, 1221 cm^{-1} and the IR bands that are observed at 762 and 720cm^{-1} could indicate B-O-B bending vibrations. The FTIR band at 627cm^{-1} has been attributed to the T^1 (Ca) translatory lattice modes. The vibrational mode of Gd/Tm/Er-O band has been found at 480cm^{-1}. The above assignments have been made for Tm^{3+} and Er^{3-} doped $Ca_4GdO(BO_3)_3$ phosphors and these results thus obtained are found to be quite clearly comparable with the results that are earlier reported in literature [22,23] and the assignment details are documented in above table.

Thulium doped phosphors have attracted substantial attention in recent years because Tm^{3+} ions provide blue luminescence with potential applications in screens and displays.

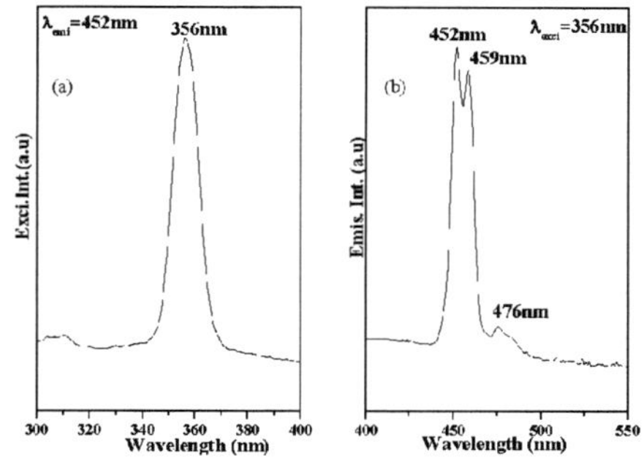

FIGURE 4(a) Excitation and (b) Emission spectra of Tm^{3+}: $Ca_4GdO(BO_3)_3$ powder phosphor

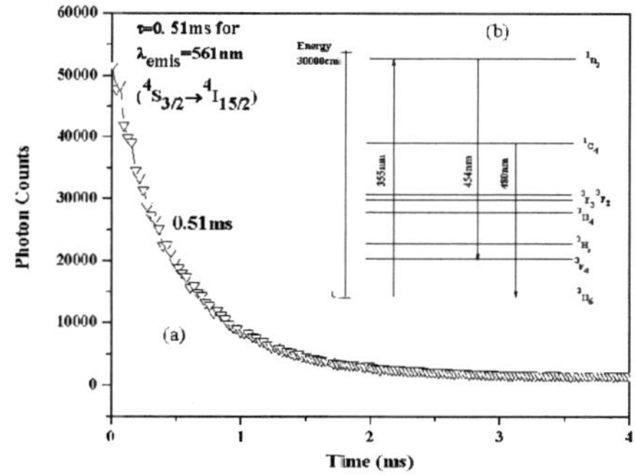

FIGURE. 5. (a) Decay curve and (b) Energy level scheme of Tm^{3+} : $Ca_4GdO(BO_3)_3$

The UV excitation spectrum of Tm^{3+}:$Ca_4GdO(BO_3)_3$ powder phosphor is shown in Fig.4(a) with excitation peak at about 356nm, which corresponds to the $^3H_6 \rightarrow {^1D_2}$ of Tm^{3+} ion. No other peaks are observed in the lower UV wavelength region (200-340nm). The emission spectrum obtained by excited at 356nm is shown in Fig.4(b), the emission peaks of Tm^{3+} were at 452nm and 476nm,which corresponds to the transition $^1D_2 \rightarrow {^3F_4}$ and $^1G_4 \rightarrow {^3H_6}$ respectively[24]. Fig5(a) presents the decay curve of the emission band $^1D_2 \rightarrow {^3F_4}$ of Tm^{3+}: $Ca_4GdO(BO_3)_3$ phosphor with an excitation at 356nm with a lifetime is τ=0.51ms, for the emission band $^1G_4 \rightarrow {^3H_5}$, lifetime was not measured due to the low intensity. An energy level scheme is given in Fig5(b) to explain the excitation and emission process in powder phosphor.

Fig.6(a) shows an excitation spectrum of Er^{3+}: $Ca_4GdO(BO_3)_3$ powder phosphor by monitoring the emission band at 561nm showing excitation bands are assigned to the electronic transitions $^4I_{15/2} \rightarrow {^4G_{9/2}}$ (365nm), $^4I_{15/2} \rightarrow {^4G_{11/2}}$ (377nm) and $^4I_{15/2} \rightarrow (^2G,^4F)_{9/2}$ (407nm) of Er^{3+}. The emission spectrum of Er^{3+}: $Ca_4GdO(BO_3)_3$ powder phosphor in the wavelength range from 500nm to 700nm is shown in Fig.6(b).

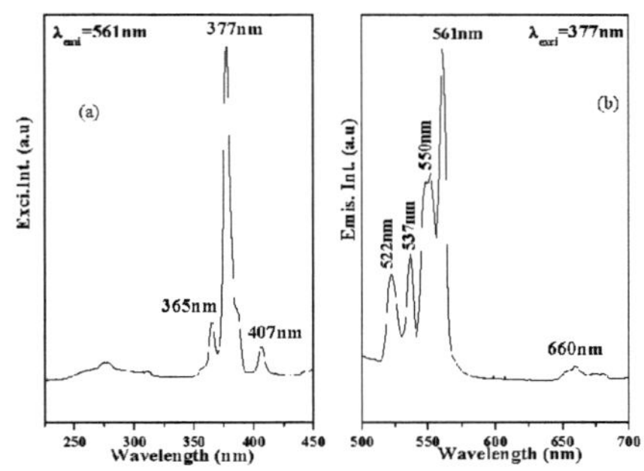

FIGURE 6. (a) Excitation and (b) Emission spectra of Er^{3+}: $Ca_4GdO(BO_3)_3$ powder phosphor

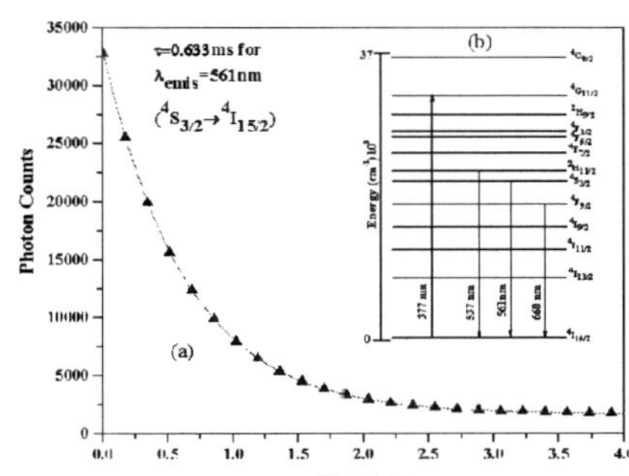

FIGURE.7.(a) Decay curve and (b) Energy level scheme of Er^{3+}: $Ca_4GdO(BO_3)_3$ powder phosphor

This emission spectra exhibited two distinct emission bands. Green emission was observed between 520-570nm corresponding to the $^2H_{11/2}\rightarrow^4I_{15/2}$ (520-540nm), $^4S_{3/2}\rightarrow^4I_{15/2}$ (540-570nm) transition, red emission was observed between 640-670nm corresponding to the $^4F_{9/2}\rightarrow^4I_{15/2}$ transition. Fig.7(a) presents the decay curve of Er^{3+}:$Ca_4GdO(BO_3)_3$ powder phosphor of the emission band $^4S_{3/2}\rightarrow^4I_{15/2}$ with an excitation wavelength at 377nm and the lifetime is 0.633ms, the decay curves of the other emission bands could not be measured due to its low intensity situation [25]. An energy level scheme has been given in Fig.7(b) to explain the emission mechanism involved from the Er^{3-}: $Ca_4GdO(BO_3)_3$ powder phosphor. The excitation wavelength 377nm fits into a relatively intense excitation structure (Fig.6(a)) and excites directly Er^{3+} ion. The 377nm light elevates an electron from the $^4I_{15/2}$ ground level up to the $^4G_{11/2}$ state, from where it may easily fall down to the $^4S_{3/2}$ through non-radiative, multi phonon relaxation.

CONCLUSION

Tm^{3+} & Er^{3+} doped $Ca_4GdO(BO_3)_3$ phosphors were synthesized by the method of solid-state reaction. From the measured emission spectra of these phosphors, bright blue emission (for Tm^{3+}) and red emission (for Er^{3+}) bands have been observed. Besides recognizing its optical properties and other related properties through XRD, SEM and FTIR have been understood in the present work. Based on the results obtained in the present work, these luminescence phosphors could be identified as novel optical materials for certain electronic display applications.

REFERENCES

1. F.S. Liu, Q.L. Liu, J.K. Liang, J. Luo, L.T. Yang, G.B. Song, Y. Zhang, L.X. Wang, J.N. Yao and G.H. Rao, J. Lumin. 111, 61 (2005)
2. L.Sirbu, V.V. Ursaki , I.M. Tiginyanu, K. Dolgalev and Robert W.Boy, Phys. stat. sol. 1, 13 (2007)
3. Y. Inaguma, T.T. suchiya, T. Katsumata, J. Solid State Chem. 180, 1678 (2007)
4. K.M. Lin, C.C. Lin, C.Y. Hsiao, Y.Y. Li, J. Lumin. 127(2), 561 (2007)
5. C.T. Chen, L. Bai, Z.Z. Wang and R.K. Li, J. Cryst. Growth, 292 (2), 169 (2006).
6. D.S. Thakare, S.K. Omanwar, S.V. Moharil, S.M. Dhopte, P.L.Muthal and V.K. Kondawar, Opt. Mater. 29, 1731 (2007)
7. C. Duan, J. Ynan and J. Zhao, J. Solid State Chem. 178, 3698 (2005).
8. Y.Wang, K.Uheda, H.Takizawa, U.Mizomoto and T.endo, J. Electro chem. Soc. 148(8), G430 (2001).
9. M. Iwai, T. Kobayashi, H.Fraya, Y.Wari and T.Sasaki, Jpn. Appl. Phys. 36, L276 (1997).
10. G.Aka, A.Kahn-Harari, D.Vivien and J.M.Benitez, J. Solid State Inorg. Chem. 33, 727 (1996).
11. C.J. Duan, W.F. Li, H.H. Chen, X.X. Yang and J.T. Zhao, Chin. Phys. Lett. 22, 711 (2005).
12. H.W. Song, J.W. Wang, B.J. Chen, H.S. Peng and S.Z. Lu, Chem. Phys.Lett. 376, 1 (2003).
13. F. Druon, F. Auge, F. Balembois, P. Georges, A. Brun, A. Aron, F. Mougel and G.Aka, D. Vivien, J. Opt. Soc. Am. B **17,** 18 (2000).
14. Y. Zhang and Y. Li, J. Lumin 113, 45, (2005).
15. B.VengalaRao,U.Rambabu and S.Buddhudu, Physica B 382, 86, (2006).
16. B.Vengala Rao, U.Rambabu and S. Buddhudu, Physica B 391, 339, (2007).
17. B.Vengala Rao, U.Rambabu and S.Buddhudu, Mater. Lett. 61, 2868 (2007)

18. A.B. Iyukhin, B.F. Dzhuriukii, Russ. J. Inorg. Chem. 36, 847 (1993).
19. Y.Wang, T.Endo, E.Xie, D.He and B.Liu, Microelectron. J. 35, 357 (2004).
20. I.V.Kityk and A.Mefleh, Physica B 262, 170, (1999).
21. G. Liu, G. Hong, J. Wang and X. Dong, Nanotechnol. 17, 3134 (2006).
22. V.Krishnakumar and R.Nagalakshmi, Spectrochim. Acta.Part A 60 2733 (2004).
23. G.D.Dzik, W.Rybaomnowski, S. Golab, L. Macalik, J. Hanuza and A. Pajazkowska, J. Mol. Struct. 555, 213 (2000).
24. H. Lai, B. Chen, W. Xu, X. Wang, Y. Yang, Q. Meng, J. Alloys compd. 395, 181 (2005).
25. C.W. Jia, J.G. Zhao, H.G. Duan, E.Q. Xie Mater. Lett. 61, 4389 (2007).

Mechanical Property of CdS Coated PET and Mica Films

Sandhya Gupta*, Vinodini Shaktawat , Kananbala Sharma., Narendra.S. Saxena* and Thaneshwar P.Sharma

Semi-conductor & Polymer Science Laboratory, 5-6, Vigyan Bhawan, University of Rajasthan, Jaipur-302004. INDIA
*E-mail: n_s_saxena@rediffmail.com,1982sanman@gmail.com

Abstract. Thick film of CdS is coated on polyethylene terephthalate (PET) and mica films (both are commercially available) using screen-printing method. The variation of complex modulus and Tan δ has been studied by Dynamic Mechanical Analyser (DMA). Complex modulus decreases with increase in temperature due to softening of the PET film. The Tan δ initially increases with temperature up to a certain temperature and then decreases with increase in temperature. The T_g of PET is found to be 107^0C. The Complex modulus of PET film is increased due to coating of CdS and the peak of Tan δ shifts to a higher temperature i.e. 112^0C indicating increase in mechanical strength. Similarly, the complex modulus of mica also decreases with temperature due to softening and Tan δ remains nearly constant. After coating CdS on mica, the complex modulus is increased due to the effect of CdS particles and there is a small peak in Tan δ due to the structural changes in CdS at interface.

Keywords: Polyethylene terephthalate, Mica film, CdS, Complex modulus, Transition temperature.
PACS: 68.60.Bs

INTRODUCTION

Among compound semiconductors, II-VI binary compound such as CdS has proven to be one of the important materials, which exhibit the markable optical and electrical properties. CdS exists in nature as two different minerals, greenockite and hawleyite. Greenockite forms hexagonal crystals with the wurtzite structure and Hawleyite has the zinc blend structure [1,2]. CdS is a direct band gap semiconductor, therefore it is used in CdS-CdTe and CdS-Cu(InGa)Se_2 thin film solar cells [3,4]. One simple use is as a photoresistor whose electrical resistance changes with incident light levels. The CdS semiconductor has n-type conductivity due to the donor centers [5], which form in CdS during growth therefore it is used in manufacturing flexible schottky diodes [6] and thin film transistor. There are many substrates, which are flexible, but PET and Mica sheets exhibit good mechanical properties so that CdS can be printed on them for getting mechanical viable CdS coating on flexible substrate.

Polyethylene terephthalate (PET) is hard, stiff, strong and dimensionally stable material. It maintains its mechanical stability up to 150°c. It has good tensile strength (55-75 mpa) and 50-150 % elongation at break [7]. Due to these properties, PET is widely used for packaging, laminating, in making thermally stable transparent sheets.

CP1004, *Thermophysical Properties of Materials and Devices-NCTP '07*,
edited by P. Predeep, S. Prasanth, and A. S. Prasad

The mica group of sheet silicate minerals includes several closely related materials having highly perfect basal cleavage. The highly perfect cleavage is explained by the hexagonal sheet-like arrangement of its atoms [8]. Mica sheets show mechanical stability up to 300-400°C. Mica sheet is highly insulating material therefore mica films are widely used in electronic press, manufacturing capacitors for radio frequency applications. Very scanty efforts have been made to study mechanical properties and its temperature variation for CdS coated PET and Mica sheets.

In the present work, an effort has been made to study the effect of CdS coating on mechanical properties like complex modulus, glass transition temperature of PET and mica films using Dynamic Mechanical Analyser (DMA) [9]. Thick film of CdS has been printed on PET and mica films using screen-printing technique.

MATERIAL PREPARATION

Commercially available polyethylene terephthalate and mica films (thickness of the order of about 20-50 μm) have been used as substrate .CdS has been prepared by chemical method using $CdCl_2$ & H_2S [10, 11]. In this method an appropriate ratio of cadmium chloride is taken in solution form. This solution is mixed thoroughly using magnetic stirrer for 20 minutes. H_2S gas produced by heating thiourea is then pass through this solution. The yellow precipitate obtained in the above process was filtered. This precipitate was then dried in open atmosphere. When precipitate was completely dried, it was then crushed to fine powder by grinding process.

CdS has been printed on PET and mica films using screen-printing technique. The paste of CdS for screen-printing was prepared by using cadmium chloride as an adhesive source and ethylene glycol as a binder. The weight of cadmium chloride was only 10% of the weight of CdS. The paste was screen printed on PET and mica substrate. CdS printed on PET film was then sintered in an oven at 100^0C for ten minutes and CdS printed mica film was sintered at 400^0C for ten minutes for ensuring a good adhesion.

RESULTS

In fig.1 it is very clearly seen that the value of complex modulus E^* decreases from 5.2×10^9 to 0.2×10^9 Pa as temperature increases where as in fig.2 the value of Tan δ initially increases, reaches at maximum value and then decreases with increasing temperature. Temperature corresponding to peak of Tan δ is known as glass transition temperature. Hence T_g of PET is 107^0C.

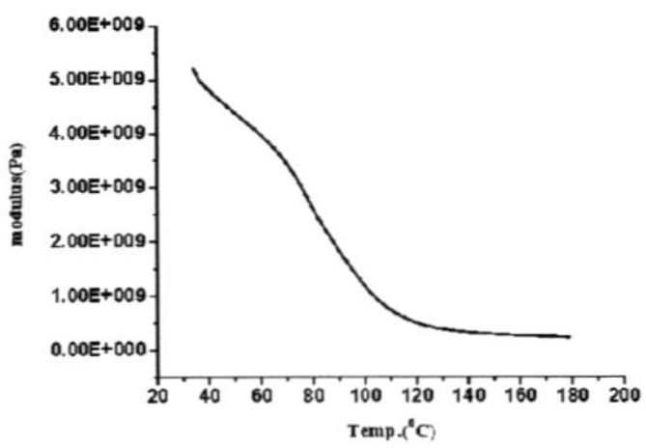

FIGURE 1. Complex modulus versus temperature for PET sheet

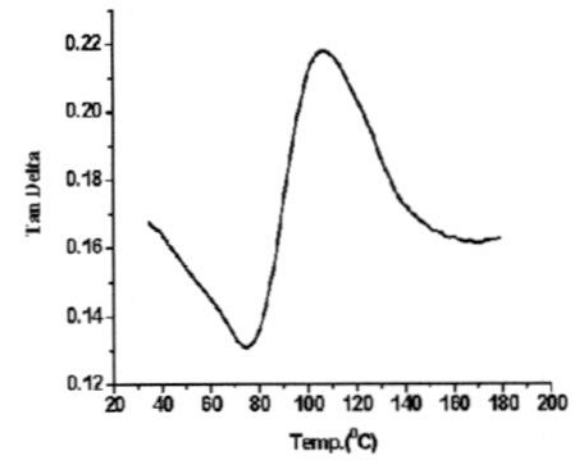

FIGURE 2. Tan δ versus temperature for PET sheet

Complex modulus and Tan δ versus temperature for mica sheet are shown in fig.3 and 4.

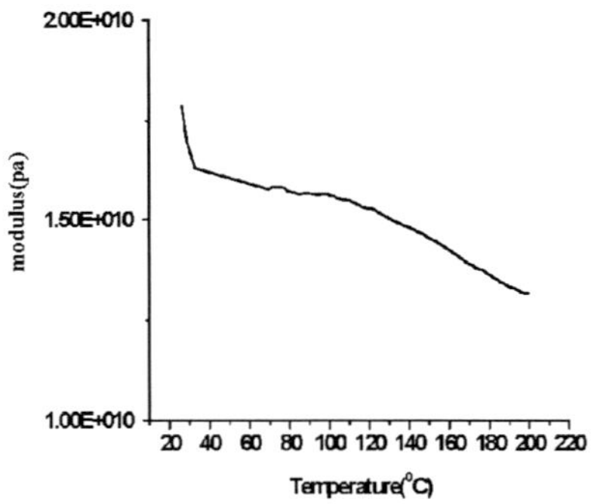

FIGURE 3. Complex modulus versus temperature for mica sheet

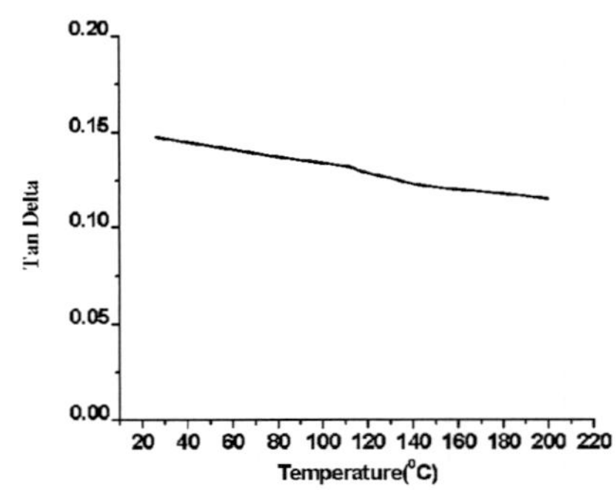

FIGURE 4. Tan δ versus temperature for mica sheet

The complex modulus decreases from 1.78 x 10^{10} Pa to 1.3 x 10^{10} Pa as temperature increases from room temperature to 200^0C. There is no specific peak in fig.4, which indicates that T_g for mica does not exist in that range of temperature.

Complex modulus and Tan δ versus temperature for CdS coated PET sheet are shown in Fig.5 and 6.

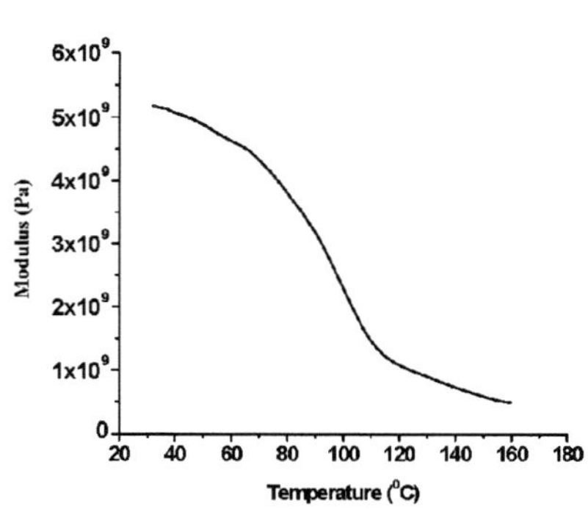

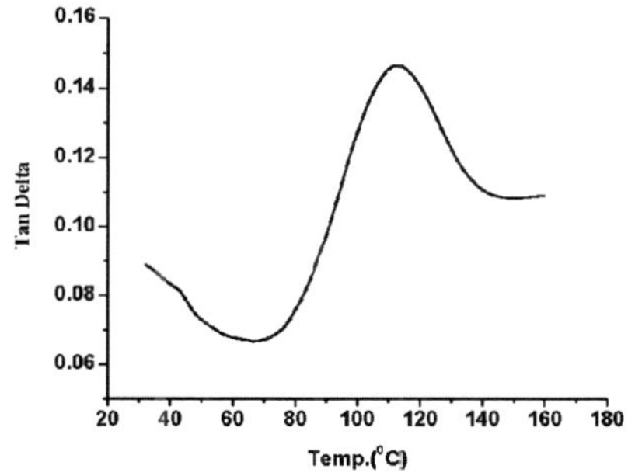

FIGURE 5. Complex modulus versus temperature for CdS coated PET sheet

FIGURE 6. Tan δ versus temperature for CdS coated PET sheet

These figs. indicate the improvement in value of complex modulus which varies from 5.16×10^{9} to 0.49×10^{9}Pa and T_g shifts towards the higher temperature (112^0C).

Complex modulus and Tan δ versus temperature for CdS coated mica film are shown in fig. 7 & 8.

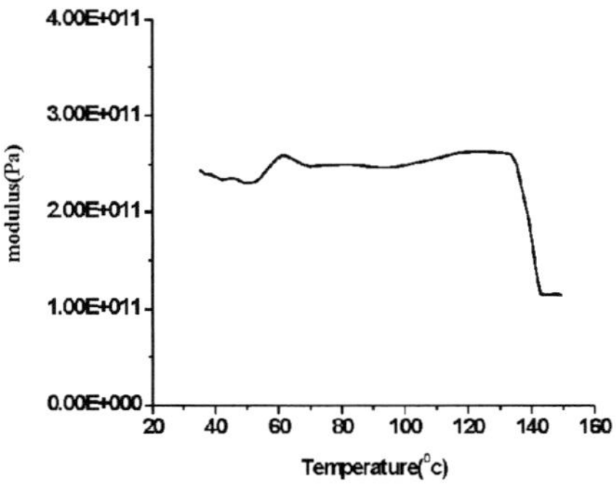

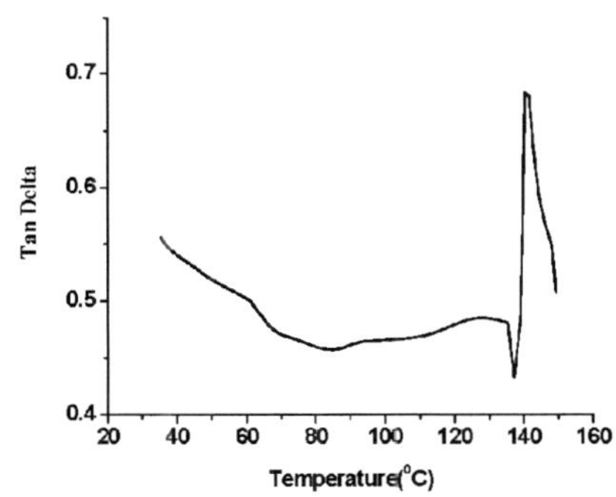

FIGURE 7. Complex modulus versus temperature for CdS coated mica sheet

FIGURE 8. Tan δ versus temperature for CdS coated mica sheet

From fig.7 it is clear that complex modulus increases with respect to the result of pure mica sheet. The value of complex modulus (2.48×10^{11} Pa) initially remains almost constant with increasing temperature up to 134^0C after that there is a fall in the values. The decreased values of complex modulus 1.15×10^{11}Pa further remain constant with increasing temperature up to 150^0C. We see that there is a small peak of Tan δ at 140^0C for CdS coated mica sheet.

DISCUSSION

The decrease in complex modulus with increasing temperature is due to the softening of PET and peak of Tan δ at 107^0C also indicates the phase transformation of PET from solid phase to rubbery phase. The value of complex modulus increases after CdS coating on PET sheet. The molecules of CdS are attached to the polymer chain of PET through local resonance, which results into more compact structure and therefore according Crankshaft model free volume decreases. The lack of free volume restricts the possibility of motion in various directions and hence they are unable to respond to the application of load/stress to which the sample is subjected. In other words we can say coating of CdS film increases stiffness of PET sheet. Then CdS coated PET sheet takes more heat to transform from solid phase to rubbery phase and glass transition temperature shifts towards the high temperature (112 ^{0}C).

The complex modulus of mica also decreases but the decrease is very slow with increasing temperature from 40^0C to 200^0C because mica is structurally more stable with respect of PET in this temperature range. Almost constant value of Tan δ in that range also supports structural stability of mica. CdS coated mica sheet shows improved value of complex modulus. The variation of complex modulus with increasing temperature for CdS coated mica sheet can be explained by understanding the structure of CdS and mica. CdS consists both wurtzite structure and zinc blend structure. But wurtzite structure is not the favored one in the bulk crystals. Wurtzite structure can be favored in some thin films and nanocrystal forms of the material. Wurtzite structure is a member of the hexagonal crystal system and consists of tetrahedrally coordinated cadmium and sulphar atoms that are stacked in an ABABAB pattern. Mica have hexagonal structure therefore at interface the CdS particles of wurtzite structure try to adjust themselves in the structure of mica on increasing temperature and at a particular temperature (140^0C) almost all the CdS particles of wurtzite structure convert in to the hexagonal structure at interface. Therefore, complex modulus shows minima whereas Tan δ shows maxima (peak) at the same temperature (140^0C).

CONCLUSION

The mechanical strength of PET and mica sheets are improved by coating thick CdS film on them using screen-printing method. The increased values of complex modulus of PET and mica sheets after CdS coating indicate that the stiffness of PET and mica sheet increases. Hence CdS coated PET and mica sheets can be used in fabrication of flexible semiconducting [12] and optical devices.

ACKNOWLEDGMENTS

Authors are thankful to Dr. Dinesh patidar, Ms. Manasvi Dixit, Ms. Deepika Chaudhary, Mrs. Rashmi Saxena, Mr. Kuldeep S. Rathore, Mr. Praveen Kumar Jain for their helps in various ways during the course of this work.

REFERENCES

1. L. V. Azaraff, "*Introduction to Solids*", Tata McGraw-Hill, New York (1989).
2. C. Kittel, "*Introduction to Solid State Physics*",John Wiley & Sons, U.S. (1963).
3. T L Chu and S S Chu, *Solid-State Electron* **38**, 533-549 (1995).
4. A. Rockett , *J. Appl. Phys.* **70**, R 81-R 97 (1991).
5. T D Dzhafarov, M Altunbas, A I Kopya, V Novrwzov and E Bacaksiz, *J. Phys. D; Appl. Phy.* **32**, 2125-2128 (1999).
6. H.Koczuka, A.Tsumura, *Synthetic Metals* **28**, 753-756 (1989).
7. J. R. Fried, "*Polymer Science and Technology*", Prentice Hall of India Private Limited, New Delhi(1999).
8. Mica, http://en.wikipedia.org/wiki/Mica
9. K. P. Menard, "*Dynamic Mechanical Analysis- A practical introduction*", Boca raton: CRC press LLC (1999).
10. D. Patidar, N. Jain, N.S.Saxena, kananbala Sharma, and T.P.Sharma, *Brazilian Journal of physics* **36**, 1210-1212 (2006).
11. L Kumar, B.P.Singh, A.Mishra, S.C.K.Misra, and T.P.Sharma, *Physica B* **363**, 102-109 (2005).
12. H. Gleskova, I-Chun Cheng, Sigurd Wagner,James C. Sturm and Zhigang Suo, *Solar Energy* **80**, 687-693 (2006).

Low Temperature Specific heat of $La_{1-x}Sr_xMnO_3$ Manganites

Archana Srivastava [a], Renu Choithrani [b] and N.K. Gaur [b]

[a] Department of Physics, Sri Sathya Sai College for Women, Bhopal 462024, India
[b] Superconductivity Laboratory, Department of Physics, Barkatullah University, Bhopal 462026, India

Abstract. Perovskite manganites $La_{1-x}Sr_xMnO_3$ (x=0.075, 0.10, 0.125, 0.15, 0.17, 0.185, 0.30, 0.50, 0.60, 0.75) have been studied as a function of temperature (1K≤T≤20K) by means of a Rigid Ion Model (RIM) for the determination of specific heat , Debye temperature, cohesive energy, restrsahlen frequency and Grüneisen parameter. The specific heat values revealed by us are in reasonably good agreement with the available experimental data at low temperature The Debye temperatures obtained by us from the lattice contributions are found to be in closer agreement with the experimental data

Keywords: Specific Heat, Thermodynamic Properties, Debye temperature, Thermal Properties, Manganites.
PACS: 65.40.b, 65.40.Ba, 67.80.Gb, 78.30.Hv

1. INTRODUCTION

Mixed valence manganites with the perovskite structure $R_{1-x}A_xMnO_3$ (where, e.g.,R=La, Nd, Pr and A=Ba, Ca, Sr, Pb) have been a popular subject of contemporary research because of their interesting physical properties such as metal-insulator transitions and colossal magnetoresistance effect [1] which makes these systems promising for magnetic sensor and reading head device applications. A number of works has been devoted to study the interplay between structure and transport properties of these manganese perovskites [2-7].

Electron-phonon interaction is one of the most relevant contributions in determining the conduction mechanism in these materials. The strong coupling between the electron and the lattice phonons brings about a large change in the physical properties of these compounds by a small change in the ratio between trivalent and divalent ions at the R site. A similar change occurs due to the external effects like the application of the magnetic field, the hydrostatic pressure or the temperature. Enormous amount of the experimental work on the specific heat, Debye temperature and other thermodynamic properties have been reported on these compounds but scant attention has been paid by theoretical workers to explain these properties. Thus, the study of the lattice thermal properties of these materials may be taken as a starting point for the consistent understanding of the more complex physical properties of the doped manganites. We have used Rigid Ion Model (RIM) to study the

CP1004, *Thermophysical Properties of Materials and Devices-NCTP '07*,
edited by P. Predeep, S. Prasanth, and A. S. Prasad

thermal properties of $La_{1-x}Sr_xMnO_3$ perovskite manganites. The essential formalism of the RIM and the results obtained are presented in the subsequent section

2. FORMALISM OF RIM

We have formulated Rigid Ion Model (RIM) by including the attractive long-range (LR) Coulomb and the short range (SR) Hafemeister Flygare [8] (HF) type overlap repulsive interactions. The potential describing the formalism of RIM is expressed as:

$$\phi = \phi_{kk'}^{C}(r) + \phi_{kk'}^{R}(r) + \phi_{kk'}^{vdw} \tag{1}$$

where the first term is the LR Coulomb attraction potential expressed as:

$$\phi_{kk'}^{C}(r) = -\frac{e^2}{2} \sum_{kk'} Z_k Z_{k'} r_{kk'}^{-1} \tag{2}$$

with $r_{kk'}$ as the separation between the the two atoms designated by k and k′ and k is the composite index indicating the type of ion and unit cell to which it belongs. The overlap repulsive energy $\phi_{kk'}^{R}(r)$ according to Hafemeister-Flygare (HF) [8] type interaction extended upto the second neighbour is expressed as:

$$\phi_{kk'}^{R}(r) = \sum_i n_i b_i \beta_{kk'} \exp\{(r_k + r_{k'} - r_{kk'})/\rho_i\} + \frac{n_i'}{2} b_i \left[\beta_{kk} \exp\{(2r_k - r_{kk})/\rho_i\}\right] + \beta_{k'k'} \exp\{(2r_{k'} - r_{k'k'})/\rho_i\} \tag{3}$$

Here $r_{kk'}$ appearing in the first term on the right represents separation between the nearest neighbours while r_{kk} and $r_{k'k'}$ appearing in the following two terms are the second neighbour seperation. $r_k(r_{k'})$ is the ionic radii of k (k') ion. n (n') is the number of nearest (next nearest neighbour) ions. b_i and ρ_i are the hardness and range parameters for the ith cation-anion pair, respectively and $\beta_{kk'}$ is the Pauling coefficient[9]. Z_k ($Z_{k'}$) and N_k ($N_{k'}$) are the valence and the number of electrons in the outermost orbits respectively. The contributions of van der Waal's (vdw) forces for the dipole-dipole interactions is determined by using the Slater- Kirkwood Variational (SKV) method [10] and are defined in our earlier papers [11-14]. The model parameters, hardness (b) and range (ρ) parameters are determined from the equilibrium condition:

$$\left[\frac{d\phi}{dr}\right]_{r=r_O} = 0 \tag{4}$$

and the bulk modulus:

$$B = \frac{1}{9Kr_O}\left[\frac{d^2\phi}{dr^2}\right]_{r=r_O} \tag{5}$$

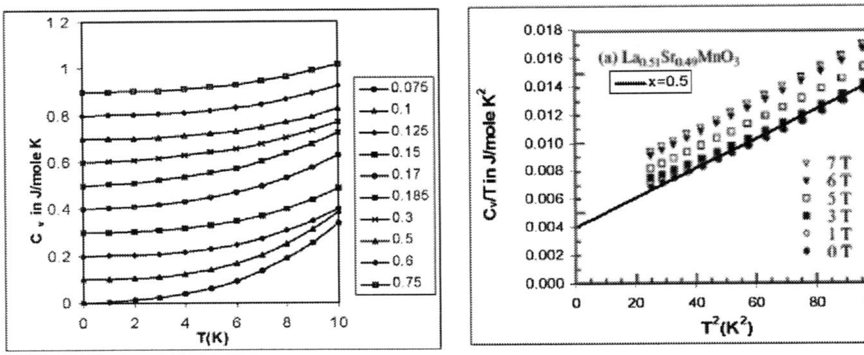

FIGURE 1. Variation of Specific heat of La1-xSrxMnO3 manganites within the temperature range $0 \leq T \leq 10K$. All the curves have been shifted upward by 0.1J/mol K each by the preceding curve for clarity in the left figure. Right figure shows calculated specific heat ($C\text{-}C_{spin}$) of LSMO at x=0.5 (—) and its match with the experimental curves [27] shown by symbols..

where K is the crystal structure constant dependent on the crystal structure and r_o is the equilibrium nearest neighbor distance. The expressions for calculating the thermodynamic properties are taken from our earlier papers [11-14]

3. RESULTS AND DISCUSSIONS

Using the input data at lower temperature from ref.[15-22] for various doping concentration of $la_{1-x}sr_xmno_3$ and vdw coefficient $c=69.8\text{x}10^{-60}$ ergcm^{-6} calculated using skv method, we have calculated the model parameters (b_1, b_2, ρ_1, ρ_2) at various compositions ($0.075 \leq x \leq 0.75$) with the help of eqn. (4,5), as a function of temperature and used them to calculate the thermal properties of these doped manganites. The values of model parameters b_1 and $\rho_1$are for the ion pairs $Mn^{3+}/Mn^{4+}\text{-}O^{2-}$ and those of b_2 and ρ_2 are for the ion pairs $La^{3+}/Sr^{2+}\text{-}O^{2-}$. To test the validity of our model, we have calculated the cohesive energy of the compounds using the eqn. (2) and found that cohesive energy depicts a quasi-linear behavior. The negative value of cohesive energy shows the stability of these compounds. The experimental values of cohesive energy of $la_{1-x}sr_xmno_3$ are not available but our calculated values are close to the reported value -140.52ev for the parent compound $lamno_3$ [23]. The calculated values of various thermal properties corresponding to various concentrations (x) are reported in table 1. The restsrahlen frequency of these compounds is close to reported value of 8.58thz for $lamno_3$ [24] the grüneisen parameter is almost near the reported limit of 2-3 for manganites [25] but our comments on reliability of these values are restricted until the report of experimental data on these compounds. In the observed doping range Debye temperature increase and the lattice specific heat decrease with increasing doping (x). The similar anomalous softening of the lattice is observed by Okuda et al. [26] over the region $0.1<x<0.4$ but we observed this trend over a wider region of doping $0.075 \leq x \leq 0.75$

Our results on Debye temperature are in good agreement with the available experimental data (see table 1). The lattice contribution to the specific heat for $La_{1-x}Sr_xMnO_3$ perovskite manganites is calculated in the low temperature region, $1k \leq t \leq$

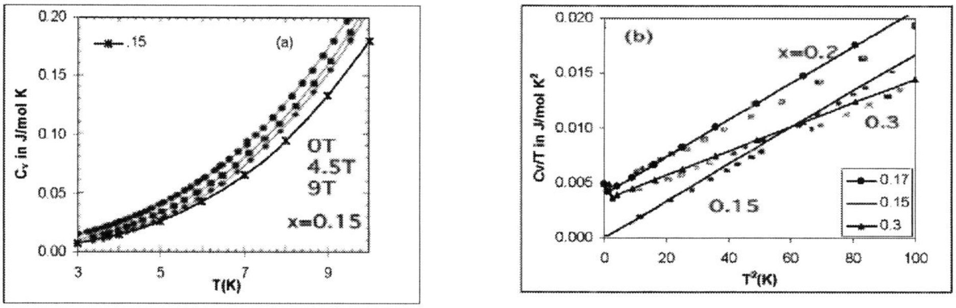

FIGURE 2. Specific heat of LSMO (x=0.15, 0.17, 0.3) (solid line) and their match with experimental curves [26] (symbol) for (0.15, 0.2, 0.3).

10k and a constant offset value has been added to the calculated values to account for the electronic and ferromagnetic spin wave contribution and the resulted values (fig.) have reasonably good agreement with the available low temperature experimental values (see fig 1,2). In order to facilitate the visualization, the curves for each doping concentration have been displaced 0.1 j/mol k upside while the curve for x=0.75 is undisplaced in fig. 1.

It can be inferred from Fig.1(right) that our specific heat values without the magnetic contributions are similar to experimentally reported values and follows the same trend as of Moritomo et al [27]. Fig. 2 is depicting the fair agreement of Cv/T and Cv values for doping levels 0.15, 0.17, 0.30 and 0.15 respectively with the experimental curves of Okuda et al [26]. It can be inferred from Fig.2(a) that higher order lattice terms are expected to contribute above 5K. It can be concluded from the above analysis that the specific heat decreases, as the doping concentration of the divalent Sr^{2+} increases from 0.075 to 0.75. The studies of cohesive properties of $La_{1-x}Sr_xMnO_3$ indicate that distortions of the lattice increase with higher doping concentrations and the stability of the heavily doped compounds reduces substantially.

4.ACKNOWLEDGMENTS

TABLE 1. The model parameters and thermal properties of $La_{1-x}Sr_xMnO_3$ system at room temperature.

Doping x	$b_1 x 10^{-19}$ (J)	$b_2 x 10^{-19}$ (J)	ρ_1 (Å)	ρ_2 (Å)	Φ in ev	f in N/m	v in THz	Θ_D in K		γ
								This work	Exp.	
0.075	2.14	1.87	0.38	0.63	-146.3 -140.52[a]	17.92	6.84 8.58[b]	327.8	302[f] (x=0)	2.0 2-3[c]
0.100	2.08	1.84	0.37	0.62	-145.7	17.90	6.93	332.0		2.0
0.125	1.62	1.62	0.31	0.53	-150.0	23.92	7.95	380.0	360[d,g]	2.4
0.150	1.52	1.58	0.30	0.51	-149.1	23.90	7.97	384.7	386[d,g]	2.4
0.170	1.05	1.41	0.23	0.41	-155.6	24.30	8.09	387.6	388[d,g]	2.1
0.185	1.46	1.54	0.29	0.50	-149.2	25.80	8.31	398.0	402[d,g]	2.5
0.300	1.10	1.34	0.25	0.43	-147.3	31.60	9.31	445.8	450[d,g]	2.9
0.500	0.92	1.24	0.23	0.42	-137.1	31.40	9.48	453.9	474[e]	3.0
0.600	0.84	1.12	0.22	0.40	-131.8	31.30	9.58	458.7		3.1
0.750	0.73	1.01	0.21	0.39	-123.8	31.20	9.72	465.5		3.2

[a] ref[22]; [b]ref[23]; [c] ref.[25]; [d] ref.[26]; [e] ref[27]; [f] ref.[18]; [g] ref.[28]

The authors are thankful to the University Grants Commission (UGC), New Delhi for providing the financial support.

5. REFERENCES

1. S. Jin, T.H Tiefel, M. McCormack, R.A. Fastnacht, R. Ramesh and L.H. Chen, *Science* **264**, 413-415, (1994); Josep Fontcuberta, Physics world Feb., **33-39,** 1999;A. J. Millis, *Nature* **392**, 147-152, (1999).Y. Tokura (ed) *"Colossal Magnetoresistive Oxides"* Gordon and Breach Publishers, London (2000).
2. J. Volger, *Physica* **20,** 49-66, (1954).
3. G.H. Jonker and J. H. van Santen, *Physica* **16,** 337-349, (1950); G.H. Jonker and J. H. van Santen, *Physica* **19,** 120-130, (1953).
4. H.L. Ju, C. Kwon, Q. Li, R.L. Green and T. Venkatesan, *Appl. Phys. Lett.* **65,** 2108-2110, (1994).
5. J. Hemberger, A. Krimme, T. Kurz, H.-A. Krug von Nidda, V. Yu. Ivanov, A. A. Mukhin, A. M. Balbashov, and A. Loidl, *Phys. Rev. B* **66,** 094410-094417, (2002); A. Szewczyk, M. Gutowska, and B. Dabrowski, *Phys. Rev. B* **72,** 224429-224436, (2005).
6. E. O. Wollan and W. C. Koehler, *Phys. Rev.* **100**, 545-563, (1955).
7. D.W. Hafemiester and W.H. Flygare, *J. Chem.phys.* **43,** 795, (1965).
8. Linus Pauling- *Nature of Chemical Bonds*, (Ithaca: Cornell University Press), 1945; L. Pauling, *Z. Kristallogr.* 67, 377, (1928).
9. J. C. Slater, K. G. Kirkwood, *Phys. Rev.* **37**, 682-697, (1931)
5. E. G. Rini, N.K. Gaur and R. K. Singh *J. Supercond.* **15**, 583-585, (2002); Nupinderjeet Kaur, N.K. Gaur and R.K. Singh, *Mod. Phy. Lett. B* **21,** 885-891, (2007).
11. N.Kaur, R. Mohan, N. K. Gaur and R. K. Singh, *Physica C* **451,** 24-30, (2007).
12. N. Kaur, N. K. Gaur, R. Mohan, and R. K. Singh, *J. Phys. Chem. Of Solids* (In Press) (2007).
13. Renu Choithrani and N. K. Gaur, *J. of Magn. & Mag. Mat.* (In Press) (2007).
14. F. Mayr, Ch. Hartinger, and A. Loidl, *Phys. Rev. B* **72,** 024425-24431, (2005);
15. A. E. Nikiforov, S.E. Popov and S. Shashkin, *Physica B* **276-278**, 772-773, (2000).
16. O. Chmaissem, B. Dabrowski, S. Kolesnik, J. Mais, J. D. Jorgensen, and S. Short, *Phys. Rev. B* **67,** 094431-094444, (2003); S. Uhlenbruck, R. Teipen, R. Klingeler, B. Büchner, O. Friedt, M. Hücker, H. Kierspel, T. Niemöller, L. Pinsard, A.Revcolevschi, and R. Gross, *Phys. Rev. Lett.* **82**, 185-188, (1999).
17. B. F. Woodfield, M. L. Wilson, and J. M. Byers, *Phys. Rev. Lett.***78,** 3201-3204, (1997).
18. Y. Moritomo T. Akimoto and A. Nakamura K. Ohoyama and M. Ohashi, *Phys. Rev. B* **58,** 5544-5549, (1998).
19. A. M. De Le´on-Guevara, P. Berthet, J. Berthon, F. Millot, and A. Revcolevschi A. Anane, C. Dupas, K. Le Dang, J. P. Renard, and P. Veillet, *Phys. Rev. B* **56,** 6031-6035, (1997).
20. D. Kim, B. L. Zink, and F. Hellman, J. M. D. Coey, *Phys. Rev. B* **65,** 214424-214430, (2002).
21. L. Vasiliu-Doloca and J. W. Lynn, Y. M. Mukovskii, A. A. Arsenov, and D. A. Shulyatev, *J. App. Phys.* **83,** 7342-7344, (1998).
22. N.N. Kovalova, J.L.Gavartin, A.L.Shluger, and A. M. Stoneham, *Physica B* **312-313,** 734-736, (2002); Roger a. deSouza, M. Saiful Islam and Ellen Ivers-Tiffe, *J. Mater. Chem.* **9,** 1621-1628, (1999).
23. I. Fedorov, J. Lorenzana, P. Dore, G. De Marzi, P. Maselli, and P. Calvani, S.-W. Cheong, S. Koval and R. Migoni, *Phys.Rev. B* **60,** 11 875-11878, (1999).
24. R.M. Nicklow and R.A. Young, *Phys. Rev.* **152,** 591-596, (1966); P. Dai, Jiandi Zhang, H. A. Mook, S.-H. Liou and P. A. Dowben, E. W. Plummer, *Phys. Rev. B* **54** , R3694-3697 ; P. G. Radaelli and S.-W. Cheong, *Phys. Rev. B* **66,** 094408-94416, (2002).
25. T. Okuda, A. Asamitsu, Y. Tomioka, T. Kimura, Y. Taguchi, and Y. Tokura, *Phys. Rev. Lett.* **81,** 3203-3206, (1999).
26. Y. Moritomo T. Akimoto H. Fujishiro A. Nakamura, *Phys. Rev. B* **64,** 064404-64407, (2001).
27. M. B. Salamon and Myron Jaime, *Rev. Mod. Phys.* **73,** 583-628, (2001).

PARTICIPANTS OF NCTP07 FROM OUTSIDE KERALA

Sr.No.	Name	Address
1	**Prof. Ranganathan**	Saha Institute of Nuclear Physics Kolkota, India
2	**Dr. J. Govindarajan**	Institute for Plasma Research, Gandhinagar - 382 428, Gujarat, India
3	**S. Raju**	IGCAR Kalpakkam, Tamil Nadu 603102, India
4	**Prof. T. K. Dey**	IIT, Kharagpur West Bengal, India
5	**Prof. JAE Desa**	Goa University Goa, India

6	**Dr. Mahesh Roy**	DRDO Jodhpur, India
7	**Dr. Mani Veloor**	CMET Hydrabad
8	**Prof. T.P. Sharma**	University of Rajasthan Jaipur
9	**Dr. Kannanbala Sharma**	University of Rajasthan Jaipur, Rajasthan.
10	**Prof. N.S. Saxena**	University of Rajasthan Jaipur, Rajasthan.
11	**Dr. M.N. Bora**	President TPSI $ Ex. Professor Department of physics G.U, Guwahati.
12	**V.S. MuthuKumar**	Dept. of Physics Sri Sathya Sai University Puttaparthi.
13	**Ginson.P.Joseph**	Dept. of Physics Loyola College Chennai, Tamil Nadu.

14	**Ms. Vinodini Shaktawat**	University of Rajasthan Jaipur, Rajasthan.
15	**Ms. Manasvi Dixit**	University of Rajasthan Jaipur, Rajasthan.
16	**Ms. Deepika**	University of Rajasthan Jaipur, Rajasthan.
17	**Mr. K.S. Rathore**	University of Rajasthan Jaipur, Rajasthan
18	**Ms. Rashmi Saxena**	University of Rajasthan Jaipur, Rajasthan.
19	**Ms Sandhya Gupta**	University of Rajasthan Jaipur, Rajasthan.
20	**Mr. Praveen.K.Jain**	University of Rajasthan, Jaipur
21	**Mr.S. Marutha Senthil**	Sri Ramakrishna Mission Vidyalaya College of Arts & science College Coimpatore.

22	**Mr. T . Krishna Kumar**	Sri Ramakrishna Mission Vidyalaya College of Arts $ science College Coimbatore.
23	**Dr. S.Agilan**	Department of physics Coimbatore Institute of Technology Coimbatore, Tamil Nadu.
24	**Mr. T. Venkatachalam**	Department of physics Coimbatore Institute of Technology Coimbatore ,Tamil Nadu
25	**Ms.B.A Saradha**	Department of physics S.V.University Tirupati.
26	**Mr.L. Raja Mohan Reddy**	Department of physics S.V.University Tirupati.
27	**Mr. A.Surya Narayana Reddy**	Department of physics S.V.University Tirupati.
28	**Mr. B.VengulaRao**	Department of physics S.V.University Tirupati.
29	**Mr. K. Harikrishna**	Department of physics S.V.University Tirupati.

30	**Mr.S.Sambasivam**	Department of physics S.V.University Tirupati.
31	**Mr.Janaki Rami Reddy**	Department of physics S.V.University Tirupati.
32	**Mr. Omkaram**	Department of physics S.V.University Tirupati.
33	**Mr.N.Reddeppa**	Department of physics S.V.University Tirupati, Tamil Nadu.
34	**Mr . Aditya Sharma**	Department of physics S.V.College Aligarch
35	**Dr Santhosh Kumar**	Department of physics Christ Church College Kanpur
36	**Mr. N.Sharma**	Department of physics Christ Church College Kanpur
37	**Ms Runmoni Gogoi**	Cotton College Guwahati

38	**Ms.Latika Kalita**	Cotton College Guwahati
39	**Ms Yojana Janu**	Defence Lab, Jodhpur.
40	**Mr BalRaju**	J.N.V University Jodhpur
41	**Mr. M Pandiavaman**	School of Physics Madurai KamarajUniversity Madurai, Tamil Nadu.
42	**Mr . K. Neyvasayum**	Thin Film Lab Madurai KamarajUniversity Madurai, Tamil Nadu
43	**G.Krishna Bhama**	Sri Meenakshi Govt. Arts College for Women Madurai
44	**Prof K. Venkatramanan**	Department of physics Sri C S Viswa Mahavidyalaya Kanchipuram, Tamil Nadu

45	**E. Jasmine Sobha**	Nesamony Memorial Christian College Marthandom, Tamil Nadu
46	**Manila .A. Christapher**	Nesamony Memorial Christian College Marthandom, Tamil Nadu
47	**Reeja.A.J**	Nesamony Memorial Christian College Marthandom, Tamil Nadu
48	**Veena .A.P**	Nesamony Memorial Christian College Marthandom, Tamil Nadu
49	**C. Christal Vasanthi**	Nesamony Memorial. Christian College Marthandom, Tamil Nadu
50	**Ms Varsha Rani Mehto**	Barkatullah University Bhopal.
51	**Pavithra Kumar.V.S**	Nesamony Memorial Christian College Marthandom, Tamil Nadu

52	**Mr Ravi.K. Sukla**	Barkatullah University Bhopal.
53	**Mr Ashish Bodhaye**	Barkatullah University Bhopal.
54	**Mr R. GopalaKrishnan**	Department of physics Annamalai Uty, Tamil Nadu.
55	**Mr.Deepak Gikwad**	University of Mumbai Mumbai.
56	**Ms Alka.R. Chauhan**	University of Mumbai Mumbai.
57	**Mr N.M .Patil**	Department of Applied physics R.T.M Nagpur Uty Nagpur.
58	**Mr S.H. Shamkuwar**	Department of Applied physics R.T.M Nagpur Uty Nagpur.
59	**Mr M.Aynyas**	S.V. College Bhopal.

PARTICIPANTS OF NCTP07 FROM KERALA

Sr.No	Name	Address
1	**C.Indira Devi**	Department of Physics T.K.M.M College Nangiarkulangara.,Kerala, India
2	**Ms. Regimol C Cheriyan**	School of Pure and Applied Physics M.G University, Kottayam,Kerala, India
3	**Lijosh Thomas .T**	Department of Physics S.B College Changanacherry.,Kerala
4	**Ms. Ann Mary Sebastain**	Department of Physics S.B College Changanacherry,Kerala, India.
5	**Anumol R**	D.B College Sasthamcottah.,Kerala, India.
6	**Amba Prasad**	D.B College Sasthamcottah,Kerala, India

7	**Jisha Krishnan**	D.B College Sasthamcottah.,Kerala, India.
8	**Jyothi George**	D.B College Sasthamcottah.,Kerala, India
9	**Mahesh Kumar V**	D.B College Sasthamcottah,Kerala, India .
10	**Praveen P.G**	D.B College Sasthamcottah.,Kerala, India
11	**Praveen .M**	D.B College Sasthamcottah.,Kerala, India
12	**Radhu Krishna**	D.B College Sasthamcottah ,Kerala
13	**Saju S**	D.B College Sasthamcottah,Kerala, India.
14	**Sandhya .G**	D.B College Sasthamcottah,Kerala, India
15	**Sandhya S**	D.B College Sasthamcottah.,Kerala, India
16	**Soumya Santhosh**	D.B College Sasthamcottah, Kerala, India

17	**Vinod Kumar.K**	D.B College Sasthamcottah,Kerala, India
18	**Sreekumar. M. Pillai**	D.B College Sasthamcottah,Kerala, India
19	**Visakh. M**	D.B College Sasthamcottah,Kerala, India
20	**Divya Udayan**	D.B College Sasthamcottah,Kerala, India.
21	**Dini Raj**	D.B College Sasthamcottah.,Kerala, India
22	**Dr. Jayakumari Issac**	Department of Physics, C.M.S College, Kottayam,Kerala, India.
23	**Jyothi.R**	Amirtha School of Biotechnology Amirthapuri,Kerala, India
24	**Resmi Rajan**	N.S.S College Pandalam,Kerala, India.
25	**Manoj Kumar B**	S.N.College Kollam,Kerala, India

26	**Priya.S**	S.N.College Kollam,Kerala, India.
27	**Nimy.M.S**	S.N.College Kollam,Kerala, India
28	**Shyni.V.S**	S.N.College Kollam,Kerala, India
29	**Soumya.U**	S.N.College Kollam,Kerala, India
30	**Varsha.S.Vijayan**	S.N.College Kollam,Kerala, India
31	**Raji. R**	S.N.College Kollam,Kerala, India
32	**Veena.V.J.**	S.N.College Kollam,Kerala, India
33	**Nisha .G.H.**	S.N.College Kollam ,Kerala, India
34	**Deepthi.V.C**	S.N.College Kollam,Kerala, India

35	**Ajith.N.S**	S.N.College Kollam ,Kerala, India
36	**Ranju B.R**	S.N.College Kollam,Kerala.
37	**Jayesh.J**	S.N.College Kollam,Kerala, India
38	**Arun Kumar.A**	S.N.College Kollam,Kerala, India
39	**Ajeeshlal.M.U.**	S.N.College Kollam,Kerala., India
40	**Bijitha.A**	S.N.College Kollam,Kerala, India
41	**Leshmi.S**	S.N.College Kollam,Kerala, India
42	**Akhil.R.Nadh**	S.N.College Kollam,Kerala, India.
43	**Manju Balachandran**	S.N. College, Kollam,Kerala, India.

44	**Manoj.M**	S.N.College Kollam,Kerala, India.
45	**Aneesh.K.M.**	S.N.College Kollam,Kerala, India.
46	**Aneesh.S**	S.N.College Kollam,Kerala, India.
47	**Renjith.R**	S.N.College Kollam,Kerala, India.
48	**Neethu Asokan.**	S.N.College Kollam,Kerala, India.
49	**Simon.M**	S.N.College Kollam,Kerala, India.
50	**Suraj.S**	S.N.College Kollam,Kerala, India.
51	**Nisha Murali**	S.N.College Kollam,Kerala, India.

52	**Ullianove**	S.N.College, Kollam,Kerala, India.
53	**Sujith.V.S**	. S.N.College Kollam,Kerala, India.
54	**Lalu Prasad**	S.N.College Kollam,Kerala, India.
55	**Abdul Ofoor. A**	S.N.College Kollam,Kerala, India.
56	**Renish.R**	S.N.College Kollam,Kerala, India.
57	**Renjith.G**	S.N.College Kollam,Kerala, India.
58	**Umesh.D**	S.N.College Kollam,Kerala, India.

59	**Rinku Viswanadh**	S.N.College Kollam,Kerala, India.
60	**Nisha.N**	S.N.College Kollam,Kerala, India.
61	**Priya.O**	S.N.College Kollam,Kerala, India.
62	**Anila sathyan**	S.N.College Kollam,Kerala, India.
63	**Achitra.C**	S.N.College Kollam,Kerala, India.
64	**Ramya.B**	S.N.College Kollam,Kerala, India.
65	**Aswathy Raj**	S.N.College Kollam,Kerala, India.

66	**Gayathri**	S.N.College Kollam,Kerala, India.
67	**Roshini L Rajan**	S.N.College Kollam,Kerala, India.
68	**Ganga Nath**	S.N.College Kollam,Kerala, India.
69	**Sreekala C O**	Department of Biotechnology Amrita Institute, Karunagappally,Kerala, India.
70	**Nisha Sasi**	NIT, Calicut,Kerala, India.
71	**Anisha Mary Mathew**	CMP Lab, S N College, Kollam,Kerala, India.
72	**Neena.P**	CMP Lab, S N College, Kollam,Kerala, India.
73	**Najidha.S**	CMP Lab, S N College, Kollam,Kerala, India.

74	**Remya jayan. S**	CMP Lab, S N College, Kollam,Kerala, India
75	**Prasanth.S**	CMP Lab, S N College, Kollam ,Kerala, India.
76	**Santhosh Kumar. S**	CMP Lab, S N College,Kollam,Kerala, India.
77	**Arun. S. Prasad**	CMP Lab, S N College, Kollam,Kerala, India.
78	**Kannan,**	CMP Lab, S N College, Kollam,Kerala, India.
79	**K. Remadevi**	CMP Lab, S N College, Kollam,Kerala, India.
80	**Achama George**	St. Stephen College, Maloor, Pathanapuram,Kerala, India.
81	**Dr. H C Unnithan**	Department of Physics,D B College, Sasthamkotta,Kerala, India

82	**D. Sushama**	Department Of Physics,MSM College Kayankulam.,Kerala, India.
83	**Dr. Bhadrakumari**	Department Of physics,MSM College Kayankulam.,Kerala, India.
84	**Dr. Jayakumar**	Department Of Physics, MSM College Kayankulam.,Kerala, India.
85	**G. Santhamma**	Department of Physics, S N College(For women) Kollam,Kerala, India.
86	**Dr. Sreelatha K S**	Department of Physics, Amrita Institute,karunagappally,Kerala, India. ,Kerala, India.
87	**Vidya.V**	Department of Physics, Amrita Institute, karunagappally.,Kerala, India.
88	**Prof. P. Predeep**	NIT Calicut.,Kerala, India.

89	**Shahul Hameed**	TKM College Of Engineering Karikkodu, Kollam,Kerala, India.
90	**Dr. Jayasree**	Departmnt of Physics, TKMM, Nagyarkulankara,Kerala, India.
91	**Sukhi.S**	Department Of Physics S N College, Kollam ,Kerala, India.

Author Index

A

B

C

D

G

H

I

J

K